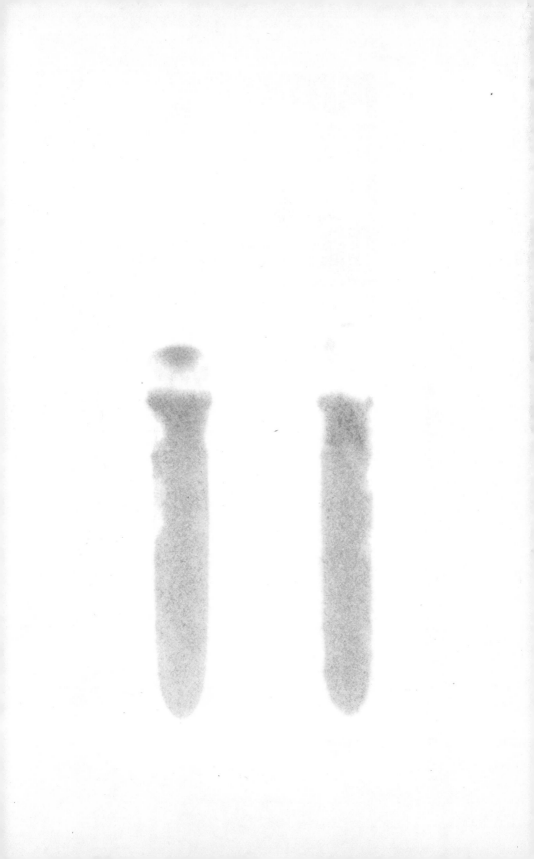

PUTNAM'S POWER FROM THE WIND

Second Edition

The 1250-kilowatt test unit at Grandpa's Knob.

PUTNAM'S POWER FROM THE WIND

Second Edition

Gerald W. Koeppl, Ph.D.

WITHDRAWN

VAN NOSTRAND REINHOLD COMPANY
NEW YORK   CINCINNATI   TORONTO   LONDON   MELBOURNE

Copyright © 1982 by Van Nostrand Reinhold Company

Library of Congress Catalog Card Number: 81-3010
ISBN: 0-442-23299-3

Manufactured in the United States of America

Published by Van Nostrand Reinhold Company
135 West 50th Street, New York, N.Y. 10020

Van Nostrand Reinhold Limited
1410 Birchmount Road
Scarborough, Ontario M1P 2E7, Canada

Van Nostrand Reinhold Australia Pty. Ltd.
17 Queen Street
Mitcham, Victoria 3132, Australia

Van Nostrand Reinhold Company Limited
Molly Millars Lane
Wokingham, Berkshire, England

15 14 13 12 11 10 9 8 7 6 5 4 3 2 1

Library of Congress Cataloging in Publication Data

Putnam, Palmer Cosslett, 1900-
    Putnam's Power from the wind.

    Bibliography: p.
    Includes index.
    1. Wind power.   I. Koeppl, Gerald W.   II. Title.
TK1541.P8   1981   621.31'2136   81-3010
ISBN 0-442-23299-3   AACR

TO
M.T.P.

To
Karen
Jacob
and
Rebecca

Foreword to Part I

In 1939 the directors of the S. Morgan Smith Company, manufacturers of hydraulic turbines, decided to explore the possibilities of large-scale wind-turbines as an additional source of power, and as a means of diversifying their product. To harness the power in the wind on a large scale required a knowledge of the habit of wind, about which science had little to say to us. To enter the field would require basic research.

T.S. Knight, Vice-President of the General Electric Company, had introduced Palmer Cosslett Putnam to us in 1939. Putnam had reviewed alternative designs of large windmills and proposed one of his own. We also reviewed the problem, and concluded that we liked Putnam's design. Encouraged by Walter Wyman, who secured the collaboration of the Central Vermont Public Service Corporation, we decided to engineer one full-scale unit of the Smith-Putnam Wind-Turbine, to determine if it was technically sound; and to undertake the necessary basic research into the habit of wind.

Under Putnam's leadership we organized a team of eminent men in the various fields of science and engineering, directed by Professor John B. Wilbur of the Massachusetts Institute of Technology, who served as Chief Engineer of the Project.

In six years of design and testing of the 175-foot, 1250-kilowatt experimental unit on Grandpa's Knob near Rutland, Vermont, in winds up to 115 miles per hour, we have satisfied ourselves that Putnam's ideas are practical, and that regulation is sufficiently smooth. We think we could now design, with confidence, 2000-kilowatt wind-turbines incorporating important improvements leading to smoother operation, simpler maintenance, and lower cost.

As soon as we learned from the test results what a production unit would look like, we began to recompute the over-all economics.

In 1939, based on 1937 prices, we had estimated that on a 4000-foot ridge in northern New England, a battery of ten 1500-kilowatt wind-turbines, 200 feet in diameter, would generate energy at about $0.0025 per kilowatt-hour.

Our 1945 estimates, based on a preproduction model similar to the test unit, showed that at the best site in Vermont the cost would be about $0.006 per kilowatt-hour. Jackson and Moreland estimated that the Central Vermont Public Service Corporation could afford to pay no more than about $125 per kilowatt for a block of wind-turbines generating energy at this cost. At 1945 prices, it would have cost us about $190 per kilowatt to install a block of such capacity. It is true that the cost savings proposed in Chapter 12 would, if realized, more than bridge this gap. But we had already spent a million and a quarter dollars, and it would cost several hundred thousand dollars more to find out whether we could actually sell in this market at a profit. Our stockholders were unwilling to make the additional investment and, accordingly, we have reluctantly abandoned the project, placing the patents in the public domain.

Although we did not publicize it, the experiment on Grandpa's Knob commanded world-wide attention, and even during the war we continued to receive inquiries from many countries. Interest in wind-power is widespread. Accordingly, we have asked Putnam to summarize our explorations in this field and to evaluate the future of large-scale wind-power.

I agree with Putnam's conclusion that large wind-turbines will be limited to applications in special regions, which, in the aggregate, however, may amount to a considerable market—a market which I hope to live to see developed.

BEAUCHAMP E. SMITH

York, Pennsylvania
November, 1946

Preface to Part I

For six years a group of eminent scientists and engineers, under the auspices of the S. Morgan Smith Company, has worked on the technical problems of the 1250-kilowatt windmill. Leading specialists in many fields have served as consultants. The ultimate economics of wind-power have been explored in a preliminary way.

This book is a summary of the problems which faced us in 1939, our attempts at solving them, our findings and conclusions.

It is directed to anyone interested in man's instinctive urge to subdue and harness his environment, and particularly to those in Government or Industry who are interested in eking out dwindling supplies of low-cost fuels with other sources of power.

Carl J. Wilcox, who wrote many of the original engineering reports, has prepared the drawings and organized the material for the text, which has been reviewed by all of our associates. To him warm thanks are due for help and criticism.

PALMER COSSLETT PUTNAM

Boston, June, 1947

Introduction to Part I

The great wind-turbine on a Vermont mountain proved that men could build a practical machine which would synchronously generate electricity in large quantities by means of wind-power. It proved also that the cost of electricity so produced is close to that of the more economical conventional methods. And hence it proved that at some future time homes may be illuminated and factories may be powered by this new means. Having provided these proofs, the installation has since been relinquished by the group that created it. It had served its purpose; economic factors worked against it; other responsibilities were to be met; other work was to be done.

The project that this volume discusses must not be written off with such dispatch as the preceding remarks suggest. However important were the three immediate proofs that it gave—and I regard them as highly important—this project, when viewed against a larger scene, grows still greater in significance.

The decision to undertake the design, construction, and operation of a wind-turbine to generate electricity led to a pioneering effort which aptly and fully illustrates how man utilizes intelligence in mastering his environment for the purpose of advancing his general welfare, in proceeding from ignorance to theoretical knowledge, and then reducing theoretical knowledge to practical application. Moreover, earlier harnessing of the winds to drive ships, to turn millstones, or to pump water, had resulted primarily in individual, independent applications; the Vermont project called for the correlation and integration of the installation with other installations using other types of power, and the tying of the result into a wide-flung and complex network unobtrusively serving thousands of people who knew little if anything of its existence.

Complex as are the electrical systems of which it was a part and the economic system out of which it grew, the wind-turbine is notable as the physical result of a project conceived and carried through by free enterprisers who were willing to accept the risks involved in exploring the frontiers of knowledge, in the hope of ultimate financial gain. Note that such financial gain would not have been at the cost of some other part of the economy. Large-scale wind-power will not create unemployment, for, in general, it displaces nothing, but rather draws on a new source to supplement steam and water, and can enlarge the use of the product it creates. It could, fully developed, bring light, heat, and power to regions that otherwise could not afford such services, or, in fact, because of physical difficulties, could not have them at all.

Free enterprise thus demonstrated in this project that it functions as effectively in the involved social structure of the present as it did in simpler societies in the past. So too in this project was demonstrated the ability of complex science and technology to focus a score of specialized skills on the various aspects of a problem, coordinating and collaborating in effective teamwork. That this scientific cooperation in a wide number of fields was carried out in this instance by a temporary organization mustered simply for the single undertaking, using people scattered from coast to coast, and nearly all on a part-time basis, is remarkable.

The project, in addition, illustrates what may be secured from a pioneering venture into new fields of knowledge. Scientific contributions of distinct importance resulted from it, particularly to knowledge of how winds behave in mountainous country, to knowledge of icing conditions, to knowledge of the behavior of metals under new and exacting conditions. Wind-energy surveys in a half-dozen portions of the globe may well be a consequence of the undertaking.

The three immediate proofs which the project gave were highly important. The four long-range demonstrations which resulted from it are a further and lasting cause for its being considered of prime significance. The story is ably recorded in this volume.

V. BUSH

December, 1946

Foreword to Part II

In the past few months, having been inactive in wind power since the first publication of *Power From The Wind* in 1948, I have attended a number of conferences and workshops on wind power. The experience has left me with two overwhelming impressions.

My first impression concerns the great momentum that now exists behind the idea of wind power. Governmental contracts with academia, think tanks, and industry have funded surveys of the national wind resource and research into parametric analyses of wind turbines. Since 1973 there has been federal funding for the wind programs of NASA and the Department of Energy to the extent of a quarter of a billion dollars, which has taken us into the third generation of design concepts. The federal government and some state governments have been supporting wind power research, development, and application by favorable tax and other legislation.

In the industrial sector, such giants as Alcoa, Bendix, Boeing, GE, Hamilton Standard, Westinghouse, and others (including some excellent small ones) have responded to the new economics of wind power by coming forward with prototype designs for domestic and foreign markets.

Among the utilities, Pacific Gas and Electric, Southern California Edison, and Bonneville Power each has put its first experimental wind turbine on stream. There are now nine multi-megawatt WTGs operating in the United States as of September 1981.

In the entrepreneurial sector, WINDFARMS, Ltd. has signed a contract to sell to the Hawaii Electric Company the energy from the first large wind farm in history. Twenty sites have been obtained on the northeast corner of Oahu, facing the trade winds. There, on Kahuku Point, twenty two-bladed, horizontal-axis, four-megawatt WTGs are scheduled to start replacing oil by 1983. U.S. Windfarms, Inc., having installed an experimental wind farm of twenty 25-KW units on Crotched Mountain, N.H., in December 1980, now has a contract with Pacific Gas and Electric to install up to 2,000 50-KW units in Altamont Pass, east of San Francisco. And many others are in the planning stage, here and abroad.

In short, thanks largely to OPEC, there is general agreement that the time for large-scale wind power has arrived, vindicating the vision that Beauchamp Smith and I shared in October 1939.

My second impression is the large fraction of the total national output of electricity that those in authority are publicly estimating will be wind-generated by the year 2000. Thirty-five years ago, discouraged by our inability to generate for less than four mils per kilowatt hour, I estimated that someday, when fossil fuels became scarcer, wind might possibly generate four percent of the national total. Today we are being given estimates for the year 2000 of eight to nineteen percent.

Late in 1980, the National Science Foundation estimated that in the year 2000 wind could generate from seven to thirteen percent of the total electricity produced in the U.S., by means of 25,000 to 43,000 4-MW units. The NSF report concluded that, to make this goal "achievable," 1,250 units should be installed by mid-1988. The goal requires an annual doubling over the next seven years.

Based on estimates by the California Energy Commission, 25,000 4-MW units would reduce emissions of SO_x, that now cause acid rain on New England, by one million tons yearly.

No other renewable resource today has these potentials.

Because so many recognize what a substantial contribution economically competitive wind power will make towards energy independence while reducing chemical pollution, and because so many are trying to learn how to supply reliable long-lived turbines and to site them properly, it has seemed appropriate to the publisher to review the state of the art and to note what progress has taken place between the end of our experiment on Grandpa's Knob in 1945 and the design for wind farms 35 years later.

To me, it is fascinating that my pioneering parametric analyses, including their optimizations, have now been generally confirmed in studies by GE and Boeing, as noted by Dr. Koeppl. Advances have been made in blade material, to lengthen life and avoid interference with TV; in computerized structural analysis; in hub design, to profit by helicopter experience; and in the use of electronic sensors and minicomputers, to smooth out the operation and increase reliability.

Dr. Gerald Koeppl, in updating my original *Power From The Wind,* has done a stunning job of digesting and presenting the enormous mass of new data. His approach has been different from mine in an essential way. My treatment was narrative, with a minimum of formal mathematics. Dr. Koeppl, dealing with a maturing art, is rigorously and formally mathematical, as is proper.

He describes the available kinetic energy in the wind, favorable and unfavorable site characteristics, the foreign and domestic wind turbines under design and test, the new, OPEC-generated economics of wind power, and the legal and environmental questions—a comprehensive survey.

A seascape or a landscape without a work of man in the middle distance is often thought to be not worth photographing or painting. An expanse of mere ocean does not say much. Put a laboring vessel in the middle distance, and there is a point of interest—dramatic value. A distant mountain range is just there. Add a forest ranger's tower: the composition begins to say something. How much more will it say when the slender tower is seen to support two blades that rotate slowly, gracefully, silently—evidence that man is once again and at last using his environment benignly!

Atascadero, CA   PALMER COSSLETT PUTNAM

Preface to Part II

Work which culminated in the new edition of *Power from the Wind* began in 1978 while I was on sabbatical at the University of Vermont. The production of electrical energy using wind power technology was not discussed during the University's Energy Forum, a series of lectures and panel sessions concerned with Vermont's future energy requirements and resources. I had encountered brief descriptions of the large wind turbine generator operated on Grandpa's Knob during the 1940s in popular journals, and knew that the Department of Energy (DOE) was developing both small and large-scale wind machines. I decided to study the potential of Vermont's wind energy resource and current efforts to produce power from the wind. I was aided by a friend who informed me that the Vermont wind turbine project had been described in a book.

I was astonished to learn that almost forty years ago, a group of eminent scientists and engineers, working under the auspices of a private company, had attempted to commercialize large-scale wind turbines to conserve our dwindling supplies of low-cost fossil fuels. Putnam's remarkable book provides a lively and interesting description of pioneering work involving wind site prospecting, large-scale machine design and performance, and the economics of wind generation. His account of the Smith-Putnam project has proved to be a valuable resource for the present Federal Wind Energy Program.

An associate, John Zimmerman, informed me that utility companies in Vermont did not participate in the federal competition in 1976 for wind turbine candidate sites, and that another opportunity to compete would be announced soon. Since the wind survey program described in Putnam's book indicated that Vermont had many promising sites for economic wind generation, we decided to encourage utility companies to join the competition, and help them prepare proposals. Carl J. Wilcox, an engineer who played an important role in the Smith-Putnam project and in the development of materials for the original edition, furnished project records which were useful for the preparation of this edition and proposals for the DOE competition. In 1979, the site which Putnam referred to as "the best large capacity site in Vermont," the Lincoln Ridge, and another site, Stratton Mountain, were selected as candidate sites for DOE/NASA wind turbines.

Putnam's book is more than a historical record of the 1940s project. It provides a definitive description of the important concepts of large-scale wind generation. Consequently, changes in the original material have been limited to typographical errors. Extensive additional material has been added to update Putnam's description of the wind resource and its potential for electrical generation, wind characteristics, wind turbine performance, the economics of wind generation, efforts to develop large-scale wind machines, and relevant legal-institutional and environmental issues. References were not cited in the text of the original edition. The new material contains extensive references to the recent, and rapidly expanding literature of wind energy. The reader may wish to read the entire original edition first, and may be compelled to do this by Putnam's lively account of the Smith-Putnam project. Alternatively, the reader may choose to read the updated material after it is introduced by the material in the text of the original edition.

The new material demonstrates that large-scale wind machines can supply a significant amount of the U.S. demand for electrical energy. The technical and economic problems can be solved in the near term, large quantities of oil can be saved for important future uses, and our environment will benefit from the utilization of a non-polluting resource. The United States government could effect the cost reduction which accompanies mass production by being the first purchaser of hundreds of large-scale wind turbines for the augmentation of existing hydroelectric facilities. Additional applications in the utility sector could then commence on a sound economic basis. However, at the present time, it appears that the economic incentives for the development of large-scale wind turbines may cause cooperative efforts involving wind farm developers, manufacturers, and utility companies to play the leading role in the commercialization of this technology.

Dimitrios Panagopoulos and Eric Spieler helped me with calculations involving wind characteristics and wind system performance. Thomas C. Boucher of the Green Mountain Power Corporation helped me understand the aspects of large-scale wind power technology which are important to a utility planner. I am grateful to Palmer C. Putnam for preparing the Foreword of the new edition and for his helpful comments and criticism concerning the new material.

I am also grateful for assistance and encouragement from: Ronald Allbee; Daniel F. Ancona III; Carl Aspliden; L.A.W. Bedford; John Bohn; Brewster E. Buxton; Richard Cambio; F.J.P. Clarke; Louis V. Divone; Stanton D. Dornbirer; Dennis L. Elliott; Wayne Foster; Joseph M. Gainza; Elmer Gaden; George Gantz; John Gilbert; John C. Glascow; H. Grastrup; Alfred J. Gross; Margaret Gross; Frederick Gross; William E. Heronemus; Stanley J. Hightower; Robert E. Howland; Sven Hugosson; David R. Inglis; James M. Jeffords; Finn Jørgensen; Anne Just; C.G. Justus; Stephen Keel; Gordon Kraft; Patrick J. Leahy; John E. Lowe; Bill Maclay; Don Mayer; Richard A. Mixer; John Norton; Donald Ogden; William Pickens; John Pillsbury; Gerard Pregent; Charles Racine; David S. Renee; Steve Rice; Jim Sanford; Richard H. Saudek; Robert L. Scheffler; Ferrall Seiler; David Sellers; Alan T. Smith; Richard Snelling; Robert T. Stafford; George Tennyson; George G. Walker; L.L. Wendell; Brendan Whittaker; Rolf Windheim; and Valerie Zimmerman.

Thanks are also due to the personnel of the word processing center at Queens College of the City University of New York, Robert Wurman, Robin Hall, and Terry Hurwitch for help in preparing the manuscript.

G.W.K.

Contents

PUTNAM'S POWER FROM THE WIND

Second Edition

Part 1

1
Personalities and History of the Smith-Putnam Wind-Turbine Project, 1934-1935

In the fall of 1941 something new had been added to the generating system of the Central Vermont Public Service Corporation. Motorists in central Vermont saw, from 25 miles away, a giant windmill (frontispiece), its polished sunlit blades flashing on top of the 2000-foot Grandpa's Knob,* 12 miles west of Rutland and overlooking the Champlain Valley. This was the experimental Smith-Putnam Wind-Turbine, undergoing its first tests. The unit was rated at 1250 kilowatts, enough to light a town, and was feeding power into the utility company's system, permitting water to be stored behind the dams when the wind blew more than 17 miles an hour.

The synchronous generator, operating in phase with the other generators on the system, was driven, through gears and a hydraulic coupling, by a two-bladed stainless-steel turbine 175 feet in diameter, whose hub was at a height of 120 feet above the summit of the smoothly glaciated Knob. The turbine responded to changes in wind direction by rotation in yaw. Power output was regulated hydraulically by controlling the pitch of the blades with a speed governor which responded to changes in rotational speed of the turbine shaft. The blades were free to cone down-wind, moving like the ribs of an umbrella. For the first time, wind had been harnessed to drive a synchronous generator feeding directly into the highline of a utility network.

The experiment is another proof that the spirit of exploration and adventure had not yet died out in those ancient citadels of capitalism, New England and Pennsylvania. This chapter briefly describes the development of the project, backed by a group of Down-east Yankees, and free enterprisers from York, Pennsylvania.

In 1934 I had built a house on Cape Cod and had found both the winds and the electric rates surprisingly high. It occurred to me that a windmill to generate alternating current might reduce the power bill, provided the power company would maintain stand-by service when the wind failed, and would also permit me to feed back into its system as dump power the excess energy generated by the windmill.

---

*This peak had not been distinguished on maps by a separate name. It was bought from a Vermont farmer whose family always referred to it as "Grandpa's." Because of this, and its shape, we christened it Grandpa's Knob.

PALMER COSSLETT PUTNAM

Manager of the Smith-Putnam Wind-Turbine
Project.

But when I came to compute the size of windmill needed to carry the peak load of the all-electrical house in the prevailing wind, it was clear that none of the small units, widely used for farm lighting, would be large enough. A much larger unit would be necessary.

Aaron Davis, a friend and neighbor, called attention to the design of a Finn, Savonius, who had just been commissioned by Colonel Henry Huddleston Rogers to put up three Savonius-type rotors on Colonel Rogers' estate at Southampton, Long Island. A review of this design made it clear that it was inherently inefficient per unit of weight, since all of the area swept was occupied by metal. Two narrow rapidly moving blades extract more energy from the wind, per unit of area swept.

This study brought me into touch with Elisha Fales, who had been among the first to apply the aerodynamic lessons of the First World War to the problem of the windmill. In order to obtain a preliminary indication of the strength of the wind at an average site on Cape Cod, Fales lent me one of his small two-bladed direct current test units, which, mounted on a 60-foot-pipe, served briefly and intermittently as a sort of anemometer.

The results of these measurements were inconclusive, but stimulated me to survey the previous work as reported in the literature and to study the designs for large, wind-driven, direct-current and induction generators that had been proposed by Flettner, Mádaras, Kumme, Darrieus, the Russians, and Honnef (Chapter 6).

My conclusion was that, if an economically attractive solution to the problem existed, it lay in the direct generation of alternating current by a very large, two-bladed, high-speed windmill, feeding into the lines of an existing hydro, or steam and hydro, system. Thus, existing hydro-storage would provide the capacity to tide over periods of no wind. When the wind blew, the dispatcher could shut down the hydro units and accumulate additional stored water in the reservoirs. From the point of view of a hydro system the energy in the wind then became merely increased stream flow. To assure utmost wind velocity, the site should have airfoil characteristics and should be capable of speeding up the wind over its summit.

Another friend and neighbor, Harold Sawyer, made a useful suggestion, which, however, did not become technically practical for another ten years. Sawyer proposed that there be incorporated between the wind-turbine and the generator a torque-limiting device that would be incapable of passing along to the generator a torque in excess of a predetermined value, regardless of those gust-generated power

VANNEVAR BUSH

Electrical engineer, President of the Carnegie Institution of Washington, formerly Director of the Office of Scientific Research and Development, now Chairman of the Joint Research and Development Board of the Department of National Defense.

surges from the turbine, too sudden to be prevented by the speed governor. The mechanical solution proposed by Sawyer did not prove feasible, but the electric coupling now available seems most promising. After a few months Sawyer left to join the staff of the McMath-Hurlburt Observatory in Michigan, assigning his interest in the project to me.

I continued to develop the design features which later became incorporated in the test unit, and made some rough cost estimates. These looked promising to Dr. Vannevar Bush, then Dean of Engineering at the Massachusetts Institute of Technology, and in 1937 Bush referred me to Thomas S. Knight, the Commercial Vice-President of the General Electric Company in New England. Knight, a member of the Cruising Club of America, had sailed Down-east all his life and was immediately attracted by a proposal to use recent developments in aerodynamics and other fields to harness the wind on a large scale. He offered to assist me in developing my ideas to the point where more accurate cost studies could be made. Under this arrangement, the aerodynamic outputs were computed from data secured in conferences with Dr. Theodor von Kármán, Director of the Guggenheim Aeronautical Laboratories at the California Institute of Technology. The wind regime was selected after conferences with Dr. Sverre Petterssen, Director of the Department of Meteorology of the Massachusetts Institute of Technology. Based on these assumptions, I considered and rejected many alternative designs and details, in the end arriving at the schematic layout that was finally built; and at a first approximation of the best dimensions for a large-scale wind-turbine of this design.

Professor John B. Wilbur, of the Massachusetts Institute of Technology (now head of the Department of Civil Engineering), also found his imagination stirred, and I retained him to carry out preliminary stress analyses. Having "frozen" the schematic layout, and Wilbur having determined the sizes of the members within

THOMAS S. KNIGHT

Commercial Vice President of General Electric Company in Boston since 1931.

SVERRE PETTERSSEN

One time Director of the Weather Bureau at Bergen. One time Head of the Department of Meteorology, Massachusetts Institute of Technology. With British Air Ministry 1942-45, and now Chief of the Weather Forecasting Service in Norway.

JOHN B. WILBUR

Head of the Department of Civil and Sanitary Engineering, Massachusetts Institute of Technology, and Chief Engineer of the Wind-Turbine Project.

limits, I retained Jackson and Moreland, Consulting Engineers of Boston, to make layout sketches and a brief report analyzing the economics of the project.

Now, at this time there were two tactical problems that had to be solved before the wind-power project could become a going concern. First, it was necessary to find someone to put up the money for a full-scale test unit. Second, a manufacturer had to be found with the experience and prestige to qualify him for this project. Knight had granted me temporary office space in the Boston Offices of the General Electric Company and under his patronage I discussed the project, with Knight's consent, with various members of his staff including Alan Goodwin, his hydroelectric specialist. Out of the discussions with the latter came the suggestion that Goodwin see Walter Wyman, President of the New England Public Service Corporation, at Augusta, Maine, and clearly the one man in New England who had both the authority and the vision to push such a project as large-scale wind-power.

Goodwin, in connection with the various hydroelectric developments in which he was interested, had been dealing with Frank Mason, Chief Engineer of the New England Public Service Corporation. Goodwin knew that Mason was an enthusiastic believer in water-power and felt that he could be interested in wind-power if it could be demonstrated to him that wind-power would enhance the value of hydroelectric units.

Goodwin went to Augusta and talked to Mason, who recognized the great possibilities in a combination of wind and water. Mason looked out of his window, saw that Wyman's automobile was parked in its usual place, and so knew that Wyman was in his office. Mason also noted that the wind was one of those strong southwesterlies that frequently come during the hot summer spells of dry weather. Mason and Goodwin went to Wyman's office. Wyman, like Knight, was a Downeaster; he owned a farm on a windy ridge near enough the coast to come under the diurnal sea breeze in summer. The three looked out at the waving trees. It was a hot day in August and the water in the Utility Company's reservoirs was low. Goodwin said, "Mr. Wyman, just look at the way the wind is blowing those trees. There is a lot of force there and it is all going to waste. Man has used the wind for centuries to blow himself around the oceans but he has never harnessed it for power on a large scale. Wouldn't it be wonderful if you could use some of the power of the wind right now when it is hot and dry and windy and your water is running low?"

This made sense to Wyman, who also knew that Goodwin never talked without a purpose. He asked Goodwin what he had in mind and the upshot was that Goodwin returned to Boston able to tell his Chief, Knight, that Wyman wanted to discuss with Knight, at his earliest convenience, the purchase of a wind-power development for the New England Public Service System.

This conference had taken care of problem number one.

Problem number two was solved this way. During the preceding fifteen years Goodwin had been promoting hydroelectric developments in collaboration with representatives of water-wheel manufacturers. The most active New England representative was Howard Mayo of the S. Morgan Smith Company. Goodwin and Mayo realized that there was just so much water and, while they had not come to the end of their water in New England, they were interested in any means of enhancing the existing or the future water-power developments of New England. It seemed logical to them that the S. Morgan Smith Company, preeminent in the field of controllable-pitch hydraulic turbines, should take on the manufacture of a wind-power unit.

Furthermore, Mayo knew that S. Morgan Smith Company was seeking to diversify its product. He took Goodwin to York, Pennsylvania, to see S. Fahs Smith, the President of the Company. The next week the two Vice-Presidents of the Company, Beauchamp Smith and Burwell Smith, came to Boston to see Knight and me.

There in Knight's office the project was born, in October, 1939, in the expectation of finding wind-power sites in New England where secondary but predictable power could be generated at the rate of 4400 kilowatt-hours per kilowatt per year, at a cost of $0.0025 per kilowatt-hour at the foot of the tower.

BEAUCHAMP E. SMITH

President, S. Morgan Smith Company.

BURWELL B. SMITH

Vice-President and Treasurer, S. Morgan Smith Company.

S. Morgan Smith Company decided that first they would redetermine the most economical dimensions and review my cost estimates. This work, described in Chapter 9, was carried out by a special staff of the Budd Company, in Philadelphia. The wind regime was re-specified by Petterssen. The aerodynamic outputs were recomputed by von Kármán. Some of the weights were estimated by Wilbur. Costs were obtained from vendors. The annual charges were set by Jackson and Moreland. The computations were carried out by Carl J. Wilcox, on loan to the S. Morgan Smith Company by the Budd Company. Variables considered included tower height, turbine diameter, turbine speed, generator rating, and generator speed.

The best design was found to be a 1500-kilowatt unit, driven at 600 revolutions per minute, by a geared-up turbine 200 feet in diameter, turning at about 25 revolutions per minute, on a tower about 150 feet high. The energy cost at the switchboard at the foot of the tower was estimated to be $0.0016 per kilowatt hour.

Satisfied with this estimate, but aware that it was based on many assumptions which remained to be tested, the S. Morgan Smith Company agreed to engineer and build a test unit to my designs. General Electric had agreed to develop and furnish the electrical equipment at cost, and Walter Wyman had arranged that Central Vermont Public Service Corporation, a subsidiary of the New England Public Service Company, should be the guinea pig, to provide the site and tie-in facilities, and should operate, and ultimately purchase, the test unit.

Fortunately both Albert A. Cree, President of the Central Vermont Public Service Corporation, and his chief engineer, Harold L. Durgin, were taken with the idea. Their system had no steam capacity. The hydro capacity was insufficient to carry peak load, and their excess demand was met under a power-purchase contract with the Bellows Falls Hydro Electric Corporation. If the test unit of the Smith-Putnam Wind-Turbine proved successful, Wyman and Cree hoped to install a block of wind capacity to supplement their hydro capacity and reduce over-all power production and purchase costs.

In the late fall of 1939, S. Morgan Smith Company realized that they were not organized either to design the test unit of this new product or to fabricate it, because of the rapidly increasing backlog of hydraulic turbine orders, and that it would be

ALBERT A. CREE

President, Central Vermont Public Service Corporation. President, Windpower, Inc.

HAROLD DURGIN

Vice-President and Chief Engineer, Central Vermont Public Service Corporation.

necessary to sublet both the engineering and fabrication. Admittedly this was not a very satisfactory arrangement; but there was no alternative, and Beauchamp Smith asked me to find somebody to handle it. Several firms with outstanding reputations as successful designers of heavy precision rotating equipment were approached.

Alfred E. Gibson, President of the Wellman Engineering Company of Cleveland, Ohio, agreed to undertake the work on a cost plus contract. The Wellman Company are designers and manufacturers of large mining and material handling equipment.

At Beauchamp Smith's request, I became Project Manager. It was my responsibility to organize the project and conduct its external relations, in accordance with the general policies laid down by Smith. He encouraged gathering together as consultants the leading authorities in the various related fields, to make possible a coordinated attack upon those problems in meteorology, ecology, aerodynamics, vibration analysis, electrical engineering, and the economics of power generation, which in 1939 remained to be solved before a large wind-turbine could be successfully and profitably operated on the lines of a utility system.

George A. Jessop, Chief Engineer of the S. Morgan Smith Company, was active in general supervision of the project, and contributed to the solution of many of the novel problems encountered.

Dr. John B. Wilbur, who later became Chief Engineer of the Project, heading the engineering and development and reporting to Jessop, was retained as structural consultant. The forces coming into the structure from the turbine were estimated from data supplied by von Kármán, who was retained as a general consultant, specifically responsible for the aerodynmaic design.

It was under von Kármán's supervision that Dr. Elliott G. Reid, of Stanford University, took charge of wind-tunnel tests of models of the turbine blades. The models had been constructed, some in California, some by the S. Morgan Smith Company in York, Pennsylvania, but all under the supervision of Mr. John Haines, now Director of Research of the Aeroproducts Division of General Motors. Haines served the project as a mechanical engineer in charge of the problems of pitch control and coning.

Professor J. S. Newell, of the Aeronautical Engineering Department of Massachusetts Institute of Technology, was consulted concerning the structural analysis of the turbine blades, whose Cor-Ten spars were designed by the American Bridge

GEORGE A. JESSOP

Chief Engineer, S. Morgan Smith Company.
Member A.S.C.E. and A.I.E.E.

HURD CURTIS WILLETT

Professor of Meteorology at the Massachusetts Institute of Technology.

CHARLES F. BROOKS

Professor of Meteorology, Harvard University, and Director, Blue Hill Observatory.

Company. The stainless-steel skin and ribs of the blades were designed by the Budd Company, noted for their pioneer work with stainless-steel trains. Budd fabricated the stainless-steel parts and assembled them around the spars received from the American Bridge Company.

The mechanical design, from the blade roots to the tower cap, was carried out by the Wellman Engineering Company under the direction of R. W. Valls, while the tower was designed and erected by the American Bridge Company.

Vibration analysis of the entire structure was placed in charge of Dr. J. P. den Hartog, at that time of Harvard University, now at Massachusetts Institute of Technology.

According to the agreement, the electrical equipment was supplied by the General Electric Company.

The meteorological program was directed by Petterssen, in collaboration with Dr. Hurd Willett, also of Massachusetts Institute of Technology; Dr. C. F. Brooks, Director of Harvard's Blue Hill Observatory; and Dr. K. O. Lange, also of Harvard.

KARL O. LANGE

One time Research Associate at Massachusetts Institute of Technology, and Research Meteorologist at the Blue Hill Meteorological Observatory. Now Professor of Physics, University of Kentucky.

ROBERT F. GRIGGS

Of George Washington University, Chairman of the Division of Biology and Agriculture, National Research Council.

Wind-tunnel tests of models of mountains were carried out by Dr. Th. Troller of the Guggenheim Aeronautical Institute, at Akron, Ohio, under the direction of von Kármán.

The ecological studies were carried out by Dr. Robert F. Griggs of George Washington University.

Jackson and Moreland served as application engineers, and from time to time carried out various studies of the economics of wind-power, in collaboration with the Central Vermont Public Service Corporation.

Beauchamp Smith had insisted on gathering together the leading men in each field; without a doubt he had succeeded. Each in turn came under the peculiar spell of the project, which roused the enthusiasm of all of us. The tone set by the Smiths was a little unusual. The realism with which they approached the project was garnished with dash—even gaiety. This spirit soon infected all who were engaged on the project, and produced an esprit de corps which went far toward counterbalancing the disadvantages that all were not housed under one roof, and indeed were not all working on the project full time. The project was administered by means of frequent gatherings in various cities, interspersed with telephone conferences. It was no easy task for Dr. Wilbur to weld together into one functioning whole all of these components, when the key personnel were scattered from coast to coast, in many cases with their attentions partly diverted to other consulting problems.

One of the critical aspects of the project was the control of the machine in all phases of operation. It was necessary to select items from among the mechanical, hydraulic, and electrical control apparatus already in use in other fields, principally in connection with hydroelectric units, and to adapt the most suitable to our needs. The selection, adaptation, and interconnection of the various control elements required careful consideration. Various alternative methods of control were devised before the final control details were determined. To head up and centralize all control problems, Grant H. Voaden, of the S. Morgan Smith Company, was appointed Chief Test Engineer.

Voaden, who had had considerable experience in this class of work in the hydraulic turbine field, worked in collaboration with Irl Martin and Walter Thorell of the Woodward Governor Company and George Jump and Herman Bany of the General Electric Company.

In May, 1940, we began ordering steel forgings. In June, Grandpa's Knob was selected as the test site, because, having an elevation of only 2000 feet, it was, we hoped, not high enough to encounter destructive ice storms. The 2-mile road connecting the site with the Vermont highway network was built that summer, under the direction of the Central Vermont Public Service Corporation, and the tower was erected by the American Bridge Company.

Stanton D. Dornbirer of the Erection Department of S. Morgan Smith Company was placed in charge of the field erection of the wind-turbine.

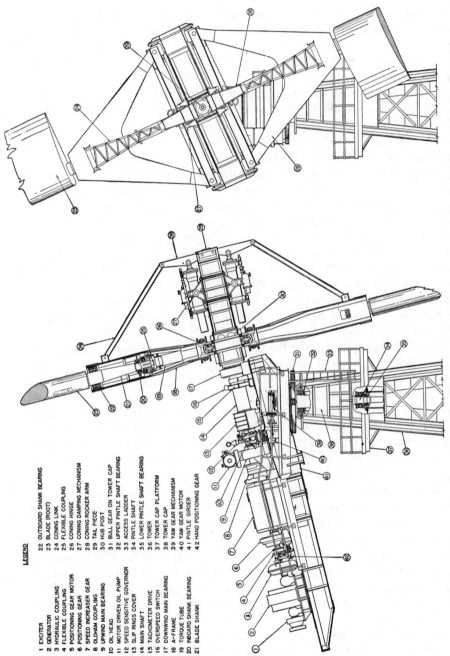

Fig. 1. Schematic detail aloft of the 1250-kilowatt test unit of the Smith-Putnam Wind-Turbine erected on Grandpa's Knob.

Erection of equipment aloft continued throughout the winter, as weather permitted. Low temperatures and high winds made rough work of handling heavy steel, but we had only one accident. On a particularly bitter, windy sub-zero day, the 40-ton pintle girder (Fig. 1) with the 24-inch main shaft and its two main bearings in place, was being trucked to the summit, on a heavy trailer drawn by the truck unit and two Caterpillar tractors in tandem. At the hairpin turn below the summit the girder rolled off the trailer-bed and turned upside-down into a deep snowbank, the 48-inch main-bearing housings having found the only opening between rocks! After recovery, inspection showed no damage to the bearings or shaft, and only minor damage to some of the plates of the girder.

Erection was completed in August, 1941, when the blades were put in place.

To facilitate carrying out the test program a field office was set up in Rutland in July, 1941. Activities of this office were under the direction of Dr. Wilbur and his assistant, G. H. Voaden. The office was staffed with three test engineers and a secretarial force.

In June, 1940, I had gone to Washington, ultimately to join Dr. Bush and the Office of Scientific Research and Development, and was not again active in the wind-turbine project until the spring of 1945.

When the turbine was assembled, William Bagley, of the General Electric Company, in collaboration with Voaden, checked the controls and carried out a number of test runs at no load, beginning at low speed and progressively altering the pitch until, after a few weeks, the turbine was run at the full speed of 28.7 revolutions per minute, without load.

Finally, on the night of Sunday, October 19, 1941, in the presence of the top management of the Central Vermont Public Service Corporation and many of the Staff of the S. Morgan Smith Company, and with the Smiths listening in by long-distance telephone, Bagley, having completed his adjustments and made his final inspections aloft, phased-in the unit to the lines of the utility company, in a gusty 25-mile wind from the northeast.

There was no difficulty. Operation was smooth. Regulation was good. After 20 minutes at "speed-no-load," the blade pitch was adjusted until output reached 700 kilowatts. For the first time anywhere, power from the wind was being fed synchronously to the high-line of a utility system.

Then began months of gathering experience, and of making adjustments leading to smoother operation. A comprehensive test program was put on a routine basis, in charge of Grant Voaden, reporting to Wilbur.

In the course of this program the test unit was operated in winds of 70 miles an hour, generating as much as 1500 kilowatts, and it was exposed, while not operating, to winds of 115 miles an hour. Regulation was found to be satisfactory. Only light icing was encountered during operation. Under these conditions the stainless-steel blades shed ice well. Under certain conditions there was some vibration, which was satisfactorily reduced by adjustments to the coning-damping system and the yaw-damping system, and largely eliminated in the plans for redesign described in Chapter 10.

The first major mishap occurred when, in February, 1943, the 24-inch down-wind main bearing failed, for reasons which are still obscure, although apparently not related to anything peculiar to wind-turbines. Because of the war it took more than twenty-four months to obtain and install a new one. But these months were not wholly lost. The skeleton staff were busy digesting observational data, and designing simplifications and improvements, many of which were carried out when the new bearing went on.

One of the most important studies made in this period was a review of loadings, by Wilcox, in collaboration with Holley, under the direction of Wilbur. They concluded that some of the loadings had been over-estimated and some under-estimated. As a net result, they felt they could now detail the design of a large wind-turbine, with conventional factors of safety throughout, and which would weigh less per kilowatt than the test unit, and cost less per pound. But, of immediate concern, they concluded that the stresses in the root sections of the blades of the test unit were higher than had been realized. Accordingly, Wilbur, early in 1945, recommended that the unit should not be operated after it had completed its test purposes, and he hoped for an early completion of this program.

With the new bearing in place, the Smith-Putnam Wind-Turbine went back on the line on March 3, 1945. Wind permitting, it was continuously operated as a routine generating station on the lines of the Central Vermont Public Service Corporation for twenty-three days. Operation was satisfactory. There was no trouble.

At 5 A.M. on March 26, 1945, my telephone started ringing. It wouldn't stop. It was Wilbur, "Put, we've had an accident. It could be worse. We've lost a blade, but no one is hurt, and the structure is still standing." I got Beauchamp Smith out of bed in York and reported the accident.

This is what had happened. In 1942, after a period of operation, the blade skins had begun to break near the roots. It was necessary to reinforce the blades at the root section. But the blade shank spar connection was already a source of anxiety since it was known to be a very highly stressed member, and stiffening the blade skin near the root would throw still greater stresses into the spar section at its weakest point. It was a Hobson's choice, and the decision was to carry out the modifications in the field. Otherwise it would have been necessary to discontinue the field test program until such time as postwar blades could be fabricated, delivered, and erected; and that choice, as it turned out, would have meant a delay of some four years. Subsequent study has shown that the field welding associated with these repairs to the blades had, as had been feared, in effect cut deep notches in the blade spars in the very zone where they were known to be most highly stressed. Stress concentrations, resulting from these inadvertent notches, had caused the spars to crack progressively. But the cracks had occurred just behind a bulkhead which made inspection impossible. At 3:10 on the morning of March 26 the turbine

Fig. 2. The blade that failed.

was operating in a smooth, steady, southwest wind of about 25 miles per hour. The spars now had hidden cracks across more than 90 percent of the cross-sectional area. The tension of centrifugal force, amounting to several hundred thousand pounds, was being withstood by only a few square inches of Cor-Ten steel.

Harold Perry, who had been the erection foreman, and was a powerful man, was on duty aloft. Suddenly he found himself on his face on the floor, jammed against one wall of the control room. He got to his knees and was straightening up to start for the control panel, when he was again thrown to the floor. He collected himself, got off the floor, hurled his solid 225 pounds over the rotating 24-inch main shaft, reached the controls, and brought the unit to a full stop in about 10 seconds by rapidly feathering what was found to be the remaining blade of the turbine. He estimates that it took him about 5 seconds to get to the controls after the first shock.

One of the 8-ton blades had let go when in about the 7 o'clock* position, and had been tossed 750 feet, where it landed on its tip (Fig. 2).

Following the failure, Beauchamp Smith asked me to review the entire project, and to estimate the future of large-scale wind-power.

These studies were carried out by Carl J. Wilcox and Stanton Dornbirer, in collaboration with Jackson and Moreland, and in consultation with George A. Jessop, John B. Wilbur, and Myle J. Holley.

---

*Subsequently calculated by von Kármán.

If the test unit, modified and "cleaned-up," but not redesigned for production, had been put into limited production in the fall of 1945, a block of 6 could have been installed on Lincoln Ridge (4000 feet), in Vermont, at a cost of about $190 per kilowatt. However, the worth of this block of 9000 kilowatts to the Central Vermont Public Service Corporation, as estimated by Jackson and Moreland, but not confirmed by the Corporation, was about $125 per kilowatt.

Several means for bridging the gap between $190 and $125 were suggested and are discussed in Chapter 12. It would cost several hundred thousand dollars to test these suggestions, however, and S. Morgan Smith Company found that the scientific research and development already carried out had cost over a million and a quarter dollars—far more than had been originally estimated by me or them. Being a small company, the Directors felt it would be imprudent to lay out further substantial funds, with no reasonable guarantee of economic success in the end. Accordingly, in November, 1945, they decided, with the concurrence of Wilbur and Jessop, to abandon the project.

In the wake of this decision, Beauchamp E. Smith and Burwell B. Smith asked me to write this account of how a small company undertook a risky and expensive exploration beyond the frontiers of knowledge—of how they succeeded technically but could not afford to find out if they had succeeded economically. The Smiths wish to make available to the world the knowledge that has been gained, in the hope that someone else, perhaps in windy Scotland, or Ireland, or New Zealand, or southern Chile, or any place where winds are strong and fuels scarce, may see fit to continue from where, regretfully, they have been forced to halt.

2
How Does Wind Behave?
Our Assumptions of 1939

INTRODUCTION

The density of the air and the average speed of the wind are low compared with other commercial propelling fluids. The kinetic energy of a unit volume of moving air is less than that of falling water or high pressure steam, for example. Accordingly, the projected area of turbine blades to be driven by the wind must be greater, per kilowatt of output, than in the case of a hydraulic turbine or a steam turbine. The structure to support this great turbine, when designed to withstand ice and wind storms, runs to great weight, and therefore to high installed cost, per kilowatt. For example, the test unit of the Smith-Putnam Wind-Turbine weighed about 500 pounds per kilowatt.

Thus the low density and speed of the wind are handicaps which stand in the way of its use as an economical prime mover. It is not feasible in this case to improve the density by supercharging, but nature can be made to assist in the matter of the speed. There are mountain ridges which act like airfoils, and which speed up the normal flow of the wind. Other types of mountain retard the flow.

An intimate knowledge of the habits of wind-flow will permit one to select a site for a turbine where the free-air speed has been increased by 20 percent or more. The power in the wind varies as the cube of the speed. For example, the power in a 20-mile wind is 8 times greater than the power in a 10-mile wind. Obviously, it is important to find a site which speeds up the wind by a factor of, say, 1.20, rather than one which may retard it.

The compression of the streamlines over a good airfoil tends to damp out turbulence and provides a further incentive to learn in accurate detail the habits of wind-flow, particularly in mountainous country.

Such accurate and detailed knowledge is necessary if the designers are to select a site where the handicaps inherent in this slow-moving and tenuous propelling medium are reduced so far as may be, and if they are to develop a design which can work on advantageous terms in economic harness with water-power and steam stations.

It will be seen shortly how far we were in 1939 from possessing a knowledge of the wind adequate for site selection, or for the economical design and operation of a large wind-turbine.

MAJOR CIRCULATIONS OF THE WIND

The sun is the ultimate source of the energy in the wind. The atmosphere enveloping the earth is a rotating, regenerative thermal engine, stoked by radiant energy from the sun. The complex dynamics of this system result in the types of major circulation shown in Figs. 3 though 6, comprising the doldrums at the equator, the northeast trade winds in the tropics, the calms of the horse latitudes, and the prevailing westerlies of the "roaring forties." This information, supplemented by 50

Fig. 3. Prevailing winds over the oceans, January-February, after W. Köppen. Width of arrow indicates strength of wind.

------>   Less than 10 miles an hour.
———>   From 10 to 15 miles an hour.
———➤   From 15 to 30 miles an hour.
━━━➤   Over 30 miles an hour.

Length of arrow indicates steadiness of wind.

years of U.S. Weather Bureau records from ships at sea, has been analyzed by Petterssen (Ref. 1-B), as set out in Table 1 and summarized in Fig. 7, the wind speeds being converted to potential power outputs at various oceanic sites.

Where land masses are absent, as in the Antarctic seas, the wind circulation is strong and fairly uniform. But, where land masses are prevalent, the atmospheric envelope is called upon to act as a heat exchanger between the thermally nearly static oceanic areas and the alternately winter-chilled and summer-heated land areas,

Fig. 4. Prevailing winds over the oceans, January-February, after W. Köppen. Width of arrow indicates strength of wind.

- - - - →   Less than 10 miles an hour.
————→   From 10 to 15 miles an hour.
————▶   From 15 to 30 miles an hour.
━━━━▶   Over 30 miles an hour.

Length of arrow indicates steadiness of wind.

producing a more erratic general circulation, which is further and profoundly influenced by local topography.

Various continental regions are known to be windy. In general, they lie between the latitudes of 25 degrees and 60 degrees and are mountainous. These areas are listed in Table 2. Information is generally lacking on which to base reliable estimates of annual output in these regions.

Fig. 5. Prevailing winds over the oceans, July-August, after W. Köppen. Width of arrow indicates strength of wind.

    ----->   Less than 10 miles an hour.
    ——>   From 10 to 15 miles an hour.
    ——▸   From 15 to 30 miles an hour.
    ━━▶   Over 30 miles an hour.

Length of arrow indicates steadiness of wind.

LIMITS OF OUR KNOWLEDGE OF AVAILABLE WIND-POWER, IN 1939

This gross survey of oceanic sites, with some indications of the windiness of certain coast lines, represents about the limit to which in 1939 we could develop a detailed knowledge of wind-power.

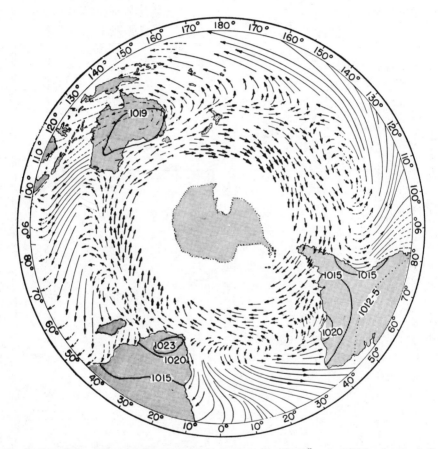

Fig. 6. Prevailing winds over the oceans, July-August, after W. Köppen. Width of arrow indicates strength of wind.

     ----→   Less than 10 miles an hour.
     ——→   From 10 to 15 miles an hour.
     ——→   From 15 to 30 miles an hour.
     ➤   Over 30 miles an hour.

Length of arrow indicates steadiness of wind.

TABLE 1. WINDY ISLANDS AND COASTS

Classified by Annual Output in Kilowatt-Hours per Kilowatt*

Region	Output in Kwh./Kw.								
	Over 6000	5000 to 6000	4500 to 5000	4000 to 4500	3500 to 4000	3000 to 3500	2500 to 3000	2000 to 2500	Less Than 2000
North Atlantic Ocean		Iceland	Faeroes Ireland Scotland Shetlands Norway	Azores Bermuda	Devonshire Wales	Cape Verde Island Martinique Porto Rico	Trinidad Aruba St. Martin	Cape Hatteras Madeira Nantucket Newfoundland Nova Scotia Santo Domingo	Bahamas Cuba Haiti Jamaica St. Martins Colon
South Atlantic	Sandwich Is. South Georgia	Falkland Is. Gough	Tristan da Cunha			Ascension St. Helena	Fern. de Noronha So. Africa	St. Paul	
North Pacific Ocean			Aleutians	Kamchatka Kuriles	Japanese Island	Formosa Palmyra		Hawaii Philippines	Guam
South Pacific Ocean	Auckland Campbell Macquarii	Bounty	Tasmania Ware-kauri	Chile New Zealand	Australia Lord Howe	Norfolk Ryukas	Kirmadec New Caledonia	Easter Fiji Gilberts Juan Fernández Luisiades Malden Marguesas Marshalls New Hebrides Pitcairn St. Felix Sala-y-Gomez Tonga	Samoa Society Is. Santa Cruz
Indian Ocean		St. Paul		Mauritius	Cocos		Madagascar	Ceylon Socotra	Andamans Seychelles Tachago

*Based on a 1250-kilowatt turbine, 175 feet in diameter. If a turbine operated continuously at 100 percent of capacity, the annual output would be 8760 kilowatt-hours per kilowatt (24 hours per day for 365 days).

1. The Free-air Wind Speed at Mountain-top Height, in Vermont

It is not possible to estimate retardation or speed-up of air-flow by a mountain top unless one knows, or can assume, a value for the undisturbed free-air speed at the elevation of the mountain top. The difficulty is that it may be impossible to find a truly undisturbed free-air speed at any point up-wind of the ridge, at the elevation of the ridge, since the free flow will be affected by the broken country lying underneath it.

Increase of Wind Speed with Height above the Ground. Even over such flat country as the level, baked-clay surface of the Great Australian Desert, the flow of the free-air is retarded by the friction of the ground. Since the frictional force decreases with elevation, the wind speed will increase as one ascends through the atmosphere.

The Height at Which the Wind Blows Unimpeded by Ground Effects—that is, the Gradient Wind Level. The rate at which speed increases with height is greatest near the ground, decreasing as one leaves the ground. It has been common practice in the literature of meteorology to assume that over flat country that rate of change of speed with elevation that is due to decreasing frictional effects will normally become negligible at a height of about 1000 meters above the ground. For practical purposes this height is known as the gradient wind level; but, over rough country, the thickness of this friction layer is normally greater. When the wind speed is high and the temperature decreases strongly with altitude, there is much turbulent mixing, and therefore good mechanical interlocking between the ground and the air mass, and the influence of surface friction will extend upward indefinitely. In these circumstances the concept of a definite height (gradient wind level), at which the wind blows unimpeded by ground effects, vanishes.

Estimating the Vertical Distribution of Speed between the Gradient Wind Level and the Ground. The wind speed at the gradient wind level can be computed from the isobars * on the daily surface weather map. Knowing the wind speed so computed at gradient wind level, and making some arbitrary allowance for the frictional effect due to a particular terrain, one can estimate the distribution of speed between the ground and the gradient wind level.

Whatever the height of the gradient wind level may be, it was assumed that the "free-air" speed just up-wind of a mountain top varied somehow with the elevation above sea level, and also with the relative elevation above the surrounding country. The true situation was found to be far more complex.

---

*Isobars are lines of equal barometric pressure.

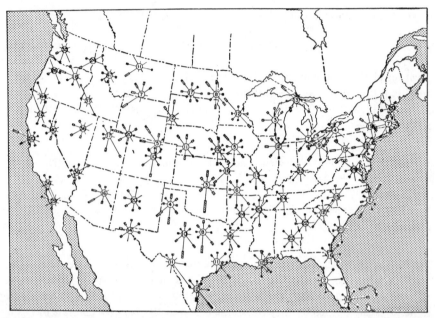

Fig. 8. Annual surface wind roses. United States Department of Commerce, Weather Bureau. The numbers in the circles refer to the percentage of calms (0-3 miles per hour). The length of the spokes indicates the steadiness of the wind.

General Lack of Suitable Wind-velocity Data. Information was not available in 1939 with which to prepare a map showing the strength and constancy of winds over the summits of potential wind-power sites in New England; nor was it possible to estimate these factors by extrapolation from U.S. Weather Bureau anemometer readings, pilot-balloon observations, or pressure maps, except in the roughest sort of first approximation.

Lacking an actual wind-energy survey, various interested persons had from time to time attempted to work up the U.S. Weather Bureau chart of surface wind velocities (Fig. 8). This chart is an accurate summary of the mean annual speeds and directions, observed at the various U.S. Weather Bureau Stations (Ref. 1-A). But isovents † passing through these points are meaningless guides to wind-power, because U.S. Weather Bureau stations are rarely located on potential wind-power sites, which are exposed high ridges lying across the prevailing wind. Weather Bureau stations are usually located either on buildings in cities or at airports. Airports are not located on mountain tops. The velocity readings at stations in cities are influenced by the turbulence and frictional drag of buildings. Readings at

---

† Lines of equal mean annual wind speed.

many stations are affected by the erection of new buildings, the growth of the cities, and the incorrect location of the anemometer in the wind shadow of sloping roofs or of other instruments. The velocity readings, therefore, are valid only for extremely localized conditions, and it is not possible, for example, to estimate the velocities at 140 feet over Mt. Abraham, in the Green Mountains, by extrapolation from the velocities recorded by the Weather Bureau in the city of Albany, 110 miles away.

The Wind Velocity Data Available. The extent of our knowledge concerning the particular wind regime prevailing over the mountains of New England was limited to anemometry from two mountain stations operated under the direction of C. F. Brooks—Mount Washington Observatory (6288 feet) in northern New Hampshire, and Blue Hill Observatory (635 feet), 10 miles south of Boston; and to upper air data from the only two pilot-balloon stations in New England—Boston and Burlington. These sets of data, combined with certain assumptions, had been used to develop a point of departure, as follows.

Our First Assumptions about Wind in the Mountains of New England. We established the respective wind speeds at anemometer height above the summits of Mount Washington and Blue Hill by taking the average speed over a 5-year period of anemometry at each station. We estimated the free-air speed at the same elevations from the data of Boston pilot-balloon observations and pressure maps, and, to establish an end point, by anemometry records of East Boston Airport (35 feet). This work was carried out by C. F. Brooks. In Fig. 9 the mean wind speeds thus obtained are plotted against elevation above sea level.

Brooks concluded that the speed-up factor at Blue Hill was close to unity (1.0), that is, that the wind passing over Blue Hill at anemometer height had been neither speeded up nor retarded. Accordingly, a smooth curve was drawn connecting the three "fixes" of the free-air speed. This (lower) curve of Fig. 9 represents the assumed variation of the free-air speed with elevation above sea level in New England. Mountain-top wind speeds falling on this lower curve would correspond to speed-up factors of unity (no speed-up or retardation). Points falling below this curve would characterize the tops of mountains whose masses had retarded the free-air, and where the speed-up factors were therefore less than unity. Points falling above this line would characterize mountain tops which had speeded up the wind (speed-up factor greater than unity).

Now, Brooks had found on Mt. Washington that the speed-up factor increased as the free-air speed increased. So we assumed that, for aerodynamically identical mountain crests, the speed-up factor would vary somehow with wind speed and, accordingly, a crude limiting envelope was suggested by drawing the smooth upper curve, which tended to converge with the lower at the lower wind speeds. Our assumption was that, in any 100 square miles of the Green Mountains, we would not

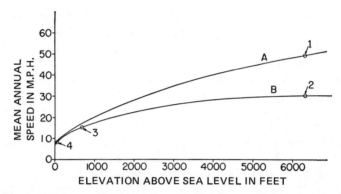

Fig. 9. Assumed variation of mean annual wind speed with variation in elevation above sea level, in New England.

Curve A represents speeds as accelerated by mountain-tops the example being the value reported in 1935 for Mt. Washington (6300 feet), of about 49 miles an hour, since found to be high by about 5 miles an hour.

Curve B represents free-air speeds.

Point 1   Long-term mean speed as reported in 1935 by Mt. Washington Observatory.

Point 2   Free-air mean speed at elevation of Mt. Washington, computed by Brooks.

Point 3   Long-term mean speed at Blue Hill Observatory, reported by Brooks in 1935.

Point 4   Long-term mean speed at East Boston Airport (sea level), reported by Brooks in 1935.

fail to find some high ridge, the wind speed on whose summit would fall between these two curves. Specifically, we estimated that we could find such ridges at 4000 feet with speed-up factors of at least 1.20. This we have probably done (Lincoln Ridge).

The Uncertainty about the Burlington Data. The analysis of the Boston and Burlington pilot-balloon observations, that led to the free-air speeds entered in Fig. 9, presents a vivid picture of the difficulties and uncertainties encountered in using such observations for our purpose.

The accumulated data from pilot-balloon observations over Boston showed the expected continual increase of wind speed with elevation. The pilot-balloon observations over Buffalo, which were used as a check, showed essentially the same characteristic. We realized that pilot-balloon measurements are selective measurements. Before the advent of radar, determination of the wind by pilot-balloons eliminated all data with cases of weather in which the balloons disappeared in fog or clouds. Such weather, more often than not, is windy. Thus, windy data are eliminated from the records, more and more as the elevation increases. It was,

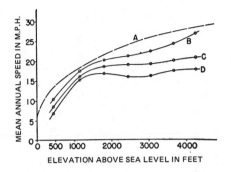

Fig. 10. Actual variation of mean annual wind speed with variation in elevation above sea level as recorded by United States Weather Bureau pilot-balloon data at Burlington, Vermont.

Curve A represents the original assumption (Curve B of Fig. 9).

Curve B shows the actual distribution in winter afternoons.

Curve C shows the mean distribution for the year.

Curve D shows the actual distribution for summer mornings.

therefore, felt that the upper-air wind speeds indicated by pilot-balloon material were in general lower than the actual and could be used safely in our speed estimates.

The Burlington pilot-balloon data, however, indicated a distribution of speed with elevation quite different from that found over Boston. Instead of a steady increase of speed with elevation, the Burlington data indicated that the wind speed increased up to about 1500 feet and then remained almost constant up to about 3000 feet (Fig. 10). By comparison with Boston it was obvious that the Burlington pilot-balloons represented a particular local condition. But how should it be interpreted? Was it characteristic of wind speeds over foothills up-wind of mountainous terrain or was the mountainous terrain responsible for a systematic error of measurement?

Now an inherent weakness of the standard pilot-balloon method * is that the speed data become unreliable whenever the balloon rises through air with strong vertical motion, in the sense that the indicated wind speed is too low when the balloon rises faster than usual. There obviously exists a field of up-currents to the windward of the Green Mountains. Pilot balloons released at Burlington into the prevailing westerly wind would float directly into this up-current field. In this manner the wind speeds indicated by Burlington pilot balloons would become too low after a few minutes, that is, after the balloon had risen to a few thousand feet.

The Interpretation on Which the Project was Based. The consensus of the consulting meteorologists was that this systematic error in the pilot-balloon evaluations at Burlington was responsible for part, if not all, of the peculiar characteristics of the Burlington wind curve. It was, therefore, decided to base the free-air speeds of Fig. 9 on the Boston pilot-balloon data alone.

The curves of Fig. 9 were used as the working hypothesis by which we estimated that the mean annual speed at the 2000-foot test site, Grandpa's Knob, would be 24 miles an hour. It proved to be about 17 miles an hour.

---

*Following the balloons with two theodolites, one at each end of a base line.

It was not until 1945 that Lange, in view of the unexpectedly low output at Grandpa's Knob, suggested and carried out a new and detailed analysis of the Burlington pilot-balloon material, which showed that the Green Mountain Range actually does exert its influence on the wind far out to windward, over the foothills. Lange's study is summarized in Chapter 4, and explains in part the failure to obtain the originally predicted wind speeds on Grandpa's Knob—a failure which resulted in an output there amounting to only 27 percent of what we had predicted, based on Fig. 9.

2. The Effect of the Geometry of a Mountain upon the Retardation or Speed-up of the Wind-Flow over Its Summit

Prior work, especially in gliding and in certain wind-tunnel programs, indicated approaches to this problem.

It was assumed that a north-south * ridge with a suitable profile would speed up the flow of free-air over its summit, by a factor of at least 1.20. (The factor at Mount Washington was thought to be in excess of 1.50.)

Consider a ridge of uniform cross-section: If the ridge does not lie parallel to the wind, it will act as a baffle and tend to deflect the wind.

As the angle increases, the effect on the wind direction increases. When the angle becomes 90 degrees and the ridge lies athwart the wind, like a dam, some of the wind will flow over the ridge, but some will stream around either end. The distribution will depend in part on how much more work is required to deflect a parcel of wind vertically than to deflect it horizontally. And this depends on the thermodynamic characteristics of the air mass. This reasoning led us to suppose that such a ridge as the Green Mountains (160 miles long from north to south) would cause significant differences in the prevailing wind direction at the northern and southern ends. As stated in Chapter 4, the effects actually found were larger than anticipated, and of great importance in site selection.

It was assumed that wind flowing over a ridge would tend to behave as shown schematically in Fig. 11. But how critical for a wind-turbine is the zone of rupture on the lee side? No answer was available. This was something we had to determine.

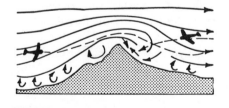

Fig. 11. Assumed behavior of wind flowing over a ridge. From Petterssen's "Introduction to Meteorology," Fig. 78, page 118.

*We assumed a prevailing west wind in Vermont.

Von Kármán showed that, on ridges with steep western faces (Fig. 12), the speed-up of the wind over the summit would be greater when the ridge is high compared to the east-west width of the base. This is summarized in Table 3 and Fig. 13.

He thought that the maximum value of the wind speed would be found just at the crest of the ridge, but it was not certain that the turbulent flow found

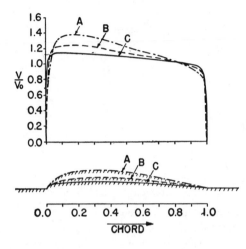

Fig. 12. Distribution of speed, expressed as a percentage of the free-air speed, over three airfoils. After von Kármán.

The maximum thicknesses of the three air-foils expressed as percentages of the respective chord lengths are:

Airfoil	Thickness
A	12.0%
B	7.5%
C	4.5%

TABLE 3. EFFECT OF THE GEOMETRY OF A RIDGE UPON THE ACCELERATION OF WIND FLOW

Height of ridge as per cent of east-west width of base of ridge	$\dfrac{V_{\text{accelerated}}}{V_{\text{free-air}}}$
4.5%	1.15
7.5%	1.23
12.0%	1.37
17.5%	1.6

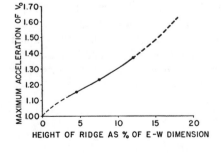

Fig. 13. Variation in the maximum acceleration of the wind-flow over a ridge, with variation in the profile of the ridge.

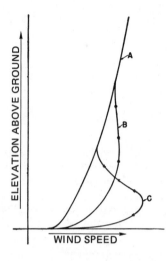

ELEVATION ABOVE GROUND

WIND SPEED

Fig. 14. Assumed speed distribution above different types of terrain.

Curve A   Flat country.
Curve B   A rounded ridge, corresponding roughly to Curve A in Fig. 12.
Curve C   A sharp ridge.

in nature over wooded and broken ridges would conform to the regular flow over smooth models that had of necessity been the basis of his estimates.

We had no reliable information, based on sustained measurements, of the way in which speed varied with height over a ridge. We did not know the height at which the maximum speed would be found, nor the variation, if any, of this height with speed. We thought that a sharp ridge would produce a high value of the maximum speed at a low elevation; while a less sharp ridge would produce a lower value of the maximum speed at a higher elevation. These assumptions are shown schematically by the dimensionless curves of Fig. 14.

In sum, we felt that the higher and sharper the ridge, the greater was likely to be the acceleration of the wind-flow over the summit; but we also felt that such a ridge would induce heavy and rapidly shifting turbulence on the lee side, which might even catch a turbine aback and wreck it. We lacked criteria for evaluating these and other factors and for making a rigorous selection of a wind-turbine site on a ridge.

As will be described, these qualitative assumptions have been confirmed; however, our inability in 1939 to make quantitative interpretations of the effect of topography upon wind-flow has also been confirmed by six years of experience. We selected as the test site a ridge which we though possessed a speed-up or acceleration factor of at least 1.2. Our records indicate that this ridge apparently retarded the flow, giving a speed-up factor of only about 0.9. We have found no criteria by which to make an economically useful quantitative prediction of the effects of topography upon wind-flow.

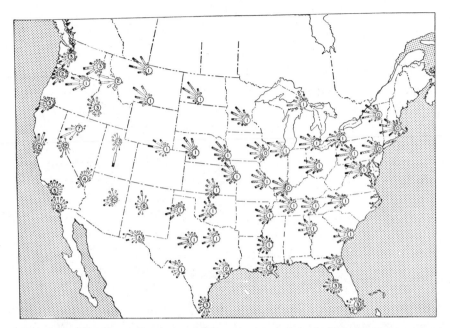

Fig. 15. Upper-air wind roses for December, January, February, 2000 meters above sea level, United States Department of Commerce Weather Bureau. Numbers in circles refer to the percentage of calm (0-3 meters per second). The length of the spokes indicates the steadiness of the wind.

3. Prevailing Wind Direction at 2000 Feet in the Western Foothills of the Green Mountains

Upper-air data indicated a prevailing wind direction of W by N in New England at 2000 meters (Fig. 15). This was confirmed at Mount Washington at 6300 feet (Fig. 16A), and in the free-air over Burlington at a little over 4000 feet (Fig. 16B). It was therefore assumed that the prevailing direction at Grandpa's Knob at 2000 feet would be about W by S, being shifted about 20 degrees to the left from the direction observed at Burlington at 4000 feet, in conformity with the Ekman spiral (Fig. 17). The error in this assumption, amounting to some 28 degrees (Chapter 4), presumably contributed to the error in the predicted output at Grandpa's Knob.

As will be described in more detail in Chapter 4, these first three assumptions of 1939 were in error and, although we have since located a 4000-foot ridge in Vermont where the originally predicted output of 4400 kilowatt-hours per kilowatt per year could probably be obtained, at the 2000-foot test site we averaged only 1200 kilowatt-hours per kilowatt per year, or 27 percent of the output originally predicted for this elevation.

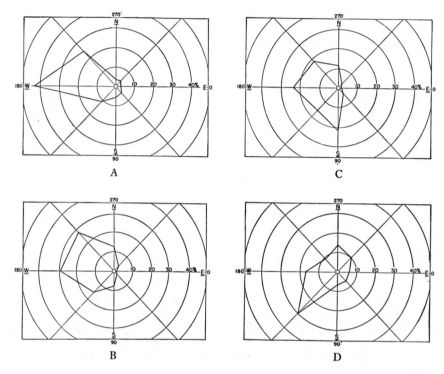

Fig. 16. Directional distribution of wind in New England. Five year averages.

A–Mt. Washington anemoscope,	6300 feet
B–Burlington pilot-balloon data,	4270 feet
C–Burlington pilot-balloon data,	2440 feet
D–Grandpa's Knob aneomoscope,	2000 feet

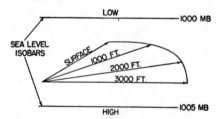

Fig. 17. Directional shift of wind in the Ekman spiral. From Petterssen's "Introduction to Meteorology," Fig. 67, page 108.

4. Influence of the Structure of the Wind on Design

We assumed that, if the aerodynamics of a crest were such as to accelerate the wind-flow, the work done in compressing the streamlines would tend to damp out turbulence. Conversely, if the crest retarded the wind-flow, turbulence might be increased. We thought we were safe in designing the test unit, which was to be located on a site with an acceleration factor of not less than 1.20, according to the gustiness factors established over fairly flat country, as at Lakehurst, N.J., and later confirmed and amplified by Sherlock (as noted by Petterssen in Ref. 2-A). Our conclusions, based on this work, were that on well-exposed, fairly flat sites, the net speed integrated over a disc area of 175 feet in diameter might change at the rate of 50 percent each second, throughout an interval of 1.6 seconds, while the net direction might simultaneously change at the rate of 90 degrees in 1.0 seconds (Ref. 3-B). As will be explained, our experience with these two assumptions was inconclusive, but we have found no reason to question them.

5. Influence of the Structure of the Wind on Estimates of Output

The power in the wind varies as the cube of the wind speed (Chapter 5). An anemometer records speed. Therefore, to compute potential turbine outputs from anemometer records, it is necessary to deal with the cube of the values recorded by the anemometer. But anemometer records are frequently given in terms of total miles of wind per hour, and the wind does not blow steadily at this average value throughout the 60 minutes. It will, on the contrary, have been fluctuating, as shown by the gustiness patterns reported by Sherlock. The cube root of the mean of a series of cubed terms is greater than the mean of the series. For example, the cube root of the mean of the cubes of the series 20.0, 30.0, 40.0, is 32.1, while the mean of the series is 30.0. The ratio 32.1/30.0 we have called the cube factor. It is always greater than 1.0.

In Fig. 18 five examples of the cube factor are plotted against the logarithm of the time range of 6 seconds to 1 hour. They were derived from records taken at Blue Hill, selected as representative by C. F. Brooks. The range of mean hourly wind speeds covered is from 17 miles an hour to 47 miles an hour. By extrapolation of the arbitrary envelopes it is seen that the annual average cube factor in the range of 1 second to 1 hour lies within the range of 1.02 to 1.14, perhaps not far from 1.08, while the average for the range of 0.1 second to 1 hour is probably a little greater, perhaps 1.10.

We did not know what minimum unit of time was characteristic of the response of the wind-turbine, and accordingly we ignored the cube factor, treating it as a safety factor in estimating outputs from mean hourly wind speeds.

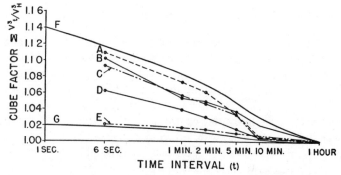

Fig. 18. The cube factor. From 5 hourly records obtained at Blue Hill Observatory, selected as representative by C. F. Brooks.

	Average Hourly Speed	*Stability Conditions*
Curve A	Speed 29.0 miles per hour	Very unstable
Curve B	Speed 18.7 miles per hour	Unstable
Curve C	Speed 39.1 miles per hour	Unstable
Curve D	Speed 47.0 miles per hour	Unstable
Curve E	Speed 17.8 miles per hour	Stable

6. Influence of Atmospheric Density on Estimates of Output

Output in kilowatts varies directly with the density (Chapter 5), which in turn varies with the temperature and the elevation above sea level. The average sea-level density in New England in January is more than 10 percent higher than in July. To obtain a value for the mean annual density at sea level, the average for each month was computed from the 5-year average temperature for each month, and from these monthly densities the annual average was computed by weighting the density for each month according to the computed output for that month (Ref. 4-B).

Density also varies with elevation (Fig. 19), the annual average value in New England at 10,000 feet being nearly 30 percent less than at sea level (Ref. 5-B).

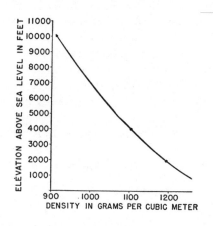

Fig. 19. Variation of atmospheric density with elevation. Annual values weighted for mean monthly temperatures in New England.

From these considerations, we estimated that the average annual weighted density at Grandpa's (2000 feet) would be 1195 grams per cubic meter. This was a good estimate. The actual weighting, based on our records, yields a 5-year average of 1201 grams per cubic meter, or half a percent higher.

7. Influence of Estimates of Icing on Design and on Site Selection

Brooks surveyed available data over a 35-year period and prepared charts showing how the maximum icing would vary with latitude and elevation above sea level (Figs. 20 and 21) (Ref. 6-B).

Brooks recognized that the deposit of an "ice storm" ranges in habit and density from light coatings of hoarfrost through rime and porous ice to clear, solid ice. Using the last as a standard, he showed that in Vermont the maximum thickness of solid 56-pound ice which might accumulate on a stationary structure, increased rapidly with elevation above sea level, from 5 inches at 2000 feet to 12 inches at 4000 feet.

No authority consulted would predict how ice would form on our stainless-steel blades, under conditions of maximum icing. We took the view that the test unit must not fail structurally; that if it did, the project would die.

This philosophy, in conjunction with our fears of heavy ice, became a strong incentive to create a design with much excess strength and to locate the test unit on the lowest site possessing wind adequate for the test program.

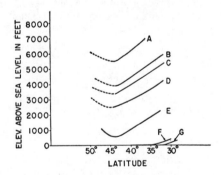

Fig. 20. The maximum thickness of 35-pound ice to be expected on a stationary structure in various latitudes in U. S. A.

Curve A	36.0 inches
Curve B	18.0 inches
Curve C	13.5 inches
Curve D	9.0 inches
Curve E	4.5 inches
Curve F	2.5 inches
Curve G	1.0 inches

Fig. 21. The maximum thickness of ice to be expected on a stationary structure in New England. Derived from Fig. 20 by Brooks.

Curve A	Latitude 58° 00'
Curve B	Latitude 43° 45'
Curve C	Latitude 47° 00'

FIRST ATTEMPTS AT ANSWERING THE QUESTIONS AND
TESTING THE ASSUMPTIONS

Question 1. Of the two practical questions to which we sought answers, the first, being the selection of the test site, was the most pressing. Certain requirements had been set up. The site should have a clear sweep up-wind (thought to be W by S), and a clean exit or tailrace to the east. It should have a bold and rounded profile. It should be located near the load center of the Central Vermont Public Service Corporation. It could not be located within the boundaries of the Green Mountain National Forest, and, most important of all, in order to avoid heavy icing, it should not be more than 2000 feet above sea level (Figs. 20 and 21).

It was necessary to select the test site in the spring of 1940. During the preceding winter, we had made map studies of Vermont, and, using certain criteria since found to have been unreliable, we had identified some 50 summits which we then though were promising wind-turbine sites. Concurrently with the map studies we had made field trips, which failed to shed much light on the problem. Following a suggestion of Brooks, S. Morgan Smith Company authorized us to invite Griggs, the ecologist, to accompany us. Griggs could find little evidence of wind below 2500 feet (Ref. 7-B) but the meteorologists concluded that the wooded summits acting as airfoils would show compressed streamlines above treetop height.

This view received some support when, during the early part of 1940, the results began to come out of the wind-tunnel at Akron (Ref. 8-B), showing that the models of Pond Mountain and Glastenbury, for example, would yield acceleration factors of 1.20, with the maximum compression of the streamlines occurring at higher elevations above the summits as the roughness increased.

Grandpa's Knob met the requirements of elevation, exposure, and profile, and was close to Rutland (12 miles), requiring only 2.2 miles of new road to connect it to a highway network. Accordingly, although the trees showed no deformation by wind and although there had not been time to put a model of Grandpa's into the wind-tunnel, it was selected as the test site in June, 1940, without benefit of anemometry.

An anemometer mast had been put up a few days before this decision was made and anemometry had begun immediately. The wind speed records throughout the summer of 1940 gave computed monthly outputs which averaged only 10 percent to 30 percent of the predicted values. Unwilling to believe that these low values were characteristic of the western foothills of the Green Mountains, Wilcox studied the trend of the reported wind speed records from Blue Hill, East Boston, and Mount Washington (Ref. 9-B), and discovered that, beginning a few months earlier, the wind speed on Mount Washington had shown a substantial drop. From this we took heart and referred to the weak wind-flow at the test site as part of a temporary but regional "anomaly." It was not until the summer of 1945, in the course of the rigorous analysis that led to the abandonment of the project by S. Morgan

Smith Company, that it was learned that the "anomaly" at Mount Washington had been caused by the application of an arbitrary correction to the anemometer records. The correction had been applied by one of the observers without notification to the users of the published data. It is quite likely that we have this observer to thank for the Smith-Putnam Wind-Turbine experiment. If it had been known at the end of 1940 that not only was there no anomaly, but also little wind at those elevations below which we did not fear ice, it is likely that the experiment would have been abandoned out of hand.

With the test site selected, we were free to develop programs to answer the remaining question in a fairly orderly manner.

Question 2. The Central Vermont Public Service Corporation requested that it be supplied at an early date with a list of the best potential wind-turbine sites in Vermont, each capable of development to 20,000 kilowatts of capacity, in order that the land might be placed under option.

The very best sites were, we thought, to be found among the high ridges of the Green Mountains. As early as 1935, in a report to the General Electric Company, I had picked the 3-mile Lincoln Ridge, trending N by E to S by W, as possibly the best site in Vermont (Ref. 10-B) on the grounds that it was unusually smooth and regular, with high relative elevation, a magnificent clear sweep across the Champlain Valley 20 miles to the Andirondacks (Fig. 22), a clean tailrace down-wind and a remarkably fair profile which looked promising aerodynamically (Fig. 23). However, this and most similar sites lay in the Green Mountain National Forest. It was

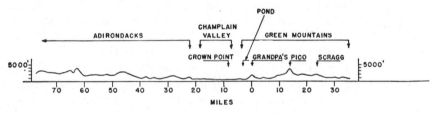

Fig. 22. East-west profile through the test site at Grandpa's Knob, 30 miles down-wind and 70 miles up-wind. Verticle scale exaggerated about 5 times.

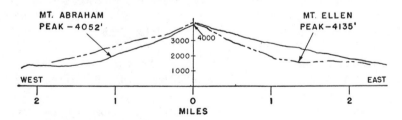

Fig. 23. West-east profiles through Mt. Abraham, the southern shoulder of Lincoln Ridge, and Mt. Ellen, the northern shoulder.

several years before permission was obtained to consider sites in the Forest and before we had learned not to fear the heavier icing to be expected at 4000 feet. Accordingly, these sites were not surveyed in 1940–1941.

Lange thought that his experience in gliding would enable him to make a preliminary selection of ridges with good aerodynamic characteristics and capable of speeding-up the free-air velocity by a factor of at least 1.20. However, he lacked quantitative criteria for selection by this method alone, and particularly he lacked knowledge of the exact elevation over the summit at which the maximum acceleration of wind flow would occur. Accordingly it was decided to check these tentative selections by testing in the wind-tunnel models of four selected mountains.

THE WIND-TUNNEL TESTS

Topographic survey sheets of Mount Washington (6288 feet), Mount Glastenbury (3840 feet), East Mountain (2200 feet) and Pond Mountain (1407 feet) were blown up by pantograph and, from the "blow-ups," wooden models were made, layer upon layer, each conforming to a contour line. The outer corner of each layer was smoothed off and the model faired.

The models, to a scale of 1 foot to the mile, were 4 feet in diameter. They were mounted vertically on a ground-board in the vertical wind-tunnel of the Guggenheim Aeronautical Institute of Akron, Ohio, where the test program was carried out by the Director, Dr. Th. Troller, under the direction of von Kármán (Ref. 8-B). The models were rotatable through 180 degrees in steps of 45 degrees each. In this way, the direction of air-flow with respect to the model could be set from the North, Northeast, East, Southeast, South, Southwest, West, and Northwest.

A tunnel speed of 85 miles an hour was chosen, in consultation with von Kármán. Some check readings were also made at 60 and 110 miles an hour.

The vertical distribution of speed over the four mountain tops was measured at each of 80, 103, 113, and 127 points, respectively, in each of six wind directions, and at each of eight elevations above each point. Each measurement was made twice, and where the results did not coincide sufficiently closely, the measurement was made a third time.

In all, some 20,000 measurements were made twice by means of a hot wire anemometer consisting of a platinum wire 0.0006 inch in diameter, stretched and soldered to two needles. The wire was 0.25 inch long, this being the scale length of the radius of the turbine blade.

The minimum scale height at which measurements were taken was 0.075 inch or 30 feet. Above about 2.5 inches or 5000 feet, measurements showed normal velocity.

Measurements were made with the hot wire in the horizontal plane, as it was found that the apparent air-flow was not far from being normal to the length of the hot wire.

Artificial roughness was built up by cementing, with lacquer, quartz sand grains 3/32 inch in diameter, to strips of scotch tape, which were then fastened to the smooth surface of the model. It was assumed that quartz grains of this diameter would simulate full-scale forests of 40-foot trees. Coarse grains and cotton were used to simulate trees up to 100 feet in height. To simulate progressive detimbering of a site, successive strips of scotch tape were peeled off.

In subsequent tests small tufts of thread, at a height of 0.25 inch above the model's surface, corresponding to about 125 feet above the surface in nature, were secured with sealing wax to pins which had been driven into the model. The tufts, photographed in each of five directions of wind, gave some indication of the effect of topography on local flow.

The wind-tunnel tests did not reproduce findings already made in full-scale. For example, the speed-up factor measured on Mount Washington full-scale was about 1.50. But the wind-tunnel showed only 1.30 on Mount Washington (Fig. 24), while indicating 1.29 for Pond and 1.44 for Glastenbury.

Again, the balsams on Pond Mountain showed striking reverse flagging * just in the lee of the summit, indicating standing waves of turbulence in the prevailing west or southwest wind. We were not able to reproduce this effect in the wind-tunnel.

Thus we were not prepared to accept at face value the indication from the wind-tunnel that, over such bare summits as Pond and Mount Washington, the maximum speed would be found at a height of 35 feet, with a rapid decrease in speed above this height, in the case of certain wind directions. Such a distribution of speed would, of course, be disadvantageous to the operation of a great wind-turbine, whose disc would extend from 40 feet to 225 feet above the ground.

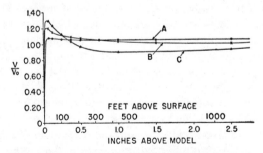

Fig. 24. The speed-up of the wind-flow at points above the summit of the model of Mt. Washington as determined in the wind-tunnel at Akron by Troller.

Curve A	Northwest Wind
Curve B	West Wind
Curve C	Southwest Wind

*Flagging is described on p. 54.

We concluded that there were two factors which made it impossible to explore for wind-power sites by means of scale models in wind-tunnels.

The first is compressibility. In the wind-tunnel, the compressibility of the air is without importance, but in the flow of air across mountains it must be taken into account, especially in connection with the stability of atmospheric stratification. It is true that the effects of compressibility can be estimated qualitatively by the principle of dynamic similarity. But our consultants came to doubt whether extrapolation to full-scale in nature could be made with quantitative certainty.

The second is that, with the exception of mountains like Mount Washington, which are high relative to the country round about, we are probably uncertain of the true free-air speed in nature. This uncertainty makes it impossible to relate with confidence a speed-up factor measured in the wind-tunnel to the factor to be found over the actual summit in question.

Because of these uncertainties we next attempted to measure the height and value of the maximum wind speed in full-scale.

We tried two methods. First, we added 30-foot extensions to the 80-foot anemometer masts on those mountains whose models were in the wind-tunnel. Unfortunately those mountains were well timbered and, although the masts extended to 110 feet, they gave us a net anemometer height above tree top of only 60 to 80 feet. Readings showed us that the height of the maximum wind speed was above this level. This result had been anticipated, and we immediately tried the second method, floating balloons.

We undertook to float a sufficient number of balloons over the summit of Pond Mountain (whose model was in the tunnel) to establish the vertical distribution of speed over the summit. This speed distribution was to be compared with the various speed distributions obtained in the wind-tunnel, at the various roughnesses of the model, in order to find that scale of roughness which corresponded to nature.

The method failed. It proved impossible in the time available to fly enough balloons exactly over the point on the crest to establish the vertical speed distribution.

A secondary objective of the floating-balloon runs was to correlate the rupture of the stream flow with the geometry of the profile. We wished to know how steep a ridge could be without producing turbulence near the summit which would interfere with the flow through the turbine disc. For this purpose, we tried to float balloons over Pond Mountain, Biddie Ridge, and other sites. We failed in this objective also. Most of the runs indicated turbulence in the lee, but this frequently caused the balloon to dip down out of sight.

THE NECESSITY FOR SPECIAL WIND RESEARCH PROGRAMS

It had now become clear that theory was inadequate and that other methods had failed either to evaluate design decisions already made or to guide us in selecting sites for ultimate development. Accordingly, the consultants recommended to

Beauchamp Smith that special wind-research programs had become essential. Smith concurred, and the coordinated meteorological and ecological studies described in Chapter 3 were launched. The findings are summarized in Chapter 4.

SUMMARY

1. A preliminary world survey of windy regions is summarized in Tables 1 and 2.

2. We lacked general wind velocity data for the mountains of New England.

3. In order to launch the Project, our consultants in 1939-1940 made working assumptions, concerning:

   a. The free-air wind speed at mountain-top height in Vermont.

   b. The effect of the geometry of a mountain upon the retardation or speed-up of wind-flow over its summit.

   c. Prevailing wind directions in the western foothills of the Green Mountains.

   d. Influence of the structure of the wind on design.

   e. Influence of the structure of the wind on estimates of output.

   f. Influence of the atmospheric density on estimates of output.

   g. Influence of estimates of icing on design and on site selection.

4. In 1940, it became necessary to select Grandpa's Knob as the test site in the light of these assumptions, and before they could be tested by field work.

5. Attempts were made to test these assumptions by wind-tunnel tests of models of mountains and by floating-balloon runs. The tests failed.

6. To learn what we had to know about the wind-flow in mountainous country, we decided to launch special wind research programs, coordinating meteorology with ecology, as described in Chapter 3.

3
The Special Wind-Research Programs, 1940-1945

INTRODUCTION

In Chapter 2 the state of our knowledge concerning the habit of wind in mountainous country was found to be meager and uncertain. The working assumptions we had had to make in 1939 were described, and it was explained why we were uneasy about most of the important ones, how the first attempts to test them, by wind-tunnel and by floating balloon, had failed, and why we felt it necessary to launch special wind-research programs, in which we would attempt to correlate meteorological and ecological measurements.

The meteorology was placed in charge of Petterssen, assisted by Lange, with Rossby, Brooks, and von Kármán in consultation. The ecological correlation was carried out by Griggs, with collaboration by myself, and both programs were at first placed under my direction, since Wilbur, as Chief Engineer of the Project, was fully occupied with design problems until after the erection of the test unit was well under way.

THE METEOROLOGICAL PROGRAM

It was decided to take speed measurements on potential sites both down-wind of the main mass of the Green Mountains and also up-wind. In order to investigate extremes, measurements were also to be taken both in the Champlain Valley at Crown Point, on Lake Champlain, and on certain 4000-foot summits in the main range of the Green Mountains. These geographical relationships are shown, partly schematically, on the profile of Fig. 22. Finally, in his report of March, 1940, Lange (Ref. 11-B) suggested equipping a number of unsuitable sites with recording anemometers. This was in a further attempt to explore the relationship between the geometry of the profile of a site and its effect upon the wind-flow.

As regards all these cases, Lange suggested that the speed-up factor could be determined by comparing the wind speed at the site, as measured by anemometer, with the gradient speed, as computed from the isobars on the daily weather maps.

In response to these various suggestions, anemometers were installed as shown in Table 4, at sites shown in Figs. 22 and 25, and selected for the various reasons itemized in Table 5.

TABLE 4. THE SMITH-PUTNAM EXPERIMENTAL ANEMOMETER STATIONS

Site	Anemometer Elevations			Period of Operation		Days of Operation
	Above Ground	Above Trees	Above Sea Level	From	To	
Pico Peak	40	10	4007	Jan. 31, 1941	March 24, 1941	52
	80	50	4047	July 2, 1940	April 30, 1941	301
	110	80	4077	Jan. 31, 1941	March 24, 1941	52
Glastenbury	80	40	3840	March 2, 1941	Sept. 30, 1941	212
Scrag	53	45	2633	May 31, 1940	Oct. 10, 1940	133
	66	38	2646	May 31, 1940	Oct. 10, 1940	133
	78	70	2658	May 31, 1940	Dec. 1, 1940	185
	108	100	2688	May 31, 1940	Oct. 10, 1940	133
Herrick	80	80	2640	June 1, 1940	March 20, 1941	292
Chittenden	80	30	2430	April 25, 1940	Nov. 30, 1940	219
East Mountain	80	40	2200	May 17, 1940	Aug. 10, 1940	85
Seward	59	13	2139	July 11, 1940	Feb. 2, 1941	206
	70	24	2150	July 11, 1940	Feb. 2, 1941	206
	80	34	2160	July 11, 1940	Feb. 2, 1941	206
	110	64	2190	July 11, 1940	Dec. 30, 1940	172
Biddie Knob Proper	75	61	2085	July 14, 1940	Dec. 1, 1940	140
Grandpa's Knob	50	50	2040	June 9, 1940	June 20, 1941	376
	64	64	2154	June 9, 1940	May 20, 1941	345
	80	80	2070	June 9, 1940	June 20, 1941	376
	110	110	2100	June 9, 1940	Dec. 18, 1940	192
	40	40	2030	Apr. 11, 1941	Dec. 31, 1945	1724
	80	80	2070	Apr. 10, 1941	Dec. 1, 1941	235
	120	120	2110	Apr. 1, 1941	Dec. 31, 1945	1731
	150	150	2140	Apr. 3, 1941	Dec. 1, 1941	242
	185	185	2175	Apr. 2, 1941	Dec. 31, 1945	1733
Biddie Knob I	50	4	1975	July 13, 1940	Dec. 31, 1940	171
	64	18	1989	July 13, 1940	Dec. 31, 1940	171
	80	34	2005	May 3, 1940	Dec. 31, 1940	242
	110	64	2035	July 13, 1940	Dec. 31, 1940	171
Moose Horn	76	46	1921	March 30, 1940	June 19, 1940	81
Middle	76	32	1836	Apr. 18, 1940	June 15, 1940	58
Pond	47	3	1407	Apr. 26, 1940	May 18, 1940	22
	64	20	1424	Apr. 26, 1940	May 18, 1940	22
	80	36	1440	Apr. 9, 1940	May 25, 1940	46
	110	66	1510	Apr. 26, 1940	May 18, 1940	22
Crown Point	151	—	240	Jan. 1, 1941	March 1, 1941	60

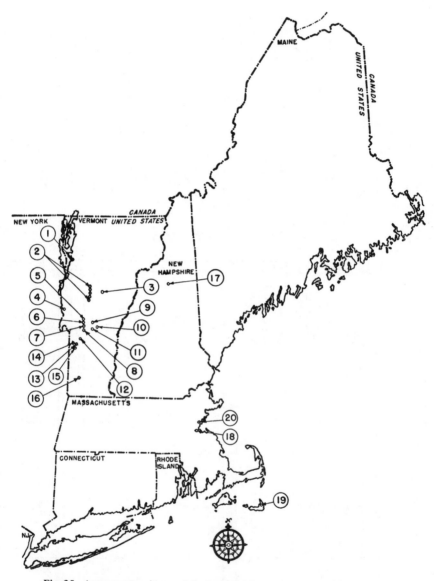

Fig. 25. Anemometer sites used in the project. See Table 4 for description.

1. Burlington
2. Lincoln Ridge
3. Scrag Mt.
4. Crown Point
5. Biddie Knob
6. Seward
7. Biddie 1
8. Grandpa's Knob
9. Chittenden
10. Pico Peak

11. East Mountain
12. Herrick
13. Moosehorn
14. Pond
15. Middle Mt.
16. Glastenbury
17. Mt. Washington
18. Boston
19. Nantucket
20. East Boston Airport

TABLE 5. THE REASONS FOR THE SELECTION OF THE VARIOUS ANEMOMETER STATIONS

Scrag

This potential wind-turbine site was interesting because it is in the wind shadow of the Green Mountains and its summit is bare. Private boundary lines restricted the location of the anemometer to a point other than that which would have been selected for a wind-turbine.

Herrick

The profile was thought to be inferior. The summit was bare and lay to the windward of the Green Mountains.

Chittenden

The site lay in the bottom of a horse-shoe, the open ends of which were directed downward and toward the west, which might, it was thought, accelerate the windflow by a funnel-effect.

East Mountain

Geographically a desirable test site, but so close to the Green Mountains (on their windward side) that the downwind exit was obstructed. The model was tested in the wind-tunnel.

Grandpa's Knob

Had been selected as the test site.

Seward
Biddie Knob Proper
Biddie Knob

Control stations on the same ridge as Grandpa's Knob.

Moose Horn
Middle
Pond

Selected to investigate the relationship between a higher peak (Moose Horn) downwind of a lower ridge (Pond), and a gap (Middle), all up-wind of the higher Green Mountains.

Crown Point

To find an inland end point on Petterssen's curve of vertical gradient of speed from the ground to the gradient wind level.

In order to measure the vertical distribution of speed above wooded summits, three stations—Pond, Biddie, and Seward—were each equipped with three anemometers at various levels, while on Grandpa's there was erected the 185-foot Christmas Tree for the purpose of measuring the vertical distribution of speed above a bare summit, the horizontal distribution of speed, and the structure of the wind. The Christmas Tree, whose location is indicated in Fig. 26, is shown in Fig. 27.

By arrangement with the Department of Meteorology at Massachusetts Institute of Technology, special daily weather maps were prepared, from which Willett, by

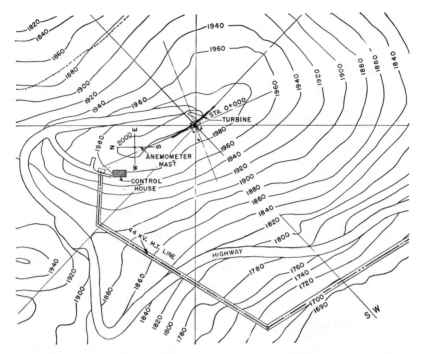

Fig. 26. The summit of the test site at Grandpa's Knob showing the access roadway, the 44-kv. high-tension line, the control house, the Christmas Tree anemometer mast, and the wind-turbine.

standard methods, computed the gradient wind for the Vermont region for three periods each day.

Two purposes were to be served by using the gradient wind. First, it was hoped that it would serve as a source of an estimate of the speed of the free-air at the elevation of each site, from which we could compute the respective speed-up factors. In this way it would serve as a reference yardstick by which the sites could be compared with each other. Second, it was hoped that predictions of gradient wind speed 30 hours in advance, with revisions 24, 18, 12, and 6 hours in advance, would make it possible to predict the output at each site, in a way which would be economically useful to a power dispatcher.

INSTRUMENTATION

The program of measuring wind speed was begun with standard unheated Robinson 2 5/8-inch, 4-cup anemometers. Too many records were lost because of ice persisting in the cups and, in the winter of 1940-1941, we designed a heated anemometer known as the rotor type (Fig. 28), based partly on similar designs developed

Fig. 27. The 185-foot Christmas tree on Grandpa's Knob.

Fig. 28. The rotor-type heated
anemometer.

in Norway and at Mt. Washington. Lacking power lines at Pico, Grandpa's, and
Glastenbury, we used gas heating in 1941. In 1943, we modified the design to
accommodate an electric heating element, and later we incorporated other improvements. Five gas-heated anemometers were built, of which three were converted
to electrical heating. After January, 1944, a gas heated and an electrically heated
anemometer were simultaneously operated at the 120-foot level at Grandpa's.

All instruments were calibrated in a 1-meter wind-tunnel at frequent intervals,
and usually showed no variation within the limits of the method (Ref. 12-B). In
these tests, it was found that the table of correction for the Robinson instruments
was essentially the standard table, which was accordingly used. In the case of the
large rotors there was some suspicion of wall interference in this small tunnel, and
accordingly each rotor instrument was calibrated by road test. The instrument was
mounted well above a truck, which, at 4:00 A.M. in a dead calm, was driven in both
directions over a 2-mile measured course. The truck was held at constant speed by
speedometer, and the actual speed was determined by a stop watch. A calibration
curve was developed for each rotor, verifying the findings in the tunnel.

In conformity with our assumptions about the cube factor (Chapter 2), it was
found that the heavy rotors tended to overrun the lighter Robinson cup anemometers. Thus, the south rotor at 120 feet gave readings that in 44 months averaged
4 percent higher than the lighter Robinson cup instrument on the west arm, at
120 feet. In Fig. 29 are plotted the ratios, north rotor/cup, and south rotor/cup,
by month respectively, for the period November, 1941–June, 1945. Because of
the interference of the Christmas Tree structure in the way of the prevailing wind,
the north rotor recorded only about 95.2 percent of the miles recorded by the
better exposed south rotor. Monthly variations in this difference of 5 percent between the two rotors are presumably due to varying directional distributions from
month to month. Variations in the ratio south rotor/cup are presumably due to

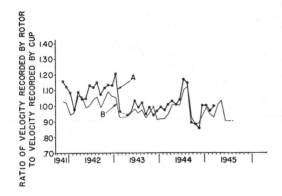

Fig. 29. The heavy rotor anemometers tend to record more miles of wind than the lighter Robinson Cup anemometers.

Curve A   The ratio between the miles of wind recorded by the North rotor and by the cup anemometer at 120 feet on the Christmas Tree.

Curve B   The ratio between the miles of wind recorded by the south rotor and by the cup anemometer at 120 feet on the Christmas Tree.

varying turbulence. The observed tendency toward a long-term trend in the two ratios, with a superimposed cyclical trend, has not been investigated.

An anemometer for use in conjunction with a wind-turbine should be dynamically similar to the turbine, as regards inertia and load. If unloaded and too light, it will presumably overrun by responding to the very short gusts of small diameter, which are without significance in turbine output; and if too heavy, it will also overrun.

Each indicated half-mile of wind was integrated and recorded by a stylus on the open scale waxed chart of a specially designed drum-type chronograph recorder (Fig. 30). The stylus was tripped by the closing of a contact in a direct current circuit powered with 6-volt dry cells. The drum was driven by clockwork. These recorders received daily servicing.

In December, 1944, at the request of the U.S. Weather Bureau, we added a USWB category-type integrating recorder at Grandpa's, on an experimental basis, and in January, 1945, we added a demand-meter type (Fig. 31A) of integrating recorder which, each 15 minutes, prints the total mileage of wind passed, in arbitrary units (Fig. 31B).

The record of the category type is an accumulation of minutes of wind in each of the ten different categories of speed shown. The limits of each category may be selected at will.

The horizontal direction of the wind was recorded at Grandpa's for the period, April 9, 1941, to December 31, 1945, by means of a standard anemoscope, the wind vane of which was 30 feet above ground, on a special mast about 40 feet north of the anemometer mast. This instrument was serviced daily. A sock was rigged to the north cross-arm of the Christmas Tree at 120 feet, and vertically above the anemoscope, to determine the vertical distribution of wind direction over this part of the turbine disc. On Pico and Glastenbury, the resident observers

Fig. 30. The drum type recorder.

tending the heated anemometers during winter months made certain visual observations of wind direction, according to a schedule.

The vertical direction of the wind was determined occasionally by measuring
with a protractor the angle to the horizontal made by rime on the anemometer
masts.

No regular aerological ascensions were made to determine the vertical distribution of temperature. The observer making the daily ascent on foot of Grandpa's
and Pico in 1940-1941 occasionally took thermometer readings at known points
and times in the ascent. These records were useful only in indicating that strong
temperature inversions (Chapter 4) did occur frequently.

The vertical distribution of speed was determined by simultaneous hourly measurements on the Christmas Tree at Grandpa's, at 40 feet, 80 feet, 120 feet, 150
feet and 185 feet, from April, 1941, to December, 1941; and for periods varying
from 1 month to 7 months during 1941, at three heights in the range 40 feet to
110 feet on Scrag, Seward, Biddie Knob, Pond; and at two heights, 35 feet and
97 feet, on Mt. Washington (Table 4) (Ref. 13-B).

Fig. 31A. The demand meter type of recorder.

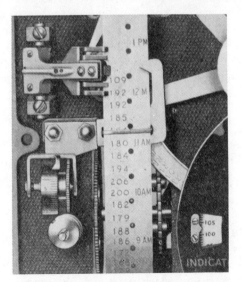

Fig. 31B. Close-up of the recording tape, demand meter type of recorder.

At Grandpa's it was found that the anemometers at 40 feet, 80 feet, and 150 feet were in the zone of influence of the mast. They were then set out on struts 10 feet long, and therefore at a distance not more than 5 effective mast-diameters away from the mast. The interference continued so markedly at the 80-foot and 150-foot instruments that they were abandoned. The interference continued at the 40-foot instrument, probably causing the 5-year average to run low by 1 or 2 percent in some wind directions.

At the other stations the masts were 5 inches in diameter and the two lower anemometers were on 3-foot struts, and thus at a distance of about 7 mast-diameters from the masts.

Instrumentation for the measurement of gust fronts—vertical and horizontal— had been developed by Sherlock who, however, informed Lange that his design was not suitable for our application. Accordingly, Lange designed a pressure-type anemometer which consisted of a pitot head faced into the wind by twin tail vanes. One of these anemometers was mounted, on a 10-foot standard, at the end of each of the four arms of the Christmas Tree at the 120-foot level; at the masthead at the 185-foot level; and at the 40-foot level; and the static head was mounted on a 10-foot strut 20 feet out from the mast on the cross-arm at the 120-foot level.

The choice of pressure instruments for the analysis of gusts was governed by the fact that we were not interested in the low speeds to which this type of instrument is insensitive.

Lange specified that the pressures were to be conducted down the mast to the 4-millimeter water manometers in the control house by 3/8-inch outside diameter tubes each 310 feet in length from the anemometer to the recorder.

The battery of manometers was mounted on the main instrument panel and photographed by two cameras. One, to provide a continuous record, operated at 8 frames per minute. The other, to analyze critical gust data, operated at 8 frames per second. Approximately 750 hours of slow film and 150 minutes of fast film were exposed and are available. They have not been investigated.

THE ECOLOGICAL PROGRAM

Deformation by Wind

Strong winds deform trees which, by the character and amount of deformation, then automatically integrate and record the force of the wind to which they have been subjected throughout their lives. Some species are more sensitive than others. Deciduous trees are poor indicators of wind velocity, showing little deformation other than brushing or stunting, probably because the amount of surface exposed to the wind varies greatly from summer to winter through the shedding of their leaves. Coniferous evergreens, which stand and take the weather the year 'round are, however, good indicators of the windiness of their habitats.

But deformation or injury caused by factors other than wind, such as temperature, insect injury, disease, lightning, or salt spray, is often mistaken by untrained observers for wind deformation. A particular caution should be given against mistaking for wind action the deformation of trees along seacoasts. Many books have striking pictures showing crippled trees exposed to the trade winds. B. W. Wells (Ref. 13-A) claims, and he supports his claim by chemical analyses, that crippling of trees along seacoasts is due largely to the salt in the spray deposited on them. The windward parts of the tree filter out the spray, leaving the leeward branches unwetted, and when the spray evaporates its salt is concentrated. Any young growth upon which it has been deposited is killed like grass on a salted tennis court.

Griggs has for years made a study of timber lines and the factors controlling their position, of which wind is one of the most important. As consultant to the S. Morgan Smith Company, he has developed his technique at various New England sites to a point where it is beginning to yield quantitative estimates of mean annual wind speed within limits narrower than those of any method other than actual long-term anemometry (Ref. 14-B).

The correlation of the degrees of deformation occurring in a certain forest, with the wind and icing regime that has prevailed throughout the life of that stand of trees, is based on our present knowledge of the physiological responses that result in tree deformation. The physiology of the flagging of balsam will illustrate the process, in which five types of response are recognized.

1. The lateral branches just below the tips of the trunk may be bent to leeward by strong winds while they are still young, and held there until their tissues harden.

2. More often, however, a young branch bends into the wind rather than before it, especially if the wind is more or less intermittent, as in New England. If young balsams on a windswept hill are examined the day after a heavy blow, most of their twigs are found turned toward the wind. This tropism is probably due to differential transpiration. More water is abstracted by the drying winds from the windward side than from the leeward side of the twig. Growth is consequently greater on the lee side, and the twig curves into the wind. The mechanism by which this curvation is later reversed, finally setting the twig in a curve to leeward, has not been worked out.

3. For a similar reason the less exposed twigs on the lee side of the tree grow longer than those to windward, which lose more water through transpiration.

4. Such twigs as manage to grow to windward during the summer are killed by winter storms, leaving the trunk bare on the up-wind side.

5. In situations where winter storms create a heavy blast of snow and ice crystals, or sand, or even pebbles, the windward side of the trunk is abraded by the flying particles. The abrasion may vary from polishing the bark to deep erosion into the wood beneath. But such abrasion may be the record of merely a few instances of very high speed, and consequently its quantitative interpretation is obscure. Often the abrasion begins only some inches above the ground, and the

base of the trunk is surrounded by a flat rosette of strong thrifty branches. Such "snow mats" indicate the permanent depth of the snow through the winter, and the wind-flow which permits snowdrifts to accumulate. Snow mats are widespread in flat lands, but in rough terrain the hollows catch most of the drifting snow and simple snow mats do not form.

However, snow is probably far more important as a protection from the severest winter storms than as an abrasive. Any thoughtful examination of trees on wind-swept situations makes clear the dependence of the trees on snowdrifts. Snowdrifts in turn are controlled by pockets of relatively slow air movement.

Balsam has been used as the example. Spruce responds similarly, but with certain important differences. Spruce requires good drainage and does not grow in places habitually covered with heavy snowdrifts, while balsam thrives under heavy snow, often recording in summer the location of winter snowdrifts by the profusion of new scrub balsam.

Correlation of Tree Deformation with Windiness

It will have become clear that our understanding of the physiology and mechanics of the deformation of trees by wind was still too imperfect to permit a rigorous quantitative correlation between observed degrees of deformation and records of wind velocity. However, the area of our uncertainty was narrowed somewhat by recognizing that, while trees may be broken, they are not deformed by isolated storms, however violent.

In the New England hurricane of September, 1938, for instance, a speed of about 200 miles per hour was reached on the summit of Mt. Washington. Great damage was done to tall trees in the forests of the lower and middle slopes; but the hurricane had no effect whatever on the deformed scrub trees with which we are dealing. This was attested by careful "before" and "after" studies in numerous locations. If the hurricane had affected timber-line trees, that effect would necessarily have been breakage or destruction. Brushing, flagging, throwing, clipping, carpets, and resurgence are adaptations gradually attained by living organisms under the long continued stress of severe conditions. The winds which produce deformation must be thought of as being not exceptional, but habitual.

Another evidence of this fact is that year after year the most severely clipped trees react in the same way to the stress of their environment. Over a period of ten years Griggs has followed the history of individual trees on Mt. Washington, by comparison of matched photographs. If their condition were due to special events, their vigor would vary from year to year as the winter was more or less severe. The almost complete uniformity of growth conditions from year to year is, however, one of the most remarkable features of these depressed trees.

Finally, it is manifest that a balsam or a spruce is very much more susceptible to injury during the few weeks of early summer, when its new twigs are soft and succulent, than after the new wood has ripened. Griggs has seen new growth completely blasted, but not deformed as described below, by spring winds of moderate intensity.

Types of Deformation Caused by Wind.

Brushing. A tree is said to be brushed when the branches are bent to leeward like the hair in a pelt which has been brushed one way. Brushing has been observed principally among deciduous trees, and is difficult to detect when they are in leaf. It is the most sensitive ecological indicator of air movement, giving clear indications of prevailing breezes too light to be of economic importance, and thus providing the lower end point of our ecological yardstick (Fig. 32).

Fig. 32. A wind-brushed oak tree near the top of Blue Hill. The monument gives the location. Interesting because the wind regime on Blue Hill is well known.

Flagging. A tree is said to be flagged when its branches have been caused by the wind to stretch out to leeward, while the trunk is bare on the windward side, like a flagpole carrying a banner flapping in the breeze. From an economic point of view the most important indications of wind in New England are given by flagged balsam and spruce.

Throwing. A tree is said to be windthrown* when the main trunk, as well as the branches, is deformed so as to lean away from the prevailing wind, as though thrown to leeward. Throw is largely produced by the same mechanism that causes flagging; that is, the wind is strong enough to modify the growth of the more vigorous upright leaders, as well as the weaker laterals, which have a lower growth potential than the leaders.

Wind clipping. Trees are said to be wind clipped when the wind has been sufficiently severe to suppress the leaders and hold the tree tops to a common, abnormally low level. Every twig which rises above that level is promptly killed, so that the upper surface is as smooth as a well-kept hedge. Sometimes such "trees" occur with the needles of spruce, pine, and fir so intricately felted together that the individual tree is completely lost in the mass, which may be so dense that one can walk on the tree tops.

Tree carpets. Where the wind is so severe as to prohibit upright growth while still allowing trees to start on the ground level, every twig that reaches more than a few inches above the ground is promptly killed. The result is a living carpet of prostrate branches spreading over the ground. The many lateral buds which remain, after the destruction of all potential uprights, are too weak to form leaders, so all tendency to form erect trunks appears lost, and the whole carpet may be held to within 2 or 3 inches of the ground. It may, however, stretch out 100 feet or more to leeward of the sheltering rock where it originally started.

It is possible to differentiate between *loose carpets,* whose mats extend to about 18 inches above the ground, and *close carpets,* which, held to within a few inches of the ground, provide the upper end point on the yardstick of tree deformation by wind.

Above this critical value of the mean speed—about 27 miles an hour at specimen height—the tree cannot survive, and the rock remains bare, even though the site is well below timber line. Such a transition is found on the Horn of Mt. Washington, 1500 feet below timber line.

---

*The term "windthrow" is also sometimes applied to more violent damage as when a single storm so throws a tree as to break its roots on the windward side without actually uprooting it, thus bringing about another type of deformity, or more often the early death of the tree.

Winter Killing and Resurgence. Situations are often found where nearly all of a tree is kept clipped to a common level, but its central leader, by reason of its greater growth potential, manages to rise above that level during the summer. With the coming of winter the body of the tree is filled with a snowdrift, leaving only the new leader exposed. Certain winter killing is the result. In the following summer, however, the same conditions obtained as before and a new leader is sent up— usually in the lee of the first. This in turn is killed the next winter. So the process goes on year after year, indefinitely. As many as half a hundred such leaders have been counted on such resurgent trees.

The power of resurgent trees to continue putting out new leaders clearly depends on the vigor of the trees, as well as on the severity of the wind. From resurgence alone, therefore, it is not possible to get good quantitative estimates of wind speed. Resurgence is, however, certain evidence that a site is subject to a heavy wind regime.

Deformation by Ice

"Ice" may form on objects exposed to wind-driven moisture, over a fairly wide temperature range on either side of the freezing point. Brooks has pointed out that the deposits of an "ice storm" may range in habit and density from coatings of hoarfrost, through rime and light-weight aerated ice, to clear, solid ice weighing 56 pounds per cubic foot.

Deciduous trees suffer characteristic damage from heavy ice storms, and the condition of a deciduous forest along a potential wind-turbine site may serve as an indicator of the frequency and severity of heavy ice storms during the growing life of that stand of trees.

Where hardwood trees, such as beech, maple, or birch, heavily laden with ice, are subjected even to relatively moderate winds, great breakage results. The following year a thick brush of small branches is put out from the broken ends. If another ice storm with wind occurs the succeeding winter, these in turn are broken off. Naturally any branch rising above the general level of the forest is more subject to breakage from wind when iced than those branches within the boundary layer. A forest which is subject to ice storms, therefore, comes to be made up of characteristic even-topped, heavy-bodied, much-branched "candelabrum trees."

Despite the fact that spruces and balsams collect more rime than the deciduous trees, they are very much less subject to ice injury. In part, at least, this is probably because the relatively short branches of the spire-shaped evergreens, when ice-laden, exert less leverage than the long limbs of the diffusely branched hardwoods.

In 1939 a quantitative ecological yardstick did not exist. It was not possible to observe a stand of deformed trees and say: "On this site, at specimen height, the frequency distribution of wind speeds has been such and the maximum thickness of solid 56-pound ice has been such, during the life of these trees."

In an attempt to develop such yardsticks, Griggs carried out a program of field observation at each of the dozen or more of stations occupied from time to time by anemometers, and also at Mt. Washington, Blue Hill, and other sites of interest, such as Lincoln Ridge in Vermont.

The ecological evidence was analyzed and compared with the meteorological evidence collected by Petterssen, Lange, and Brooks. The conclusions were compared with the principles of dynamic meteorology, in collaboration with Rossby and von Kármán.

By 1945 there had been developed a crude but useful ecological yardstick for estimating windiness. It is described in the next Chapter.

SUMMARY

1. Special wind-research programs had been shown to be necessary.

2. A meteorological program was developed, in charge of Petterssen, with Lange, Rossby, Brooks and von Kármán in consultation. On 14 mountain tops, stations were established for measuring wind speeds; at 4 of these stations, the vertical gradient of the speed was measured; and at Grandpa's, the 185-foot Christmas Tree was erected, with 100-foot cross-arms at the 120-foot level, for measuring the structure of the wind.

3. New instruments were developed.

4. An ecological program was developed by Griggs, based on criteria described in this Chapter and correlated with the meteorological program.

4
Behavior of the Wind in the Mountains of New England, 1940-1945

THE METEOROLOGICAL EVIDENCE

Introduction

The program of wind measurement described in Chapter 3 was carried out over the 5-year period, 1940-1945. Once the instruments were installed, the program was maintained by a very few individuals who were not called up for military service because, as a power plant, the project carried a double A priority rating.

However, it was not possible, because of the war, to maintain a staff adequate to evaluate all the measurements taken. The evaluation is still incomplete and the original data are available to any interested person. After V-J day, the data of most immediate interest—the direction and speed measurements—were worked up by Lange, Wilcox, and myself.

The results, which we have not had time to analyze thoroughly, have been reviewed by von Kármán, Petterssen, Rossby, Willett, Brooks, and Lange and are here summarized, in preliminary engineering form.

The Frequency-Distribution of Wind Direction

We have hourly observations of wind direction over the 5-year period, 1940-1945, from the anemoscopes on Mount Washington (Fig. 16A) and on Grandpa's (Fig. 16D). We have two series of pilot-balloon observations made four times a day at Burlington, throughout the 2-year period, 1941-1943, centered around the elevations 4270 feet and 2440 feet, respectively (Fig. 16B, C).*

The most frequent wind direction on Mount Washington (6300 feet) is a little north of west (about 285 degrees true). In the free-air over Burlington, at 4270 feet, the prevailing direction has shifted 5 degrees to the left and lies at about

---

*The total number of cases in this series was 3215, of which 3044 were used. To compute the speed at 2440 feet, about 92 percent of the 3044 were used; 77 percent were used to compute the speed at 4470 feet. The missing observations result from low cloud decks.

280 degrees. A further shift to the left at Grandpa's (2000 feet), relative to that at 4270 feet, was to be expected, in conformity with the Ekman spiral (Chapter 2), and amounting to about 20 degrees. This consideration alone indicated a value for the prevailing direction at Grandpa's of a little south of west (about 260 degrees), while the value actually found is 232 degrees (about southwest), a difference of 28 degrees.

It is possible that this differential shift of 28 degrees is caused by factors which are clearly operating in the case of the Burlington pilot-balloon data. In Fig. 16B it is seen that, while the prevailing direction in the free-air at Burlington at 4270 feet is in a rather broad band centered at about 280 degrees, the wind at 2440 feet (Fig. 16C) shows two bands of frequency accumulation, one centering at about 195 degrees, and one at about 130 degrees, with a minimum at about 235 degrees. In conformity with the Ekman spiral (Fig. 17), one would have expected to find the frequency-distribution at 2440 feet similar to that at 4270 feet, but shifted to the left and centered at about 260 degrees.

The explanation offered is that the mass of the Green Mountains, lying to the eastward of Burlington and averaging 4000 feet in height, tends to deflect the flow of the prevailing westerly winds below this elevation, so that the flow tends to be northerly and southerly around the ridge.

If this deflection is a fact, it should be reflected in the wind speeds recorded by pilot-balloon data.

Fig. 10 shows the speed plotted against elevation, based on the mean of 2 observations daily during the 5-year period, June, 1940-July, 1945. At about 2300 feet the speed is seen to decrease somewhat, dropping in summer about 4.5 percent in about 500 feet before continuing to increase (Fig. 10). In 1939 it was assumed that this speed decrease was apparent rather than real because, until the mountains had been passed, the balloon was being lifted in a rising wind, invalidating the assumption of uniform rate of ascent, and yielding, therefore, speeds lower than the true wind speeds. Accordingly, as described in Chapter 2, this decrease in speed was ignored in making our predictions of output at Grandpa's. Lange believes that part of this effect, and possibly most of it, is due to the drop in speed accompanying the change in direction.

If this is the true explanation, the effect should be most pronounced in the relatively stable light winds of July mornings and least pronounced in the strong well-mixed winds of January afternoons. This is found to be the case in Fig. 10, where the curve for July mornings shows a pronounced loss in speed, while the curve for January afternoons shows little effect.

There is no doubt that the effect of the Green Mountain range extends to the windward into the Burlington region, and causes the originally west wind to be deflected into southerly and northerly streams.

A wind rose of a similar bifurcated shape is indicated by the ecology on Mount Abraham (4100 feet), where the deformation indicates prevailing north-westerly winds, heavy enough to cause turbulence and reverse flow in the lee, both recorded by flagging of the spruces; and southwest winds, also recorded by flagged spruces, but unaccompanied by a standing turbulence severe enough to cause reverse flagging from the northeast.

The apparent deflection of the prevailing west wind at Grandpa's (28 degrees toward the south) is probably another example of the same type of influence.

The conclusion is that on the windward side of a mountain range there will be a dislocation in the distribution of wind directions as compared to that expected from the Ekman spiral; and a deficiency of wind speed on the summits of the windward foothills, as compared with the speed to be expected by interpolation in the standard vertical distribution of speed from the ground to the gradient wind level.

Had this reasoning originally been applied to the height of the test site (1990 feet), it would have resulted in a decrease of 5 miles per hour in the prediction of the mean annual speed, that is, from 24 miles per hour to 19 miles per hour. The value actually found at the test site was 16.7 miles per hour. If the value 19.0 miles an hour is the true speed of the free-air at the elevation of the test site, then Grandpa's Knob did not speed-up the wind, but rather slowed it down some 12 percent, implying a speed-up factor of about 0.90.

The Frequency-Distribution of Wind Speed

The principal use of frequency-distribution curves of wind speed is in the construction, for any turbine design, of a master curve to show the variation in annual output with variation in mean annual speed, as shown in Figs. 33 and 34.

The many hundred speed-frequency curves we have constructed have two origins, which stand in generic relation to each other. The first group was built up

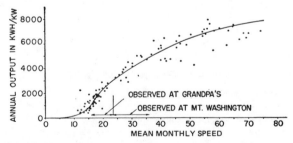

Fig. 33. The relationship between monthly output and the mean wind speed for the month. For convenience in comparison with other output data referred to in the text, each monthly output has been converted to the equivalent annual output.

Each of the 120 points on the curve represents a monthly output computed from the anemometer records for that month, plotted against the mean speed for the month. This curve applies specifically to the dimensions and operating characteristics of the test unit.

from actual hourly anemometer readings over periods varying from two to five years. Such frequency curves, for Blue Hill, Grandpa's, Nantucket, and Mount Washington, are shown in Fig. 35. The second group was built up from hypothetical curves, which can be constructed from an actual smoothed curve by using various assumed

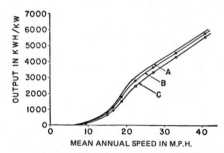

Fig. 34. Variation in annual output with mean annual speed at three elevations above sea level.

Curve A   Sea Level. Density equals 1254 grams per cubic meter.
Curve B   4000 feet above sea level. Density equals 1105 grams per cubic meter.
Curve C   10,000 feet above sea level. Density equals 911 grams per cubic meter.

These outputs were obtained by multiplying the eight speed frequency-distribution curves of Table 10 by the power-speed curve of Fig. 50.

This curve makes it possible to estimate the annual output from a turbine of this design at a site where the mean annual speed is known, provided the wind regime has about the same speed distribution as found in interior New England.

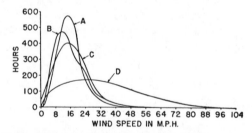

Fig. 35. Speed distribution of wind in New England; 5-year averages.

Curve A   Blue Hill, mean annual speed 18 miles an hour.
Curve B   Nantucket, mean annual speed 16 miles an hour.
Curve C   Grandpa's Knob, mean annual speed 17 miles an hour.
Curve D   Mt. Washington, mean annual speed 34 miles an hour.*

Five-year averages. The area of each curve adds up to the 8760 hours of one year. The speeds are those recorded at the respective anemometer heights.

---

*Uncorrected. See page 71; page 74; Fig. 43; and Table 7 for corrected value. The frequency distribution curve of the corrected speeds is not available.

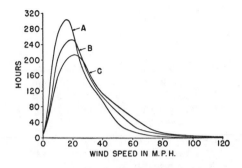

Fig. 36. Speed distribution at Mt. Glastenbury (3840 feet).

Curve A Computed from pilot-balloon observations and pressure maps. It is a free-air speed.

Curve B Results from applying a hypothetical speed-up factor of 1.20 to Curve A.

Curve C Results from applying a hypothetical speed-up factor of 1.40 to Curve A.

speed-up factors as multipliers. In Fig. 36 is shown the basic curve for Glastenbury as computed from pilot-balloon observations and pressure maps (speed-up factor = 1.00). To this curve there have been applied speed-up factors of 1.20 and 1.40 and the transformed and resmoothed curves are plotted on the same figure.

Experience gained in constructing speed-frequency curves had taught us that all of the smooth curves were of a similar type, defined by statisticians as a Pearson Type 3 function. We could, therefore, derive a frequency curve from a series of fixes such as occur when the Beaufort scale is used to report wind speed at sea, as in Figs. 3 through 7. Curves were worked up in this manner for various oceanic locations, making use of the Weather Bureau's summary of reports from ships at sea.

It is also possible to develop crude speed-frequency curves when only the mean speed is known, since the curves, at least those from the interior of New England, exhibit the following characteristics:

A. The intercept on the ordinate axis of hours is not at zero, but at some small number, usually at least 2 hours, and sometimes as high as 20 hours, representing no wind movement; that is, calms.

B. The two sides of the figure are concave upward.

C. The skewness increases with mean speed.

D. The most frequent speed is always lower than the mean speed but varies with it (Fig. 37).

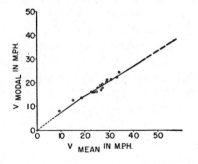

Fig. 37. Relationship between the mean speed and the most frequent speed, in 16 speed-frequency distribution curves.

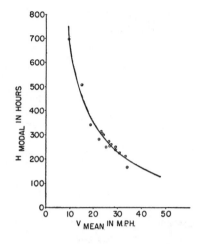

Fig. 38. The relationship between the number of hours during which the wind blows at the most frequent speed, and the mean speed, in 16 speed-frequency distribution curves.

The number of hours during which the most frequent wind blows decreases as the mean speed increases.

E. The number of hours during which the wind blows at the most frequent speed decreases as the mean speed increases (Fig. 38).

A curve sketched to meet these five requirements can be checked, since its area must total 8760* hours and the mean speed, computed graphically from it, must equal the value of the mean speed taken as the fix.

Actual curves from a coastal station would doubtless show the effect of the diurnal sea breeze; and those from tradewind stations would show a still different distribution. It is emphasized that the characteristics just described refer to curves of speed-distribution found in the interior of New England.

Vertical Distribution of the Horizontal Component of Wind Velocity

In Fig. 39 there has been plotted the average value of some 2500 hourly observations of speeds in excess of 15 miles per hour at Grandpa's, at 40 feet, 120 feet, and 185 feet, for the period, May 22, 1941, to December 1, 1941 (Ref 15-B). All values are referred to 140 feet, the height of the hub of the test unit. From a study of the variation of this vertical distribution with wind direction, it was found that 85 percent of all the hourly observations would fall within limiting curves, the band width being less than ± 3 percent at 40 feet, and ± 1 percent at 185 feet.

In Fig. 40 there are plotted the vertical distributions of speed over three wooded mountains—Pond, Seward, and Biddie 1—obtained from hourly records for the periods April 27 to May 18, July 11 to November 30, and July 15 to November 30, 1940, respectively (Ref. 16-B). The ratios of speeds at the observed heights to the speeds at the standard height above the tree tops of 35 feet are plotted against the logarithm of the height above tree top. It will be seen that the conformity

---

*The number of hours in a year is 24 × 365 = 8760.

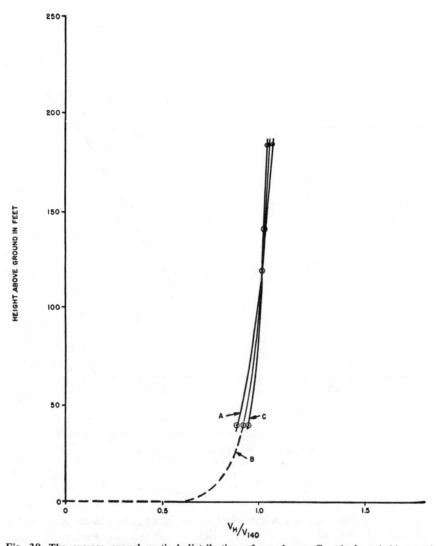

Fig. 39. The average annual vertical distribution of speed over Grandpa's, a bald summit. Curves A and C are arbitrary envelopes drawn around the mean value shown in Curve B. The envelopes include 85 percent of the 2500 hourly observations, of which Curve B is the mean. The band width at 40 feet is ± 3 percent, and at 185 feet is ± 1 percent.

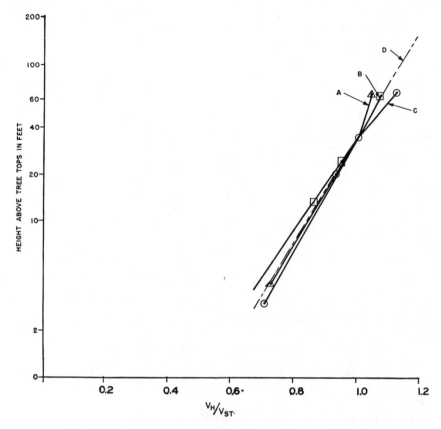

Fig. 40. Vertical distribution of speed over three wooded summits.

Curve A	Biddie I
Curve B	Seward
Curve C	Pond
Curve D	The mean gradient

of the three distributions below the reference height is good. The dispersion of the points at 110 feet above the ground is thought to be due in part to the fact that these anemometers were mounted on 30-foot extensions of 2-inch pipe, which in high winds behaved like a whip lash, introducing unknown errors into the readings of these instruments. In each case the datum plane was the assumed average tree-top level. It is probable that these three distributions could be brought into greater coincidence by small changes in the estimates of the average height of the tree tops.

In Fig. 41 the mean vertical distribution of speed over Grandpa's, a treeless summit, is compared with the mean distribution over the three wooded mountains. The vertical distribution over Mount Washington in the range of 35 feet to 97 feet,

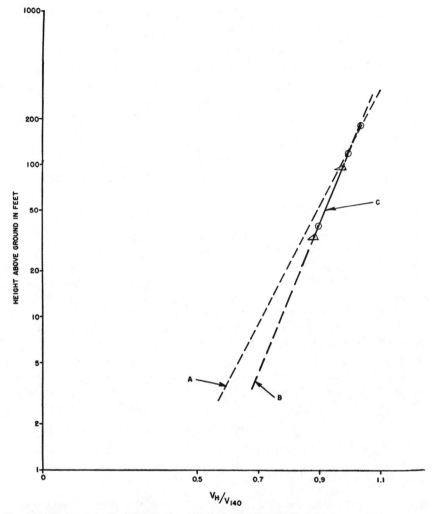

Fig. 41. The mean vertical distribution of speed over Grandpa's, a treeless summit, compared with the mean distribution over the three wooded summits of Fig. 40.

Curve A   Mean distribution represented in Fig. 40 by Curve D.
Curve B   Mean distribution over Grandpa's.
Curve C   Observed distribution over Mt. Washington.*

obtained from hourly readings in the period, August 28 to September 13, 1940 (Table 6) (Ref. 17-B), is also plotted and found to be identical in slope with this sector of the Grandpa's curve.*

*The summit of Mount Washington is cluttered with buildings, making the effective height of the two anemometers uncertain.

TABLE 6. RATIO OF SIMULTANEOUS HOURLY SPEEDS ON MT. WASHINGTON
AT YANKEE NETWORK TOWER, 97 FEET, TO THOSE AT 35 FEET ON THE 2 FT.
BY 2 FT. TOWER, EACH MEASURED BY ROBINSON 3-CUP ANEMOMETER,
AUGUST 28 THROUGH SEPTEMBER 13, 1940

Compass Direction	Ratio	Per Cent
N	1.133	3.36
NE	1.623	3.60
E	0.957	2.90
SE	1.073	4.35
S	1.141	5.54
SW	1.185	10.93
W	1.058	44.02
NW	1.068	25.01
Mean	1.155	—
All, weighted	1.097	—
		100.00

The value of V_{mean} at 35 feet of 41.35 miles per hour has been supplied by Brooks, based on anemometry during the 5-year period July, 1941, through June, 1945.

The important findings are that, on the average, the height at which the maximum speed occurs over a summit such as Grandpa's is about 200 feet; and that the difference in average speed between the lower edge of the turbine disc at 50 feet and the upper edge of the disc at 237 feet is not over 15 percent. These results, which were not available until after the turbine was on the line, came as a relief to designers and backers alike.

It is interesting that the vertical distributions of speed over such complex and varied airfoil profiles as those shown in Fig. 42 for the prevailing wind directions at these mountains, should not only be logarithmic, as predicted by Prandtl and as found by Rossby in the case of vertical distributions over flat lands, but should all have so nearly the same slope.

The Vertical Distribution of the Vertical Component of Velocity

The angle of the wind-stream, over a ridge, measured from the horizontal, might be expected to vary in value from that of the slope of the ridge to zero degrees at some point near the height where the maximum speed is found. At Mount Washington, Brooks reports a mean angle of 15 degrees in west winds, observed at a height of 35 feet. At Chittenden, I found 10 degrees at 50 feet (tree-top height being 30 feet) dropping to 3 degrees at 80 feet, in a west wind.

On Grandpa's the angle, as determined by rime on the Christmas Tree, has been observed to vary from 10 degrees at 20 feet to approximately 0 degrees at 185 feet. The average value over the disc area at Grandpa's is thus about 5 degrees, confirming the original estimates on which we had based the design angle of inclination of the turbine shaft.

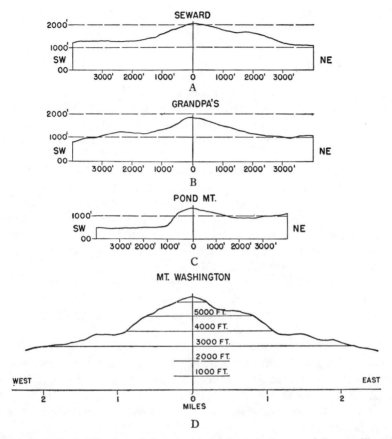

Fig. 42A, B, C, D. Profiles through four mountains drawn in the way of the prevailing wind at each mountain.

Gust Fronts

About 750 hours of film, at 8 frames per minute, and about 150 minutes at 8 frames per second, were taken of the 8 manometers recording the pressures in the pressure-type anemometers distributed over the Christmas Tree as shown in Fig 27. These records have not been analyzed, and are available to any interested person.

If our assumption is correct that a compression of the streamlines tends to damp out gustiness, and if the speed-up factor at Grandpa's is actually less than unity (about 0.9), as we have concluded (p. 76, and Fig. 43), then it should follow that the gustiness factors we experienced during the test program exceeded what we would have found on Pico (4000 feet; speed-up factor = 1.05), or on Abraham (4000 feet; speed-up factor = 1.7). But, whether the maximum gustiness factor we did experience at Grandpa's exceeded the value we had assumed in designing the test unit, and described in Chapter 2, is unknown.

Temperature Inversions

That part of the atmosphere which concerns us is warmest at the bottom, where it absorbs heat from the earth's surface by radiation, conduction, turbulent exchange, and convection. The variation of temperature with height shows great fluctuations, but the normal rate is a decrease of 1.7 degrees centigrade per 1000 feet of ascent.

Pressure decreases with elevation, and a gas, in expanding without addition or loss of heat, from one pressure to a lower one, suffers a drop in temperature. Thus a parcel of air moving upward suffers a drop in temperature as its pressure drops. Provided no heat is added or subtracted during the process of vertical movement, the temperature of the parcel of air falls almost exactly 1 degree centigrade per 100 meters (adiabatic lapse rate, or temperature gradient). This is 3.1 degrees centrigrade per 1000 feet, or nearly twice the normal rate.

When circumstances are such that the temperature increases with elevation, rather than decreases, we are dealing with an inversion, of which two kinds are of interest to us. These are the turbulence inversion and the radiational or ground inversion.

The Turbulence Inversion. The turbulence inversion occurs at the top of the layer of turbulence induced by the roughness of the surface. It is, therefore, more common and more pronounced over mountainous country. It marks the return, from the adiabatic lapse rate prevailing throughout the layer of turbulence, to the normal lapse rate above it. The wind velocity just above the inversion level tends to be higher than that just under it.

Since the inversion level is a layer of stability opposing passage of air through it, a turbulence inversion lying over the summit of a site, within some unknown critical height range, would act as one wall, though a yielding one, of a venturi tube, of which the other is a profile of the site. The streamlines that pass through the tube rather than around the ends of the ridge will be compressed, increasing the wind speed over the ridge.

Radiational or Ground Inversion. There occur principally at night time, when a layer of cold air, cooled by conduction from the cold ground, underlies warmer upper air. Two principal and opposing effects of ground inversion enter into site selection. The first tends to increase speeds, by replacing the broken surface of the ground with the smoother, more frictionless surface of the top of the lake of cold, stagnant air. The second tends to decrease or obliterate wind movement. At a site drowned in such a lake there is little or no wind.

In addition, such layers of heavy cold air cascading down the sides of a mountain under gravity are common at night, often protecting the surface from the general wind.

Since no temperature lapse-rate measurements were made, we could not expect to correlate the ratio of wind speed at Pico (4100 feet) with that of Grandpa's (2000 feet) in detail as a function of the temperature gradient.

But, the ratio of wind speed at Pico to that at Grandpa's has been determined for day and night separately over a 7-month period, and has been found to be 4.8 percent higher by night than by day, while that between Glastenbury (3700 feet) and Grandpa's, over an 18-month period, has been found to be 5.0 percent higher by night than by day.

In the absence of temperature measurements, we do not know whether this indicated nocturnal stability is caused merely by a decrease of the lapse rate, or a lowering of the level of a turbulence inversion, or a ground inversion.

The net effect of inversion levels on wind-turbine sites in Vermont remains unknown.

Icing

Although ice several inches thick was observed on the stationary structure several times, the maximum thickness observed on the rotating stainless-steel turbine blades was about 1/2 inch on the leading edge. As this skin of ice began to peel off, the unit would begin to run rough and was usually shut down for this reason.

Once or twice the unit was started up after the (stationary) blades had accumulated an unbroken sheet of ice about 1/4 inch thick. Each of these times it was shut down before the ice had been flexed off, and for some reason other than roughness.

Grandpa's Knob was probably unusually ice-free during the test period (1940-1945), but the relative inability of ice to build up on the rotating blades gave us confidence that, if we encountered a heavy ice storm, we could either continue to operate or we could motor the blades at, say, 5 revolutions per minute, enough to cause flexing and to inhibit deep ice growth.

We concluded that ice was no bar to turbine operation at 4000 feet in the Green Mountains, although only further experience would tell whether blade de-icing equipment would be economically justified.

Long-term Mean Speeds at Various Sites in New England

In Fig. 43 are plotted the 5-year mean speeds as observed and estimated at the Smith-Putnam mountain-top stations (Ref. 18-B), correlated with similar data for the same period for Mount Washington, Blue Hill, and East Boston Airport, and compared with the observed vertical distribution of the speed of the free-air in the same height range (sea level to 6300 feet). Computations are shown in Table 7.

The variation of free-air speed with elevation above sea-level, in the range 434 feet to 4270 feet, was determined by Lange, who examined the four daily pilot-balloon runs at Burlington, 75 miles to the north of Grandpa's, from 1941 to 1945.

TABLE 7. CALCULATION OF V_{140} AT ANEMOMETER STATIONS

Station	1. Length of Period of Observation	2. Tree-Height, Feet	3. Anemometer Height Above Ground, Feet	4. Hub Height A.S.L., Feet	5. Horizontal Distance from Grandpa's, Miles	6. $\frac{V_{anem.}}{V_{Gr80}}$	7. Probable Error in Ratio of Col. 6
Crown Point	2 months	44	80	240	19.0	0.655	±13%
Pond	22 days	14	75	1501	13.0	0.693	±16%
Biddie Proper	5 months	50	80	2150	4.0	0.913	±0.6%
Chittenden	7 months	0	80	2490	10.5	0.894	±0.8%
Grandpa's	60 months	46	80	2130	0.0	1.000	±0.0%
Seward	6 months	0	80	2220	2.5	0.959	±0.5%
Herrick	9 months	46	80	2700	7.0	1.044	±0.4%
Biddie I	6 months	46	80	2065	1.5	0.909	±0.5%
Glastenbury	18 months	40	80	3900	46.0	1.152	±4%
Pico Peak	7 months	30	70	4107	13.5	1.235	±1%
Mt. Washington	60 months			6428	81.0		—

TABLE 7. CALCULATION OF THE MEAN ANNUAL SPEED AT HUB HEIGHT (H = 140 FEET) AT VARIOUS ANEMOMETER STATIONS

Station	8. $V_{anem.}$ Col. 6 × $V_{Gr_{90}}$, mph.	9. Speed at 140' With Trees — Gradient Factor, from Fig. 41	10. Col. 8 × Col. 9, mph.	11. Speed at 140' Trees "Removed" — Gradient Factor, from Fig. 41	12. Col. 10 × Col. 11, mph.	13. $V_{Freeair}$, mph. from Fig. 43	14. Speed-up Factor Col. 12 ÷ Col. 13, C_z
Crown Point	10.34	0.993	10.27	1.000	10.3 ± 15%	18.0 ⎫	0.57
Pond	10.94	1.125	12.31	1.036	12.8 ± 18%	18.9 ⎪	0.71
Biddie Proper	14.41	1.089	15.69	1.010	15.9 ± 1.6%	19.0 ⎪	0.84
Chittenden	14.12	1.143	16.14	1.043	16.8 ± 1.8%	18.9 ⎪	0.89
Grandpa's	15.79	1.055	16.66	1.000	16.66 ± 0.0%	19.0 ⎬ ± 5%	0.88
Seward	15.14	1.130	17.11	1.038	17.8 ± 1.5%	19.0 ⎪	0.94
Herrick	16.48	1.055	17.39	1.000	17.4 ± 0.9%	18.8 ⎪	0.92
Biddie I	14.35	1.130	16.22	1.038	16.8 ± 1.5%	20.2 ⎪	0.90
Glastenbury	18.19	1.116	20.30	1.033	21.0 ± 5.0%	20.5 ⎭	1.04
Pico Peak	19.50	1.130	22.04	1.024	22.6 ± 1.5%	20.5	1.10
Mt. Washington					44.0 ± 7%	30.0 ± 10%	1.47

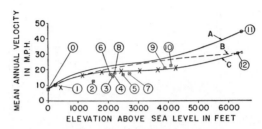

Fig. 43. The 5-year mean speeds observed and estimated at hub-height (140 feet), at the Smith-Putnam mountain-top stations, correlated with similar data for the same period at Mt. Washington, Blue Hill, and East Boston airport, and compared with the observed vertical distribution of the speed of the free-air in the same height range (sea level to 6300 feet).

Curve B represents the original assumption of Curve B of Fig. 9.
Curve C represents free-air speeds.
Curve A represents accelearated speeds.*

Point 0	East Boston	Point 6	Grandpa's
Point 1	Blue Hill	Point 7	Herrick
Point 2	Pond	Point 8	Seward
Point 3	Biddie Knob	Point 9	Glastenbury
Point 4	Biddie 1	Point 10	Mt. Washington

He estimates a probable error in his curve as applied to the Burlington region of ± 2 percent. But its application to the Grandpa region, 75 miles to the south, introduces an unknown error.

The value of the free-air speed at the elevation of Mount Washington (6288 feet) was determined by Willett. He studied all North American upper-air data from 1910 through 1941, as summarized in upper-level pressure maps, which he analyzed in conjunction with pilot-balloon soundings from Burlington and Boston for the periods January, 1924-December, 1941, and November, 1926, through October, 1931, respectively.

He estimated the mean free-air speed at the elevation of Mount Washington to be 30.0 miles per hour ± 10 percent. (Point 12, Fig. 43.)

In Fig. 43, the lower solid-line curve through Lange's seven points was projected to pass through Willett's point for the elevation of Mount Washington, and through the sea-level value for the East Boston Airport.

The dotted-line curve is the assumption of 1939, discussed in Chapter 2, and which resulted in over-estimating Grandpa's output.

---

*It is an arbitrary reference line, connecting the observed speed at Mt. Washington with that at sea level at East Boston, and conforming to the free-air distribution of Curve C.

The Long-term Mean Speeds at the Anemometer Stations

The 5-year mean speed on Mount Washington at anemometer height was supplied by Brooks. He determined a ratio between the speed at anemometer height (35 feet) and the speed at a temporary station at 97 feet on the Yankee Network Tower. This ratio is 1.10 compared with 1.08 for Grandpa's in the same height range (Fig. 39). The distribution of this ratio by direction is given in Table 6.

If this value is taken up to 140 feet by extrapolating along the Grandpa's curve of vertical distribution of speed, we get a value for the mean annual speed at 140 feet of 46.8 miles per hour. But the buildings on Mount Washington effect a net lowering of the heights of the anemometers, by an unknown amount; and there are other uncertainties. Accordingly, we arbitrarily take $V = 44.0 \pm 3.0$ as being the best judgment value for the 5-year mean speed at 140 feet above the summit of Mount Washington. (Point 11, Fig. 43.)

The 5-year mean speed at anemometer height on Blue Hill was also supplied by Brooks, and taken up to 140 feet along the curve of vertical distribution measured over Grandpa's.

The 5-year mean speed at 120 feet on Grandpa's was recorded as described in Chapter 2, and taken up to 140 feet along the measured vertical gradient of Fig. 39.

Of these three values of mean annual speed at 140 feet, that for Grandpa's is the most reliable, while the values for Mount Washington and Blue Hill are based on the assumption that the vertical distribution of speed over these two summits is identical with, or at least closely similar to, the vertical distribution over the summit of Grandpa's, at least up to 140 feet. It is true that all the vertical distributions we have measured support this assumption. Nevertheless, not enough vertical distributions above mountain tops have been measured to make the assumption secure and we must admit that the values of mean annual speed at 140 feet at Mount Washington and Blue Hill may be in error either way by as much as 15 percent.

Estimates of the 5-year mean values of speed at 140 feet over Pond, Biddie Knob, Biddie Proper, Seward, Chittenden, Herrick, Glastenbury, Pico, and Crown Point involve uncertainties of another kind.

When we began our investigations of wind speed in Vermont in 1939, we assumed that the only way in which to determine the long-term mean speed at any mountain site was to maintain there a heated and tended anemometer for a period of a year or more. But we have learned that a year's record of wind speed, standing by itself, is unlikely to be closely representative of the long-term mean speed at that station. A convenient illustration of this is contained in the records of computed output for the test unit at Grandpa's. In analyzing five years of wind speed records from Grandpa's, we find that the output computed from the speed records for a single year may depart from the 5-year average of computed output by ± 26 percent. In a 20-year period the departure of the computed output of a

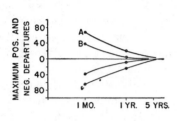

Fig. 44. Reliability of wind-power. Maximum positive and negative departures in output at two stations expressed as percentage deviations in output for the given time interval from the average output for that time interval experienced in five years of records.

Curve A   Grandpa's
Curve B   Mt. Washington

At Grandpa's, for example, the output in any single month may depart from the normal 5-year expectancy for that month by about 75 percent.

single year, from the 20-year average, would be greater—perhaps ± 30 percent, as indiciated in Fig. 44.

What is needed is an estimate of the long-term mean annual speed at the potential site. It is not practical to wait twenty years while this record accumulates.

Fortunately there appears to be a useful short cut. If the site in question is in the neighborhood of a long-established meteorological station, then the long-term mean speed at the survey station may be determined as follows:

Determine the ratio between the speed recorded at the survey station and that recorded at the control station in the same period of time. Multiply the long-term mean speed at the control station by this ratio to obtain the long-term mean speed at the survey station.

For example, let us assume that an anemometer station has been maintained for twenty years on a certain 200-foot mast. The long-term mean speed at 200 feet is known to be 25.0 miles per hour. It is desired to find the long-term mean speed at the 100-foot level on the mast. An anemometer is mounted there, on suitable struts to avoid mast shadow. It is operated for 24 hours, and the average of the 24 hourly values of the ratio between the speed at 100 feet and that at 200 feet is found to be 0.8, with a maximum of 0.9 and a minimum of 0.6. At the end of 48 hours, the 48 observed ratios are collected into eight 6-hour averages, with a mean for the 8 values of 0.82, a maximum of 0.86, and a minimum of 0.74.

At the end of seven days, the average of the seven daily ratios is found to be 0.80, with a maximum of 0.81 and a minimum of 0.79, or the mean velocity at 100 feet is 80 ± 1.2 percent of that at 200 feet = 0.80 × 25.0 = 20.0 ± 0.25 miles per hour; and so on, to any desired accuracy.

This method was used in estimating the 5-year average speeds at our various anemometer stations, whose records we ratioed in to the record at our control station at Grandpa's, where we had a continuous record for the five years, at the 120-foot level. The speed found at the 120-foot level was converted to that at hub height, 140 feet, by interpolation on the curve of vertical distribution of speed described on pp. 66-70 and in Fig. 39.

The length of the record required at the survey station in order to establish the ratio of the speed there to that at the control station (Grandpa's, 140 feet), with a stated accuracy, varies with the slant distance between the stations (Fig. 45). Thus, to obtain the ratio with an error of about 10 percent, it was necessary to operate Glastenbury, 38 miles away and 2000 feet higher than Grandpa's, for 90 days; Pico, 12 miles away and 2000 feet higher, for 30 days; Herrick, 10 miles

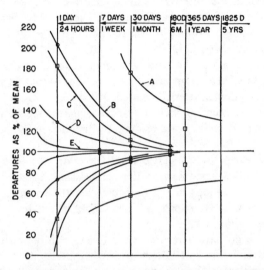

Fig. 45. A chart to indicate the minimum length of period of observation necessary to establish the ratio between the mean speed at Station X and the mean speed at the control station, with a given percentage of uncertainty.

Curve	Station
A	Departures in the ratio of the wind speed at Mt. Washington to that at Grandpa's from the mean ratio.
B	Departures in the ratio of the wind speed at Glastenbury to that at Grandpa's from the mean ratio.
C	Departures in the ratio of the wind speed at Pico to that at Grandpa's from the mean ratio.
D	Departures in the ratio of the wind speed at Herrick to that at Grandpa's from the mean ratio.
E	Departures in the ratio of the wind speed at 185 feet on the anemometer mast at Grandpa's to that at 120 feet on the same mast from the mean ratio.

Explanation: For example, over the period of observation, which was about a year, the mean speed in any 24 hours at the 185-foot station was never more than 6 percent and never less than 5 percent of the ratio between the mean speed for the whole period at 185 feet and the similar mean speed at 120 feet. But the minimum time interval required to establish a ratio with the same certainty between the wind speed at Grandpa's and that on Herrick (4 miles away from Grandpa's and 500 feet higher), is 30 days.

away and 400 feet higher, for 10 days; and Grandpa's 185-foot station, 65 feet higher on the same mast, for about 14 hours.

In Table 7 there are summarized the estimates of the 5-year mean values of speed at 140 feet at our various anemometer stations, with some indications of the probable error in each value.

These estimates were made as follows:

First, there was computed, for each month of observation, the ratio between the wind movement at the observation station and the wind movement at the control station, viz., the 140-foot level at Grandpa's. The monthly values were averaged to give an average value for the ratio over the period of observations, which varied from 6 months to 18 months.

Next, the average value of the ratio, so computed for each observation station, was multiplied by the 5-year mean value of the speed at Grandpa's at 140 feet.

All reliable observational material was used. When there was doubt concerning the condition of the instrument, particularly under icing, the record was thrown out. No attempt was made to select speed readings by direction or by speed group.

Sensitivity of this Ratio to Changes in Wind Direction. Since anemoscopes were not operated at the nine anemometer stations in question, it is not possible to make a direct analysis of variations in the ratio between the mean speed at 140 feet at Station X and that at Grandpa's, with variations in the direction of the wind. There are a few clues, however.

At Grandpa's we do have a record of the variation in the ratio between mean speed at 185 feet and that at 120 feet, with wind direction, for about 6 months (Table 8). The mean value of this ratio, summing up and weighting all directions, is 1.042. The maximum deviation from this mean occurs in winds from the northeast and amounts to about 2 percent. The records on the summit of Mount Washington show much higher variations in the ratio between the speed at 97 feet and

TABLE 8

Variation in the ratio $\dfrac{\text{Mean speed at 185 feet}}{\text{Mean speed at 120 feet}}$, with variation in the wind direction, at Grandpa's, for about six months.

Direction	Ratio V_{185}/V_{120}
North	1.061
North East	1.064
East	1.048
South East	1.032
South	1.046
South West	1.040
West	1.029
North West	1.051
ALL	1.042

that at 35 feet, as listed in Table 6. The ratio in an east wind is found to be about 59 percent of the ratio in a northeast wind. This high variation, of course, is due to the many buildings which clutter up the summit of Mount Washington.

A third indication of a different sort was somewhat dubiously afforded by the wind-tunnel tests of model mountains. In these tests, whose uncertainties are described in Chapter 2, the speed measured at a certain height above a certain point over the summit of the model was found to vary by as much as 40 percent with changes in wind direction.

Finally, a clue was afforded by comparing the ratios of the mean speed at 140 feet at the nine observation stations with that at Grandpa's for intervals of time shorter than one month. In the case of Pico, the ratio was determined for 6-hour periods and it was found that the dispersion was high (Fig. 45).

The explanation offered is that part of this dispersion is probably due to factors discussed on pp. 77-78, but that part of it is due to the directional distribution of wind speed during the periods of observation.

Sensitivity of this Ratio to Changes in Wind Velocity. It has been found that, in the case of stations separated by a substantial slant distance, the ratio is not independent of speed. Fig. 46 shows that the ratio between the mean speed at 140 feet at Station X and that at Grandpa's is some function of the velocity at Station X. This means that ratios determined during periods when the wind velocity

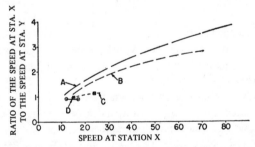

Fig. 46. The variation in the ratio between the mean speed at two stations with the speed at one of the stations. It is seen that the ratio increases as the speed increases where the two stations are separated in elevation, but that where the two stations are not separated in elevation, the ratio is nearly independent of the speed.

Curve A   Ratio between the mean speed at Mt. Washington and the mean speed at Grandpa's.

Curve B   Ratio between the mean speed at Mt. Washington and the mean speed at Glastenbury.

Curve C   Ratio between the mean speed at Glastenbury and the mean speed at Grandpa's.

Curve D   Ratio between the mean speed at Seward and the mean speed at Grandpa's.

was below the long-term average will yield a ratio that is too low, and, similarly, if the ratio was determined during periods when the wind speed exceeded the long-term average, the ratio will be too high. However, when the observation station lies at about the elevation of the control station and is exposed to a nearly identical wind regime, the uncertainty arising from this consideration becomes unimportant.

It will be noted that in Fig. 46 the ratio of the mean speed at 140 feet over Seward to that at Grandpa's is nearly independent of the value of the speed at Seward, whereas in the case of the comparison between Glastenbury and Grandpa's there is a small but definite increase in the value of the ratio, with increasing speed at Glastenbury.

A similar argument should apply to wind-speed data coinciding with abnormal temperature inversions, which might produce extreme variations in the ratio between two summits separated somewhat in elevation.

If, therefore, the ratio were determined by using only those speed records which coincided with:

A. The prevailing wind direction,
B. A wind speed not far removed from the assumed long-term mean annual speed, and
C. An absence of inversion levels,

then a convergence would be obtained substantially tighter than that of Fig. 45.

On the other hand, it must be admitted that the periods of observation which form the statistical bases for Fig. 45 were short, and that, had the experience been longer, greater aberrations would have been found, and the convergence would have been less tight.

Until more experience is gained, it is suggested that those two factors may balance each other. The convergencies of Fig. 45 may be taken as qualitatively representative of the variation in uncertainty of the ratio with the length of the period of observation, for various slant distances.

The uncertainties in the ratios so computed are tabulated in Column 7 of Table 7. Thus, the estimated long-term mean speed at anemometer height is thought to contain a probable error at Pond Mountain of ± 16 percent, at Crown Point of ± 13 percent, at Glastenbury of ± 4 percent, and at Biddie Proper, Chittenden, Seward, Herrick, Biddie 1, and Pico of ± 1 percent or less.

These ratios were first established between the respective anemometer heights. Values of the long-term mean speed were then computed and these values were taken up to 140 feet along the curves of vertical distribution of speed established in Figs. 39 and 40. In the case of bald summits (Herrick), the vertical distribution used was the average distribution measured at Grandpa's (Fig. 39). In the case of timbered summits, the distribution used was the mean of the distributions found at Pond, Seward and Biddie 1 (Fig. 40), followed by the "removal" of the trees by increasing the anemometer height by the amount of the height of the trees, along the Grandpa gradient. Thus all the speeds apply to bare slopes. Table 7 summarizes

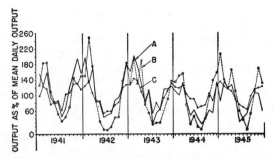

Fig. 47. Variation of the computed monthly output at three stations for 120 consecutive months, expressed as percentage deviations from the mean output for the five years.

In order to make the results comparable among months of differing lengths, the output for each month has been divided by the number of days in that month.

Curve A	Grandpa's
Curve B	Mt. Washington
Curve C	Blue Hill

these computations, and in Column 12 of Table 7 are indicated the over-all uncertainties in the speeds of Fig. 43.

From our limited experience in New England, we may conclude that observation stations which are separated from the control station by as much as 50 miles in horizontal distance and 2000 feet in vertical elevation, can be ratioed in to the control station, and the long-term mean speed established, with an error of about 10 percent, in a period of 90 days or less. If the field work is scheduled for the mean periods of April-May or September-October, and if care is used to exclude speed data coinciding with abnormal temperature inversions or directions or speeds, it seems likely that the error would be about 10 percent at the end of a 60-day period.

Where stations are more closely related, whether in horizontal distance or in elevation, the time interval for a given degree of uncertainty is correspondingly reduced. Where stations are as widely separated as Mt. Washington and Grandpa's (say, 80 miles and 4000 feet), there is no useful relationship. This is shown in Fig. 45, and also in Fig. 47, on which are plotted the sixty computed montly outputs at Mt. Washington, Blue Hill, and Grandpa's. It will be seen that there is no useful statistical relationship.

Mountains as Airfoils

The upper curve of Fig. 43 is a reference curve, and joins the speed actually observed on Mt. Washington (44 miles an hour) with that observed close to the ground

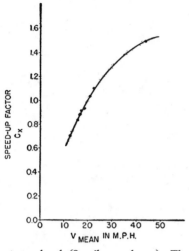

SPEED-UP FACTOR C_x

V_{MEAN} IN M.P.H.

Fig. 48. Relationship between the speed-up factor, C_x, of a ridge and the free-air speed over the ridge.

The eight values of the speed-up factor are taken from Column 14 of Table 7, and are derived from relationships shown in Fig. 43, and described on pages 76, 77, 80.

The amount by which a mountain ridge, acting as an airfoil, increases the free-air speed, increases as the mean speed increases. The low dispersion is remarkable in view of the wide differences in geometry among these mountain airfoils as shown in the profiles in Fig. 42.

at sea level (8 miles an hour). Thus the speed at any point on the upper curve is related to the speed at the underlying point on the lower curve by the speed-up factor C_x. At sea level this is unity, but on Mt. Washington it is $44/30 = 1.47$, which is the amount by which Mt. Washington, acting as an air-foil, has speeded up the free-air flow.

We had assumed in Chapter 2 that, in the case of aerodynamically identical mountain ridges, the speed-up factor C_x would tend to increase with the speed. But our measurements show (Fig. 48) that, while the speed-up factor at eight different sites does actually vary quite closely with the free-air speed, it varies hardly at all with the geometrical differences among the eight profiles. This interesting and unexpected discovery has not been investigated.

At Grandpa's, whose location is shown in plan in Fig. 25 and in elevation in Fig. 22, we find that the speed-up factor C_x has a value of about 0.88 (Table 7).

At nearby Seward, whose elevation is almost the same as Grandpa's, the value of C_x is 0.94, or about 8 percent higher. The respective WSW–ENE profiles are shown in Fig. 42. Although the Seward profile does appear to be somewhat fairer, it must be admitted that we lack criteria, based on map study alone, for predicting that the speed over the summit of Seward would be 8 percent greater than that over Grandpa's.

A spectacular exception to the relationship shown in Fig. 48 is the Horn of Mt. Washington. The Horn, which is the northern end of Chandler Ridge, lies at 4100 feet, on the east-west profile shown in Fig. 49. Brief wind-speed measurements at the Horn have confirmed the impression of the summit observers, that the mean speed on the Horn is substantially greater than that on the summit, 2200 feet higher. Specifically, the mean speed at 140 feet over the Horn was found to be 47 miles an hour, giving an apparent value for the speed-up factor there of about 2.10.

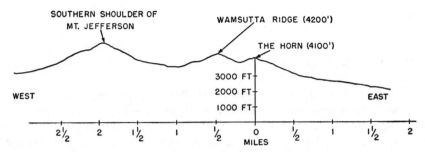

Fig. 49. West-east profile through the Horn on Mt. Washington.

It seems clear that we know very little about the effect of the geometry of a mountain-mass upon the wind-flow.

SUMMARY OF THE METEOROLOGICAL EVIDENCE

1. The Frequency-Distribution of Wind Direction

On the windward side of a mountain range, there will be a dislocation in the distribution of wind directions, as compared with the Ekman spiral; and a deficiency of wind speed over the summits of the windward foothills, as compared with the speed to be expected by interpolation in the standard vertical distribution of speed from the ground to the gradient wind level.

2. The Frequency-Distribution of Wind Speed

The use for a speed-frequency distribution curve is described.
Methods of constructing the curve are given.
Examples are shown.

3. The Vertical Distribution of the Horizontal Component of Wind Velocity

a. At Grandpa's the average height at which the maximum speed occurs is not over 250 feet. Turbine towers more than 200 feet high are uneconomical, at least on hilltops.

b. The curves of vertical distribution of speed, over all the summits measured, are logarithmic, and have nearly identical slopes.

4. The Vertical Distribution of the Vertical Component of Wind Velocity

The mean flow through the turbine disc at Grandpa's, extending from 50 feet above the ground to 237 feet, is *up* about 5 degrees from the horizontal.

5. Icing

Icing was unusually light throughout the test period. The stainless-steel blades appeared to flex off light ice easily. It is not known whether de-icers would be economically justified at 4000 feet in Vermont.

6. A method has been developed for determining the long-term mean speed at an observation station by ratioing the results of a 60-day period of observation to the simultaneous value observed at a control station, and multiplying the long-term mean speed at the control station by this ratio. This method was developed in the course of estimating the long-term mean speeds at nine mountain stations in New England.

7. On the eight "average" sites studied, the speed-up factor depends upon the free-air speed, very little upon the profile of the site.

8. A spectacular exception to this (the Horn at 4100 feet on Mt. Washington) makes it clear that we know very little about the effect of the geometry of a mountain-mass upon the wind-flow.

9. Seward (2080 feet), the best of the low stations, would have yielded 27 percent greater output than Grandpa's, the test site (1990 feet), for reasons suspected, but not yet known.

THE ECOLOGICAL EVIDENCE

The wind-power engineer is interested in the mean annual speed which he can count upon for the generation of electricity. It was made clear in Chapter 3 that considerable components of the mean speed do not contribute to the deformation of trees—the occasional very severe storms which, although of short duration, pile up considerable increments to the annual mean, and the light breezes which, because of their prevalence through many days, also contribute a large fraction to the whole.

On the other hand, it seems equally certain that for each species of tree in each habitat, there is some critical value of the mean wind speed below which deformation will not occur, no matter how constant and sustained the wind may be.

This limit is obviously difficult to establish, and today we have only a general idea of where it lies in the cases of one or two species.

But wind-turbines likewise can make little use of light wind, and hurricanes are sources of danger rather than of power to the installations. Hence, in dealing both with trees and with turbines, it would seem, at first, that one should really be concerned with something other than mean annual speed. But, as will be shown presently, this is not the case in turbine design. It has been found that, in interior New England at least, if we know the mean annual speed, then, for any turbine design, we can know the output (Fig. 34). This is because the speed-frequency distribution curves, measured in New England, are all of the same statistical type;

we are permitted, to make the tentative assumption therefore, that, in New England at least, tree deformation is likewise a function of mean annual speed. The problem, so far as the trees are concerned, is to ascertain at what value of the mean annual speed the wind begins to flag balsams and spruces and how frequently a wind of given speed must blow in order to be reflected in the shape of the trees.

The first step in correlating degrees of tree deformation with measurements of wind speed was to relate the observation stations to each other in an ascending order of measured mean annual speed, and to compare this list with a list of the same sites arranged in an ascending order of observed tree deformation. Would the lists match? At the first attempt, Griggs readily differentiated between the three groups of sites with high, moderate, and low mean annual speeds, respectively. Later work brought more refined correlation within the groups and, in fact, pointed to the necessity for a careful redetermination of the long-term mean annual speed at each site.

The second step was to determine the quantitative relationship between the measured or estimated value of the long-term mean speed at specimen height, and the degree of deformation.

In Table 9 are set forth, in a descending order of wind speeds, the types of deformation and the values of mean speed at the height of the top of the specimen. The practical range of mean speed at tree-top height over which we possess ecological indicators in New England is thus seen to be from about 10 miles per hour to about 27 miles per hour. As indicators in this range we have five progressive types of deformation (brushing, flagging, throwing, clipping, and carpet) applied to a number of species from the most sensitive (while pine and hemlock) to the least sensitive (balsam and spruce).

Balsam is a good indicator in New England, throughout a range of mean annual wind speeds which includes the economically useful range. The entire scale of deformation of balsam corresponds to a remarkably short range of mean speed at *specimen height,* running from minimal flagging at about 17 miles per hour to a 6-inch carpet at about 27 miles per hour, a range of about 10 miles per hour, for the five easily recognized stages of deformation.

It should be noted that the critical value of mean annual speed, above which the specimen cannot grow, is not a constant. As the tree top increases in height and lengthens the distance of the circulation system joining the roots to the topmost new growth, a less severe wind regime is required to induce deformation, while, as the tree is gradually suppressed in height, progressively higher values of mean annual speed are needed to produce deformation. Griggs says the explanation for the greater relative tenderness of tall specimens is doubtless to be sought in the little understood sap-pumping mechanism of a tree. In any event, the evidence summarized in Table 9 shows that a 30-foot balsam will begin to show deformation when the mean speed at specimen height is about 17 miles per hour, and that it will be held to a close carpet a few inches thick when the mean speed at a height

TABLE 9. SUMMARY OF QUANTITATIVE ECOLOGY

Station	Elevation of Specimen Above Sea Level, feet	Types of Deformation of Trees		Wind Speed V at Height x (in Feet) of Top of Specimen above Ground, mph.	Wind Speed at Hub Height V_140, mph.	Potential Annual Output, kwh./kw.
		Deciduous	Coniferous Evergreens			
Mt. Washington (The Horn)	4100		Balsam, spruce, and fir held to 1 foot.	$V_1 = 27.0 \pm 1.0$ (Obs)	44 (Extr)	6000
Abraham	4040		Balsam, spruce, and fir held to 4 feet.	$V_4 = 21.5$ (Est)	35 (Extr)	4850
Cutts	4100		Balsam thrown.	$V_{26} = 19.2$ (Inter)	28 (Extr)	3800
Nancy Hanks	3900		Balsam strongly flagged.	$V_{30} = 18.6$ (Inter)	25 (Extr)	3300
Pico	4000		Balsam flagged	$V_{30} = 17.9$ (Inter)	21.4 (Extr)	2700
Glastenbury	3790	Birch, maple, beech, and cherry held to 40 feet partly by ice.	Balsam shows minimal flagging.	$V_{40} = 17.3$ (Inter)	19.8 (Extr)	2200
Herrick	2560	Hardwoods not held to a level.	Balsam unflagged.	$V_{40} = 15.5$ (Inter)	17.3 (Extr)	1450
Grandpa's	1990		Top killing.	$((V_{46} = 14.2$ (Inter)	16.7 (Inter)	1200
Seward	2100			$(V_{46} = 13.9$ (Inter)	16.7 (Extr)	1200
Biddie I	1850				15.8 (Extr)	950
Pond	1400		Hemlock and white pine show minimal flagging.	$V_{40} = 10.6$ (Obs)	12.3 (Extr)	450

of 1 foot is about 27 miles per hour. Thus we cannot say of a specimen held, for example, to 4 feet, that the mean speed at 4 feet is also 27 miles per hour. It is doubtless something less—how much less we do not yet know.

Nevertheless, if the ecological yardstick shown in Table 9 applies generally throughout New England, then the deformation of the trees at Station X can be used directly to tell us something about the mean annual speed at *specimen height.* And if the vertical distribution of speed which was found to exist over Grandpa's, Pond, Seward, Biddie 1, and Mt. Washington is assumed to exist also over Station X, then the mean annual velocity at hub height can be directly estimated, from the deformation at tree-top height. These assumptions require much testing before they can be accepted as the basis for quantitative estimates.

It seems clear, however, that, in interior New England at least, ecology can be made to relate sites in an order of merit; indicate which sites are of submarginal economic interest; and provide direct estimates of long-term mean annual speed, with a degree of uncertainty which is not known today.

For example: The summit of Mt. Abraham, the southern shoulder of Lincoln Ridge, lies at 4040 feet, the same elevation as the Horn of Mt. Washington, and about 1500 feet below timber line. Healthy balsam which normally reaches a height of 30 feet in this habitat is held to a height of 4 feet on Mt. Abraham. Therefore we say that, at specimen height, the mean annual speed is somewhat less than that which has been measured at the specimen height of 1 foot on the Horn, viz., somewhat less than about 27 miles per hour. Let us estimate this as 21.5 miles per hour, by somewhat more than splitting the difference between the mean speed of 20 miles an hour, which causes heavy flagging of 30-foot balsams, and that of 27, which causes the 1-foot carpet.

Then, using the argument of Table 7 and Fig. 41, we estimate a value for the long-term mean annual speed at hub height of 140 feet, after the removal of the 4-foot trees, of about 35 miles per hour, with an uncertainty of perhaps 5 miles an hour either way. Entering Fig. 34 with this value for mean speed, we estimate the long-term annual output, for a certain design, at 4850 kilowatt-hours per kilowatt. Since, in this range of mean speed, the output is a nearly linear function of mean speed, it follows that the uncertainty in our estimate of output is about the same as that in our estimate of mean speed, viz., about 16 percent either way.

In another example, 30-foot balsams are "strongly flagged." The mean speed at tree height is then estimated directly from Table 9 at 20 miles per hour. Using the same arguments as before (Table 7 and Fig. 41), we estimate the mean speed at 140 feet over the bare summit, after the removal of the trees, to be 24 miles per hour, with an uncertainty of perhaps 3 miles per hour either way.

Entering Fig. 34 with this value we find an annual output of 3150 kilowatt-hours per kilowatt, with an uncertainty of about plus 16 percenr or minus 20 percent.

Whatever the quantitation may prove to be, the types of deformation enumerated above are extremely sensitive, and record, as do tufts on a model in a wind-tunnel, the turbulence of the air-stream over an irregular surface. Of great interest to the users of wind is the evidence that in such a terrain as high New England, the prevailing westerly winds flowing over a mountain top form standing waves of turbulence, recorded neatly and precisely by the coniferous evergreens, which indicate that *here* the output will be so much, while *there,* 250 feet away, it may be one tenth as much. Such evidence of sharp boundaries in the standing turbulence is found on the summits of Pond, Pico, Abraham, and the Horn, and along the carriage road on Mt. Washington just below the half-way house.

For example, the 4200-foot Wamsutta Ridge of Mt. Washington runs from S by E to N by W, and lies athwart the prevailing wind, which, 85 percent of the time, has a westerly component. Wamsutta Ridge lies 2000 feet below the summit and 1300 feet below timber line. Along most of the crest of the ridge the trees grow between 15 and 30 feet high. But in one patch a hundred yards wide, the high-speed wind-stream has been deflected downward for unknown causes, and sears the ridge. Here balsam grows only in the lee of rocks and only to a height of 1 foot. The transition from this zone of high wind to the zone of winds permitting nearly normal growth occurs in a matter of yards.

Another and a similar example occurs on Chandler Ridge, one-half mile downwind and to the east of Wamsutta Ridge, and lying 500 feet lower. Here the patch of 1-foot balsam carpet on the Horn extends a half mile from north to south, and the transition zone between the carpet and the normal 30-foot forest, while it is graduated, takes place in a matter of yards.

The ecological yardstick of long-term mean speed is admittedly still crude, and has been developed from fragmentary data, with a scale probably applicable only to New England. But we think that, at least in interior New England, it provides a useful method of taking the first step in a wind speed survey. It is hardly necessary to add that, before commercial development is undertaken at a potential wind-power site, the ecological survey should be followed by anemometry long enough to relate the site to the long-term mean speed determined at a meteorological control station, as described on pp. 73-80.

Icing

If our ecological yardstick is a crude measure of mean speed, it is cruder yet as a measure of maximum icing, for the reason that a "candelabrum" forest may have been produced by a few exceptionally heavy ice storms unassisted by wind; or it may be the result of less severe icing, but accompanied or followed by strong winds.

Quantitation of this type of deformation seems most difficult and uncertain.

However, the negative evidence may be useful. If a mature forest shows no damage due to icing, this indicates that, during the life of these trees, maximum icing has not been adequate to break branches unaided by wind.

SUMMARY OF THE ECOLOGICAL EVIDENCE

1. Occasional very severe storms do not deform trees.

2. For each species and each habitat, there is some critical minimum value of the mean wind speed below which deformation will not occur. We do not know very much about this limit.

3. Tree deformation is a sensitive indicator of the unpredictable heterogeneity of wind-flow through and over mountains. Local transitions from prevailing very high winds (which hold the balsams to a carpet), to prevailing winds so moderate that the balsams reach normal growth without deformation, occur with a matter of yards.

4. Wind-flow in mountainous country is turbulent. While the small structure of this turbulence is as yet unpredictable, it frequently consists of standing waves so nearly permanent (under the conditions of prevailing westerlies in New England) that the evidence is recorded in the deformation of the trees, particularly on down-wind slopes and shoulders, where "reverse" east-wind flagging is found, often to within 100 feet of the west-wind flagging on the summit.

5. Tree deformation is a poor yardstick of maximum icing, although absence of breakage by ice may be significant.

6. Balsam is the best indicator of mean wind speed in mountainous New England. Deformation begins when the mean speed at specimen height (say, 30 feet to 50 feet) reaches 17 miles per hour. The other end of the scale is reached when balsam is forced to grow like a carpet, at a height of 1 foot, by a mean speed of about 27 miles per hour.

7. In this range of 10 miles per hour there are five easily recognized types of progressive deformation (brushing, flagging, throwing, clipping, and carpet).

8. The ecological yardstick of mean wind speed is still of unknown accuracy, but, at least in interior New England, we believe it can be made to:

    a. Relate sites in an order of merit;

    b. Indicate which sites are of submarginal economic interest;

    c. Provide direct estimates of long-term mean annual speed, with a degree of uncertainty which we do not know, but which may be of the order of ± 20 percent.

9. The southern three-quarters of Lincoln Ridge appears to be the best large capacity site in Vermont. On Mt. Abraham (4040 feet), the southernmost shoulder of the ridge, the mean speed at hub height is perhaps about 35 miles per hour and the average annual output would therefore be about 4850 kilowatt-hours per kilowatt from a 175-foot, 1500-kilowatt unit.

5
The Power in the Wind and How to Find It

The power in the wind can be computed using the concepts of kinetics. Power is equal to energy per unit time. The energy available is the kinetic energy of the wind. The kinetic energy of any particle is equal to one-half its mass times the square of its speed, or $\frac{1}{2}MV^2$. The volume of air passing in unit time, through an area A, with speed V, is AV, and its mass M is equal to its volume multiplied by its density, ρ, or, $M = \rho AV$. Substituting this value of the mass in the expression for the kinetic energy, we obtain

$$\text{Kinetic energy} = \frac{1}{2}\rho AV \cdot V^2 = \frac{1}{2}\rho AV^3.$$

If a non-dimensional proportionality constant k is introduced to convert the energy to kilowatts, then, when

ρ = the density, in slugs* per cubic foot = $\dfrac{M}{g} \cdot \dfrac{1}{\text{ft}^3}$**;

A = the projected area swept by the turbine, in square feet = $\dfrac{\pi D^2}{4}$;

V = wind speed in miles per hour; and

k = 2.14×10^{-3};

the expression for the power in the wind becomes

$$\text{power in kilowatts } (kw.) = 2.14\rho AV^3 \times 10^{-3}.$$

For example, the power passing through the disc area of the test unit in a 30-mile wind with a density of 2.33×10^{-3} slugs per cubic foot is:

$$Kw. = 2.14 \times 2.33 \times \frac{\pi}{4} \times (175)^2 \times (30)^3 \times 10^{-6} = 3240 \text{ kilowatts.}$$

THE POWER WHICH CAN BE EXTRACTED FROM THE WIND

It is obviously impossible to convert all the power of the wind into useful power. The portion that is usable is determined by aerodynamic and mechanical efficiencies (Ref. 19-B) described in Chapter 10. The over-all efficiency varies with the wind

---

*A slug is a unit of mass equal to the weight divided by the acceleration of gravity.
**At the test site (1990 feet), the first approximation of the 5-year mean value of the density was found, by weighting the average density for each month by the proportional output during that month, to be 0.00233 slug/ft³.

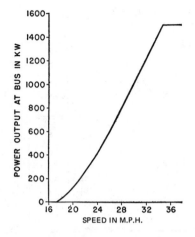

Fig. 50. The relationship between the power output at the bus and the wind speed for a 30-rpm wind-turbine rated at 1500 kilowatts, operating at an elevation of 4000 feet in a wind-stream whose mean annual density is 1126 grams per cubic meter.

speed. In Fig. 50 there is plotted the power output at the generator in kilowatts, against the wind speed in miles per hour, for a specific turbine design rated at 1500 kilowatts and turning at 30 revolutions per minute in air with a density of 1125 grams per cubic meter, corresponding to the annual average found at 4000 feet in Vermont. This design starts generating at about 17 miles an hour, and the rate of change of output with speed is quite speed in the speed range 17 miles an hour to 21 miles an hour. But in the speed range above 25 miles an hour, the relationship between output and wind speed is very nearly linear, because of the way in which the aerodynamic, mechanical, and electrical efficiency curves are combined in this particular design. For any speed-frequency distribution curve, there is some most economical combination of turbine speed and generator rating.

Knowing the output in kilowatts for any value of wind speed in miles per hour, it is possible to compute the annual output at a site whose long-term speed-frequency distribution curve is known by converting the speeds to outputs, multiplying by the hourly distribution, and summing to determine the total annual output expressed in kilowatt-hours.

A wind-turbine whose design is economical will show a maximum over-all efficiency of about 35 percent, usually at some low value of wind speed, say 18 miles an hour, and will convert to electrical energy about 6 percent of the energy in the wind which annually passes through the disc area.

The actual monthly frequency distributions of speed during 60 months at Mt. Washington and at Grandpa's have been converted into computed outputs. These outputs, with the output scale converted to annual output in kilowatt-hours per kilowatt, are plotted against the average speed for each of the 120 months, in Fig. 33 (Ref. 20-B). There is considerable dispersion, as would be expected. The mean curve is also plotted. This curve refers to the test unit design, namely, a rated capacity of 1250 kilowatts, turning at 28.7 revolutions per minute, in wind

TABLE 10. SUMMARY OF THE RELATIONSHIP BETWEEN MEAN ANNUAL SPEED AND UNIT OUTPUT

Station	Period	Yrs. of Obs.	Type	Assumed Speed-up Factor	Mean Annual Speed in mi./hr.	Output in Kwh./Kw.		
						Sea Level	4000 Ft.	10,000 Ft.
1. Boston 15	1934–1938	5	Smoothed	1.00	9.37	195	169	132
2. Nantucket	1930–1934	5	Smoothed	1.00	14.95	903	807	653
3. Grandpa's	1940–1945	5	Smoothed	1.00	16.51	1263	1133	929
4. Blue Hill	1933–1939	7	Computed	1.00	18.58	1973	1801	1523
5. Blue Hill	1933–1939	7	Computed	1.20	22.28	3026	2819	2477
6. East Mt.	1931–1939	9	Computed	1.15	26.93	3809	3633	3338
7. East Mt.	1931–1939	9	Computed	1.40	33.04	4749	4564	4274
8. Mt. Washington	1933–1940	8	Smoothed	1.00	40.79	5917	5787	5558

whose average annual density of 1195 grams per cubic meter is characteristic of an elevation of 2000 feet in Vermont.

The variation of annual output with variation in mean annual speed has also been computed for a unit rated at 1500 kilowatts and turing at 30 revolutions per minute, in wind whose density of 1125 grams per cubic meter corresponds to the annual average found at 4000 feet in Vermont (Fig. 34). This curve passes through eight points, determined by multiplying the eight speed-frequency distribution curves listed in Table 10 and some of which are shown in Figs. 35 and 36, by the power-speed curve of Fig. 50. The speed-frequency distribution curves had all been smoothed, and it will be noticed that the resultant points in Fig. 34 fall on a fair S-curve similar to that of Fig. 33, but with small dispersion.

With the curve of Fig. 50, it is possible to estimate the output of this particular design at any site (in interior New England); where the mean speed at hub height is known. A similar curve (for interior New England) could be prepared from these same speed-frequency distribution curves for any other design for which the power-speed curve was known; *and the same sort of curve could be prepared for any other region for which a suitable collection of speed-frequency distribution curves was available.*

It must be recognized that actual output over a 20-year period may depart by a few percent from that indicated by entering the smooth curve of Fig. 34 with the value for a 20-year mean speed, because of skewnesses in the monthly and annual speed-frequency distribution curves. And, of course, in the case of an individual year, the departure of actual output from that indicated by entering Fig. 34 with the observed mean speed for the year may be as much as ± 10 percent.

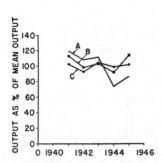

Fig. 51. Fluctuations in the annual output during a 5-year period, expressed as percentage deviations from the mean output for the five years.

Curve A   Grandpa's
Curve B   Blue Hill
Curve C   Mt. Washington

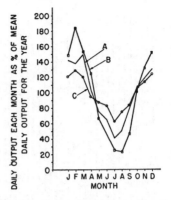

Fig. 52. Seasonal variations in output computed from five years of records.

Curve A   Grandpa's
Curve B   Blue Hill
Curve C   Mt. Washington

THE PERIODIC FLUCTUATIONS IN WIND-POWER

In Fig. 51 are plotted the computed *annual* outputs during the 5-year period, 1941-1945, at Grandpa's, Blue Hill, and Mt. Washington, each expressed as a percentage of the respective mean outputs for the five years.

Grandpa's, having the lightest of the three wind regimes, fluctuates from plus 19 percent to minus 26 percent, showing less stability than either the seacoast station or the high mountain station.

The computed *monthly* outputs for these stations during this period are shown in Fig. 47. While the general trends are similar, some interesting comparisons are seen to occur in February, 1942; November, 1942; January, 1943; March, 1944; and January, 1945. They have not been investigated.

In Fig. 52 are shown the variations in computed output for the three stations throughout the year, expressed as the percentage of the mean daily output for the

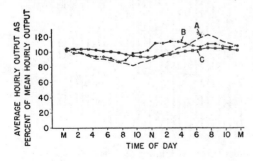

Fig. 53. Diurnal variation in output.

Curve A   Grandpa's
Curve B   Blue Hill
Curve C   Mt. Washington

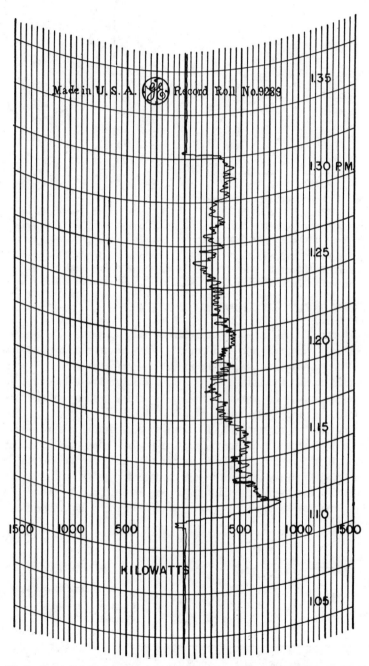

Fig. 54. Fluctuation of output during a 20-minute run on January 25, 1943, in wind varying from 19 miles per hour (output at 75 kilowatts) to 28 miles per hour (output at 870 kilowatts).

year, averaged over the five years. Thus at Grandpa's the daily computed output during February averaged 138 percent of the mean daily output for the five years; and in July, 41 percent.

It will be noted that Grandpa's *seasonal* variation in computed output stands in the middle between the more regular computed output of Mt. Washington and the less regular computed output of Blue Hill.

In Fig. 53 is shown the computed mean *diurnal* variation in output at the three stations (Ref. 21-B). The least diurnal variation occurs at Mt. Washington, the most at Grandpa's, and the sea-breeze effect is noticeable at Blue Hill. At both Mt. Washington and Grandpa's, the maximum speed occurs from 7 P.M. to 8 P.M., while the minimum occurs 2 hours later at Mt. Washington than at Grandpa's.

In Fig. 54 is reproduced a typical output strip-chart for a 20-minute period, showing the type of *short-term* fluctuation in output generated by the test unit when the wind speed was varying from about 19 miles an hour (output at 75 kilowatts) to about 28 miles an hour (output at 870 kilowatts). The period covered was from 1:12 P.M. to 1:32 P.M., January 25, 1945.

RELIABILITY OF WIND-POWER

The basis for study is the variation in computed aeroelectric output at Blue Hill, Grandpa's Knob, and Mt. Washington.

Figs. 55B, C, D have been drawn to show the maximum positive and negative departures in the daily output of each month, expressed as a percentage of the mean daily output for each month during the 5-year period, 1941-1945, for each of these four stations.

It will be seen that departures from expectancy, as established by five years of experience, tend to be least at Mt. Washington, with Grandpa's tending to be more reliable than Blue Hill.

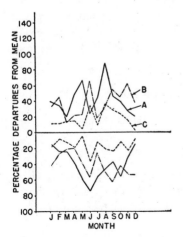

Fig. 55. Reliability of wind-power. Maximum positive and negative departures in computed monthly output from the expectancy for that month as measured over a 5-year period, for each of three stations.

Curve A   Blue Hill
Curve B   Grandpa's
Curve C   Mt. Washington

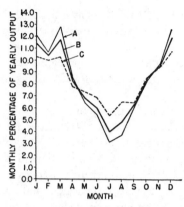

Fig. 56. Average seasonal distribution of output, computed from five years of records at two stations and interpolated for the proposed 9000-kilowatt installation at Lincoln Ridge.

Curve A   Grandpa's
Curve B   Lincoln Ridge
Curve C   Mt. Washington

As might be expected, wind output is most reliable in the fall, spring, and winter, and least so in mid-summer.

In five years of experience at Grandpa's the computed output for one month never exceeded 163 percent of the expectancy for that month, or fell below 39 percent of it. The limits for an individual year were 120 percent to 78 percent.

For the proposed 9000-kilowatt installation on Lincoln Ridge (4000 feet), we have estimated that the reliability would be greater than at Grandpa's but less than at Mt. Washington, and that the average monthly fluctuation throughout the year would be approximately as shown in Fig. 56.

SHORT-TERM PREDICTABILITY OF WIND-POWER

If utility power dispatchers on an integrated system containing some fuel burning capacity knew some hours in advance what the wind was going to do, they could at times reduce some of the "floating" and "banked" standby steam capacity, thus endowing short-term predictability with economic value.

Accordingly, Willett in Cambridge attempted to predict 30 hours in advance, with confirmations 24, 18, 12, and 6 hours in advance, which one of three conditions as listed in Table 11 would prevail during a 6-hour period at the test site at Grandpa's.

TABLE 11.   CONDITIONS TO BE PREDICTED

No output	No winds of more than 17.5 miles per hour
Some output	
Full output	No winds of less than 34 miles per hour

The results of several hundred predictions, analyzed only qualitatively (Ref. 23-B), show that the correlation factors were high enough to be useful, running from about 0.70 for the 30-hour prediction to over 0.90 for the 6-hour prediction.

HOW TO SELECT A WIND-POWER SITE

Aerodynamic Criteria

After five years of increasing familiarity with the problem of site selection, we can point to no analogy between the profiles of mountains and the profiles of airfoils by which one can predict mean wind speeds at hub height within limits which will be useful.

For example, the Horn, the 4100-foot spur lying to the north of Mt. Washington (6288), has been shown in Chapter 4 to be windier than the summit. *Yet directly up-wind of the Horn are two ridges, each higher than the Horn; and the lower and nearer of these ridges is well timbered on most of its crest, with evidences of far less wind in general than on the Horn.*

Another example is Mt. Ellen (4135 feet), the northern end of the 3.0-mile Lincoln Ridge. The southern end is Mt. Abraham (4040 feet). In Fig. 23 the profiles are compared. Each summit is a low nubbin sticking up above the mean height of the ridge (4000 feet). Each is a terminating shoulder, with a deep ravine to the north and south, respectively. Aerodynamically, there is little to choose. Yet the ecological evidence is very strong that the output at Abraham would be much larger than the output at Ellen.

A third example is that of Grandpa's (2000 feet) and Seward (2100 feet). They lie within 1½ miles of each other (Fig. 25) and their profiles are similar, yet the output at Seward, computed from anemometer records, is 30 percent greater than at Grandpa's, for reasons which we do not understand.

So far as aerodynamic criteria are concerned, we must sum up by saying that we lack the basis for arranging wind-power sites even in an order of merit.

Ecological Criteria

An ecological yardstick has been developed for use in the Green Mountains and the White Mountains of New England as described in Chapters 3 and 4. Where trees are deformed by wind, the long-term mean annual wind speed at specimen height may be estimated directly by a trained observer. The criteria of tree deformation by wind are striking, and, being easily taught, are ideal as the basis for preliminary reports by hunters, timber cruisers, fire wardens, and similar persons frequenting mountain crests.

Summary of the Criteria for the Selection of a Wind-Power Site

In sum, and repeating the lessons of Chapter 4, we believe good wind-turbine sites will be found at certain points on ridges lying athwart, or nearly athwart, the flow of the prevailing wind; that they are not likely to be found among upwind foothills, although they may be found on down-wind shoulders; and that the deformation of the trees will serve to arrange a list of potential sites in an order of merit and will indicate whether the wind-flow at a potential site is sufficiently interesting to warrant a program of anemometry.

SPECIFICATION FOR A REGIONAL WIND-POWER SURVEY

Case 1. The Region is Timbered. Until more field work is carried out, there is no reason to believe that the ecological yardstick developed in the Green Mountains and the White Mountains of New England, and described in Chapters 3 and 4, would have the same scale values in other regions. However, there is no reason to doubt that ecological criteria would permit arranging windy sites in any timbered area in an order of merit, and therefore it would seem logical that the first steps in any regional survey of wind speed in timbered regions should consist of field trips by an ecologist.

If a meteorological control station, with anemometer records for at least a 10-year period, does not exist within 50 miles in distance and within 2000 feet in elevation, of the sites to be surveyed, then it will be necessary to create such a control station, using the ratioing method described in Chapter 4.

Accordingly, we would suggest the following steps for a regional wind-power survey in timbered country:

1. An illustrated questionnaire directed to hunters, timber cruisers, fire wardens, high trappers, hikers, and others, to bring out ecological evidence of local high winds in regions thought to be generally windy, and in which occur ridges lying athwart the prevailing wind direction. The illustrations should cover brushing, throwing, flagging, clipping, and carpets.

2. Field trip by an ecologist accompanied by a meteorologist, to evaluate the evidence collected by the questionnaire, and to arrange potential sites in an order of merit.

3. If necessary, the creation of a mountain-top anemometer control station, the wind speed at which is ratioed into the long-term mean speed of the local free-air, the latter being determined in turn by ratioing the local radar pilot-balloon runs to data from the nearest upper-air station, whose 20-year average speed is taken as the base.

The exposure of the anemometer is everything. Exposures down-wind of even such open structures as the Christmas Tree result in errors of 5 percent over a long term. An anemometer should be mounted at mast head, well in the clear, and any lower anemometer should be mounted on a strut at least 10 mast-diameters in length, and pointing up into the prevailing wind, at 45 degrees to the prevailing direction.

4. The long-term mean speed, at observation stations lying within 50 miles in radius and 2000 feet in elevation of the control station, is determined by ratioing short records into the record at the control station, using selected data, as described in Chapter 4.

5. If this program is made to include the "mean" periods, usually April-May or September-October, uncertainties in ratioing the observational data to the control data will be held to a minimum, and 90 days of selected record, in conjunction with the ecological evidence, should provide estimates close enough (± 20 percent) to guide policy, since in the range of mean speed at hub height (140 feet), that is, from 25 miles an hour to 45 miles an hour, output is a nearly linear function of mean speed, and the uncertainty in the predicted output would, therefore, be about the same as the uncertainty in the predicted mean speed (Fig. 33).

6. Once the wind regime for the region is established on a quantitative basis, the ecological yardstick, which heretofore has been lacking, can be supplied with confidence, and preliminary estimates of output at many timbered sites can be made directly, without anemometry.

Case 2. The Region is not Timbered. Where the region is not timbered it will be necessary to set up many more observation stations. As explained in the previous section, we are not yet able to study a range of mountains and, based on geometry alone, arrange a group of potential sites even in an order of merit. For this reason, a wind speed survey in an untimbered region would consist solely of Steps 3, 4, and 5.

SUMMARY

1. The power in the wind in kilowatts may be given by the formula

$$\text{Power} = \text{Kw.} = 2.14\rho A V^3 \times 10^{-3}$$

where

ρ = the density, in slugs per cubic foot $= \dfrac{M}{g} \cdot \dfrac{1}{\text{ft}^3}$;

A = the projected area swept by the turbine, in square feet $= \dfrac{\pi D^2}{4}$;

V = wind speed, in miles per hour.

2. *Estimating Annual Output.* A curve has been developed which gives the annual output (in interior New England), for a specific design, when the mean annual speed is known. The method may be applied to any region for which speed data are available.

3. *Criteria of Site Selection.* The best criterion for a survey of wind speeds in windy forested country, preliminary to anemometer measurements, is the deformation of coniferous evergreens.

4. *Specifications are Suggested for a Regional Wind-power Survey,* based on experience summarized in Chapter 3.

6
Designs of Other Big
Windmills, 1920-1933

Toward the close of the First World War, and immediately following it, scientists in France, Germany, and Russia became interested in developing a modern theory of the windmill, in conjunction with war-born progress in the theory of the airscrew. Joukowsky, Drzewiecki, Krassovsky and Sabinin in Russia, Prandtl and Betz in Germany, and Constantin and Eiffel in France, were the architects of modern windmill theory. Betz was the first to show that no windmill could extract more than 16/27 (about 59.3 percent) of the energy passing through the area swept.

The war had stimulated developments, not only in propeller design, but also in light metals and in radio. Demobilized youngsters, with Yankee ingenuity, mounted the propellers from their Curtiss Jennys on the tops of barns to drive small battery-charging sets for their homemade radios; and engineers were active in designing larger windmills, in accordance with modern theory.

It is not a purpose of this chapter to attempt a definitive history of the development of the theory and practice of windmill design. In the Bibliography there is a fairly complete list of source material for such a history. Rather, I shall review the principal designs of the largest windmills, and indicate why no one of them appeared quite satisfactory.

Three paths were followed by the pioneers. A group, including Flettner and Mádaras, sought to harness the Magnus effect, the thrust exerted by a cylinder spinning in a wind-stream (Fig. 57). Flettner had crossed the ocean in the rotor ship *Baden-Baden*, propelled by this thrust. Another path was explored by Savonius, in Finland, who built S-shaped rotors, with the axis vertical. A third group, Kumme and Bilau in Germany, Darrieus in France, the staff of the Central Wind Energy Institute in Moscow, Fales in the United States, and others, had evolved designs, some small, some large, which were variations of the high-speed propeller type.

It is convenient to distinguish between windmill designs by means of the tip-speed ratio—the ratio between the peripheral speed of the tip and the true wind speed. Dutch windmills possessed a low tip-speed ratio, of 1.0 or less. "American" multi-bladed windmills had a higher ratio, of about 2.0. Modern windmills of the propeller type were high-speed, with working ratios of 5.0 and more. A low tip-speed ratio is associated with low efficiency but with the ability to start under load. A higher ratio means higher efficiency, but inability to start under load.

Fig. 57. Flettner's rotor ship "Baden Baden" which successfully crossed the Atlantic in 1925.

Flettner in 1926 built a windmill (Ref. 4-A), the four blades of which were tapered rotating cylinders, driven by electric motors. With a diameter of 65 feet 8 inches, it was rated at about 30 kilowatts in a wind of 23 miles per hour. Each cylinder was 16 feet 5 inches long, 35.5 inches in diameter at the outer end, and 27.5 inches at the inner end. The tower was 108 feet high. The windmill drove a direct current generator through a 1:100 transmission.

The disadvantages of this design seemed real enough to warrant discarding it on von Kármán's judgment, without a rigorous analysis. Von Kármán pointed out that a rotor-type turbine is considerably less efficient, per unit area swept, than the propeller-type, without compensating savings in weight; that regulation to constant torque, by control of the speed of spinning, would be made expensive by the momentum of the spinning masses; and that, finally, the higher thrust would require greater tower investment per kilowatt. Thus there seemed no inducement to explore the Flettner design further.

Mádaras proposed harnessing the Magnus effect in a different way (Ref. 5-A). On a circular track he would operate a train of flat cars. On each car would be a spinning cylinder, 90 feet hight by 28 feet in diameter, driven by an electric motor. A component of the Magnus thrust would lie parallel to the track, and would propel the train. The component of the thrust at right angles to the track would be absorbed in the heat of friction at the contact between the flange and the rail. Electric generators in the car axles would generate power, transmitted by the rails to a switchboard for distribution. The cylinders would reverse rotation twice in each circuit of the track.

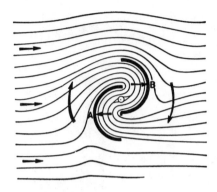

Fig. 59. Horizontal cross-section of wing rotor. Note that "A" is propelled by wind spilling from "B".

The system had a high total of aerodynamic, mechanical, and electrical losses and did not lend itself to the mountain-top locations where the wind was to be found. Further, its cylinders for extracting energy from the wind were, of necessity, mounted close to the ground, in the zone of frictional retardation, where the wind speed is low.

A single full-scale cyclinder was constructed on a concrete base at Burlington, N.J. (Fig. 58) and put under test in October, 1933. The Magnus thrust measured when the cylinder was spun in a light wind conformed to previous experience, but did not provide the economic basis for successful competition with other forms of power generation.

The *Savonius* rotor (Ref. 6-A) was a vertical cyclinder sliced in half from top to bottom, the two halves being pulled apart by about 20 percent of the diameter (Fig. 59). In principle it resembled a cup anemometer, with some recirculation of flow which in effect converted that cup which was coming backwards up-wind, into the second stage of a two-stage turbine.

The Savonius design possessed fairly high efficiency (31 percent), but was inefficient per unit weight, since all the area swept was occupied by metal (Fig. 60). Thus, in order to develop 1000 kilowatts in a wind of 30 miles per hour, a cylinder about 100 feet in diameter by 300 feet in height would be required, presenting 30,000 square feet of metal to do work which could be done by a two-bladed 175-foot turbine with about 1/30th as much metal area.

The *Kumme* design (1920) (Ref. 7-A) (Fig. 61) consisted of six blades, with rather complex guying, and was presumably operated at a moderate tip-speed ratio. The vertical generator on the ground was an attractive idea, which, however, did not stand up under economic analysis. When adequate provision was made to absorb the forces which cause the windmill to "walk" around the horizontal bevel gear aloft, throwing it out of yaw; and to keep the long flexible shaft aligned, it was found that the generator on the ground was more costly than aloft.

The *Darrieus* design was one of great refinement and elegance. In 1929 the Compagnie Electro-Mécanique erected at Bourget, France, a two-bladed unit, 20

Fig. 58. The test of the pilot model of the Mádaras rotor at Burlington, New Jersey, October 1933.

Fig. 60. The Savonius wing rotor arranged for pumping water.

Fig. 61. The Kumme wind-turbine built in Germany in 1920.

meters in diameter (Ref. 8-A) (Fig. 62). The operating tip-speed ratio was very high (10.0), probably higher than could be justified economically. It can be seen that coning was frozen, giving a built-in sweep-back, which reduced the operating bending moments at the blade roots. Since the shaft was horizontal, sweep-back also provided stability in yaw, permitting the elimination of a tail vane or a yaw mechanism.

Fig. 62. The 20-meter Darrieus wind-turbine erected at Bourget in France in 1929.

For a small windmill, without great differences in speed at different points throughout the disc area, sweep-back seems a sensible way to reduce the average working stress at the blade root. But in a disc area so large that steep gust gradients are the rule, I felt that coning would be better. As will be told, in the test unit we began operating with the coning strongly restrained, but we found that the freer the coning the smoother the operation.

Sweep-back will not, of course, give stability in yaw when the axis of rotation is inclined to the horizontal. But a horizontal shaft incurs more debits than credits.

The *Russian* design was bold and practical, if one has in mind the limitations under which they were then working. In May, 1931, after two years of wind measurement, a wind-turbine 100 feet in diameter was put in operation on a bluff near Yalta, overlooking the Black Sea, driving a 100-kilowatt, 220-volt induction generator, tied in by a 6300-volt line to the 20,000-kilowatt, peat-burning steam-station at Sevastopol, 20 miles distant (Fig. 63) (Ref. 9-A).

Regulation was by pitch control. Pitching moments were brought nearly to balance by adjustable counterweights on struts, and pitch was varied as a function of rotational speed, by means of centrifugal controls acting on the offset ailerons. Fluctuation in output in a 7-minute run ranged from about plus 20 percent to minus 15 percent.

Thrust was taken up by an inclined strut whose heel rested on a circular track on the ground. Automatic control of yaw was obtained by driving this strut around the track, by means of a 1.1-kilowatt motor responsive to a wind-direction vane aloft. Hub height was about 100 feet.

The generator and controls were aloft in a streamlined house. The axis of rotation was inclined at 12 degrees to the horizontal, but in the sense that the up-wind end was high. The blades were up-wind of the tower. The main gears were of wood. The blade skins were of roofing metal. The maximum aerodynamic efficiency at 30 revolutions per minute was 24 percent, reached at a tip-speed ratio of 4.75. The wind speed at which rated output was reached was 24.6 miles per hour.

In a wind regime characterized by a mean annual speed of 15.0 miles per hour, the annual output reported was 279,000 kilowatt-hours; generated at an average level of 48.4 kilowatts in the windy month of March; 18.0 kilowatts in the quiet month of August; and 32.0 kilowatts for the year.

After two years of experimental operation, it was planned in 1933 to add two units of 100 kilowatts each, but of varying design, as steps intermediate to the ultimate installation of the 5000-kilowatt units called for by the Second Five-Year Plan.

I felt that the principle weaknesses of the Russian design—low efficiency, crude regulation and yaw control, high weight per kilowatt, and induction generation—had been imposed upon the designers by the state of industry in Russia, where heavy forgings, large gears, and precision instruments were unavailable.

Another proposal considered was that of *Fales,* who had been among the first to become interested in high-speed windmills. He proposed a single high-speed

Fig. 63. The 100-foot, 100-kilowatt DC Russian wind-turbine erected near Yalta on the Black Sea in 1931.

Fig. 64. Design proposed by Honnef of Berlin in 1933. This wind-turbine was to stand 1000 feet high and the inventor rated it at 50,000 kilowatts.

blade, suitably counterweighted. Among the objections to this was the compelling one that, under icing conditions, the counterweight could no longer be relied upon to counterbalance the iced-up blade.

Honnef of Berlin suggested a bold design in 1933, a model of which is shown in Fig. 64 (Ref. 10-A).

Impressed by the increase of wind speed with height above the level ground, he proposed that the tower be "about 1000 feet in height, or even higher." He shows a well-designed tower, which is particularly strong in torsion. In an effort to reduce the tower cost per kilowatt, Honnef designed it to support five 250-foot turbines, in a light framework free to yaw, and supported as a unit on bearings at the head of the tower. Gears are a costly bottleneck in any design of a large windmill. Honnef solved this problem by eliminating it. His generators were built into his turbines, each of which was double, consisting of two counter-rotating members. At the rim of each of the two turbines in a pair were located suitable copper and iron, such that one member of the pair acted as the rotor and one as the stator in a direct current generator. The two speeds were 10 and 17 revolutions per minute, giving an equivalent generator speed of 27 revolutions per minute.

I felt that Honnef had exaggerated the importance of height. Knowing the rate at which wind speed increases with height and also the rate at which tower cost increases with height, it is possible to plot, for any disc area, the increase in annual output in kilowatt hours with increasing tower height; and, likewise, the increase in the total investment of the wind generating station as tower height increases. In collaboration with the American Bridge Company, and my meteorological colleagues, I had studied this relationship for a variety of tower designs and turbine sizes, and had concluded that, for a 200-foot diameter turbine on a ridge, there was no justification for a tower higher than 150 feet. It did not seem likely that the most economical height over level ground would be much greater.

Further, the Honnef scheme of a panel of five 250-foot turbines presented such formidable problems in stress analysis that it had no place in any first attempt, which should be limited to one large turbine.

Finally, I could find no one who could think of an economical structure whereby the air-gaps in the five 250-foot generators could be so maintained as to yield reasonably high electrical efficiencies when such a structure was exposed to the heterogeneous and violent stresses imposed by wind, sun, and ice.

Many other designs were considered, including several variations of the attempt to extract energy from cross-sections of the wind-stream larger than the diameter of the turbine disc. This proposal is an enticing will-o'-the-wisp, which so far has not been realized.

SUMMARY

A review of the prior art led me to propose a design of my own. S. Morgan Smith Company considered the prior art, and adopted my design as the basis for engineering studies of what became known as the Smith-Putnam Wind-Turbine.

7
Development, Fabrication, and Erection of the Test Unit, 1939-1941

SELECTION OF BASIC DESIGN

No one of the published designs of the larger wind-turbines, described in the previous chapter, appeared attractive economically.

It seemed to me that the most economical design would be a larger wind-turbine than any yet built, with two, or possibly three propeller-type blades, having a radius about three-quarters the height of the supporting tower, operating at a high tip-speed ratio, and located on a ridge which would increase the "free-air" speed of the wind.

"Three-quarters" was a rough number, which merely illustrated my conviction that, if it was unprofitable to extract energy from the frictionally-retarded winds close to the ground, it was equally uneconomical to debit the output of a small turbine with the cost of a tall tower underneath it.

Of course I had no notion of the most economical proportions of the turbine, or the proper shape of the ridge. Before the first of these two problems could be investigated, it was necessary to clarify the basic design. I will not follow all the threads of the development of the main features of the design, which can be summarized as follows:

1. Ability to Spill the Power from High Winds

A wind-turbine to produce 1000 kilowatts from a 30-mile-an-hour wind of sea-level density, and operating with an over-all efficiency of 30 percent at its rated capacity, must have a diameter of 175 feet (Chapter 5). But this great structure must be gale-proof, a need which implies the ability to reduce the turbine area somehow, when a gale impends. To accomplish this reduction one could allow the turbine disc to swing out around the vertical axis until the disc lay parallel to the wind (Fig. 65A); or one could accomplish the same result about the horizontal axis (Fig. 65B); or one could feather the blades to a position of low lift and minimum drag. The latter seemed to be the most attractive solution mechanically, and the gustiness of the wind made it preferable aerodynamically. Furthermore, the mechanism for

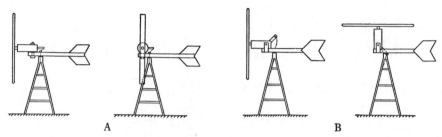

Fig. 65. A conventional means for spilling power in high winds. (A) Schematic device for rotating the turbine out of the wind about the vertical axis. (B) Schematic device for rotating the turbine out of the wind about the horizontal axis.

feathering the blades would be merely an extension of the means for controlling the pitch; and control of the pitch had been selected as the method of regulation.

2. Regulation of Torque Input

Hamilton Standard Propellers and others had developed controllable pitch to the point where it was accurate and dependable. Kaplan in Czechoslovakia had introduced the principle to the hydraulic turbine known in this country as the Smith-Kaplan turbine. In each case the pitch was controlled by the action of a centrifugal fly-ball speed governor.

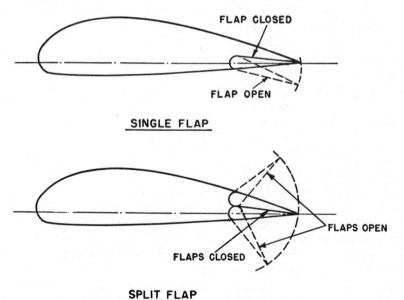

Fig. 66. Flaps, for the purpose of controlling torque.

In the hands of the Woodward Governor Company, this type of control had been developed until it would respond to a speed change of as little as 0.01 of 1 percent.

Other methods of achieving regulation were possible. Instead of altering input torque by altering the pitch of the blades, one could alter it by flaps (Fig. 66), or by spoilers. Any one of these three means of controlling input torque could be controlled by a torque-meter rather than by a speed governor.

Although von Kármán pointed out some of the advantages of flap control over pitch control, less was known about it, and Frank Caldwell, Chief Engineer of Hamilton Standard Propellers, and the Engineers of the Budd Company recommended the adoption of pitch control, at least for the early test units, since it did not then seem likely that the economic differences between these two controls would make or break the project.* This argument applied even more forcefully in the case of the torque meter, which would require a heavy development program before we could feel the same confidence in it that we felt in a Woodward Governor.

So, I decided to specify the regulation of input-torque by control of pitch through a Woodward fly-ball speed governor.

It was clear that regulation by any means would be assisted by a high inertia in the turbine—that is, the power output of a large heavy turbine would be smoother than that of a small light turbine. So, the larger the most economical unit proved to be, the better, so far as regulation was concerned.

3. Number of Blades

It was found that the increase in annual output from the addition of a third blade would be about 2 percent, based on estimates by von Kármán (Ref. 3-B), and would not justify the increase in the investment. Accordingly, the two-blade design was adopted.

4. Coning

To reduce the heavy bending moments in the root sections of the blades, I proposed to allow the blades to cone.

By coning is meant the freedom of each blade, independently of the other, to move down-wind (positive coning) or up-wind (negative coning), through the angle illustrated schematically in Fig. 67. The instantaneous value of this angle is the result of a balance between the wind pressure on the blades, the centrifugal force, the angular momentum of coning, and any coning-damping that may be applied.

The idea was not new. Autogyro rotor blades were free to cone about a vertical axis. Propellers had been built free to cone a few degrees, with a coning angle that

---

*But see Chapter 12.

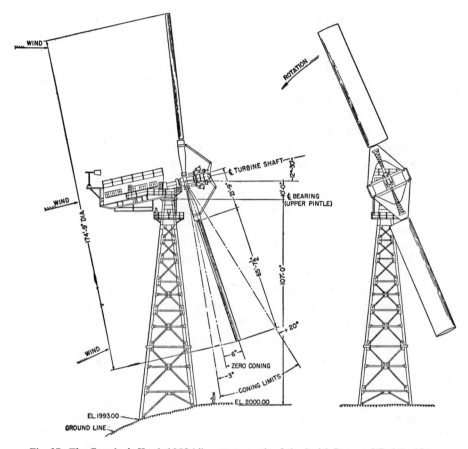

Fig. 67. The Grandpa's Knob 1250-kilowatt test unit of the Smith-Putnam Wind Turbine.

was forward. Later, propellers and wind-turbines were designed without freedom to cone, but with a built-in coning angle, in order to keep the bending moments to a minimum under working conditions. But built-in coning did not seem adequate when dealing with diameters of 175 feet, with each blade being loaded by the local wind with which it was immediately in contact. In order to allow my blades to cone, it was necessary to mount them down-wind of the tower. This provided an added advantage, by permitting inclination of the axis of rotation.

5. Inclination of the Axis of Rotation

As previously explained, I counted on finding in New England, lying athwart the prevailing westerly winds, a ridge whose aerodynamic characteristics would tend to speed up the free flow of the wind. The streamlines would be most compressed just over the crest of the ridge, so the turbine disc would be located there (Fig. 12). But at this point the streamlines would lie at some angle to the horizontal, say 10 degrees or 20 degrees (Fig. 67). Therefore, it would be best if the turbine disc could somehow be tilted out of the vertical in order to lie more nearly at right angles to the streamlines.

Mounting the blades down-wind permitted tilting the axis of rotation at will. Nothing was known about the best angle of tilt. In the first place, I did not know the average angle of inclination of the streamlines. In the second, I did not know how the output from the disc area would vary as the plane of the disc was tipped out of the wind. I assumed that on any ridge whose profile was not so abrupt as to cause severe rupture of the flow (Fig. 11), the streamline angle would be less than 20 degrees. Von Kármán assured me that the output would not be sensibly diminished by tipping the plane of the turbine disc out of the wind through a small angle (Ref. 3-B).

There was a further consideration. A main shaft with a long overhang would cause high bending moments. Minimum overhang was desirable. A study was made of tower shapes to find the most economical slope of the tower legs. The value of 8 degrees from the vertical was determined by Wilbur. Therefore, an angle of inclination of the axis of 8 degrees, with minimum overhang of the shaft, would cause the blade tips to pass close to the tower legs. To avoid this close passage and to provide for some negative coning, I chose the value of 12½ degrees for the inclination to the horizontal of the axis of rotation, feeling that this would lie within ± 5 degrees of the prevailing mean flow, and would provide adequate clearance for the blade tips (Fig. 67).

6. Yaw

Small windmills are usually faced into the wind by a tail vane. The large Dutch windmills were "walked" into the wind by a beam on a wheel rolling along a circular path on the ground. The Russians copied this idea, adding an electric motor to drive the heel of the beam around a mono-rail track.

A tail vane adequate to control a large wind-turbine in yaw must possess a great area (or the equivalent) on an arm of some length. It seemed simpler to cause the turbine to follow in yaw the indications of a yaw vane (Fig. 67), by means of suitable servo-mechanisms and a yaw motor capable of driving the turbine around a bull-gear at the head of the tower.

7. Location of the Generator

If the generator could be located on the ground, and mounted vertically at the foot of a vertical shaft driven by a system of bevel gears aloft, it would be easier to maintain, and weight aloft might be reduced. But studies of such a layout, in comparison with one in which the generator was aloft, showed an advantage for the latter design of several dollars per kilowatt. Accordingly I decided to specify that the generator be mounted aloft.

8. Type of Generator

Knight told me that most utility customers would prefer a synchronous generator, and that, if an induction generator was supplied, it would usually be necessary to back it up with some condenser capacity. Studies were made of these two generating systems, and it was found that there was little or no economic advantage in induction generation. Since a system of regulation which proved adequate for synchronous generation would also be adequate for induction generation, and since the latter seemed less generally desirable, a synchronous generator was specified.

The economics of direct-current generation, with subsequent conversion to alternating current, were prohibitive in 1934, and remain so in 1946.

9. Coupling

Once a synchronous generator is phased-in to a high-line, it will tend to stay in synchronism with the alternating current flowing through the high-line. To force it out of phase requires the application of a good deal of torque—perhaps 2 or 3 times the rated torque—such as might occur if the governor should fail.

But a centrifugal speed governor acting against such a violently surging "head" as a wind-stream, requires a generous range of speed-change so long as the generator remains phased-in. To provide a greater speed-change, or slip, we inserted between the generator and the gear a coupling capable of slipping (Fig. 67).

Two types of coupling were available—the electric induction type and the hydraulic type. The hydraulic coupling had been used in many applications. The electric coupling was of recent development and had been used principally in marine service at rather low speeds. In addition, the electric coupling was heavier and more expensive than the hydraulic coupling, which was accordingly specified for the test unit.

The electric coupling proposed for the production unit is a true torque-limiting device, which will limit input torque to the generator to not more than 110 percent of rated torque.

10. Pintle Shaft and Bearings

The simplest solution for the pintle shaft and bearings seemed to be a large vertical pipe supporting the horizontally rotatable (yawable) structure aloft, and in turn supported by radial and thrust bearings in the upper tower structure.

11. Economics of a Battery of Units

An isolated unit would require its own road for erection, its own switchgear, and its own high-line. To reduce these unit costs, a battery of 5 or 10 units was assumed to be tied together at a common switchboard.

THE BEST PROPORTIONS FOR A WIND-TURBINE

Having settled on these design features and being encouraged by Knight, I began a systematic search for the most economical size. With Knight's assistance, prices were obtained on some seventy-odd generator-gear combinations in a wide range of capacities and speeds; and from the Bethlehem Steel Corporation smooth curves were obtained from which it was possible to estimate weights and costs of four-legged towers in a wide range of heights and loadings. Rossby advanced a hypothetical vertical distribution of wind speed over a crest, and I assumed a hypothetical speed-frequency curve, with a mean annual speed of 25.0 miles an hour.

The computations are described in Chapter 9 and are summarized graphically in Fig. 68, which shows that the most economical unit is, apparently, rated at 600 kilowatts, driven by a turbine with a diameter of 226 feet, and able to produce energy at the switchboard at the foot of the tower at a cost of $0.0015 per kilowatt-hour (Ref. 10-B). When talking with others, I first added 0.50 mil to this estimate; later 1.0 mil, bringing the earliest public estimate to $0.0025 (2.5 mils) per kilowatt-hour.

As described in Chapter 9, one of the first steps taken by the S. Morgan Smith Company was to review this estimate. The work was carried out early in 1940 at the Budd Company plant (Ref. 24-B), and led to the conclusion that the best unit would have a diameter lying between 200 feet and 235 feet, a capacity lying between 1500 kilowatts and 2500 kilowatts, a generator speed of 600 revolutions per minute, and blade type 2-M, as shown in Fig. 69.

Such a unit, in the wind regime then assumed by Petterssen to have a mean annual speed of 28.8 miles an hour, would show an energy cost at the switchboard at the foot of the tower of $0.0016 (1.6 mils) per kilowatt-hour.

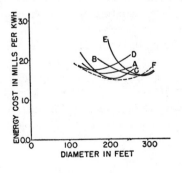

Fig. 68. First approximation of the most economical dimensions. Variation in the cost of energy with variation in diameter, for various generator ratings.

Curve A   250 kilowatts
Curve B   350 kilowatts
Curve C   500 kilowatts
Curve D   750 kilowatts
Curve E   1000 kilowatts
Curve F   The envelope of minimum
                energy costs

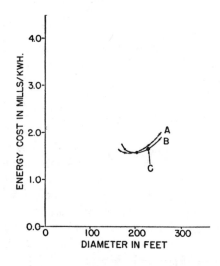

Fig. 69. Second approximation of the most economical dimensions. The variation in the cost of energy with variation in the diameter, for three turbine ratings.

Curve A   1000 kilowatts
Curve B   1500 kilowatts
Curve C   2000 kilowatts

Wilbur found that the best proportions would not vary greatly within the limits of most wind regimes which would be commercially interesting. This point has not been thoroughly investigated, but in studying a large number of wind regimes, we have found no reason to doubt that it is an acceptable assumption.

PROPORTIONS SELECTED FOR THE TEST UNIT

When it was found that the results of our second study confirmed in a general way the results of my first approximation, the S. Morgan Smith Company, on March 10, 1940, decided to move ahead with the project. The question of a small test unit, 25 feet or so in diameter, was explored. It was felt that the secret of smooth regulation lay in high inertia, which might not be provided by a small unit; the design, fabrication, and testing of which would in any case cost nearly as much as would the full-scale unit. To eliminate the risk of poor regulation and other scale effects, the full-scale test unit was decided upon, in the smallest size and on the shortest tower thought to be characteristic of the range of economical sizes. Accordingly, S. Morgan Smith Company selected a rating of 1250 kilowatts, a diameter of 175 feet, a generator speed of 600 revolutions per minute and a hub height of 125 feet, as the proportions for the smallest test unit which would be representative of the most economical size.

As soon as George Jessop had made this decision, the engineering of the test unit began. At first this work was carried out informally by Dr. Wilbur, under my nominal supervision in my capacity as Project Manager, but, in June, on my recommendation, Dr. Wilbur was formally made Chief Engineer of the Project, and this phase of the work thereafter came under his direction.

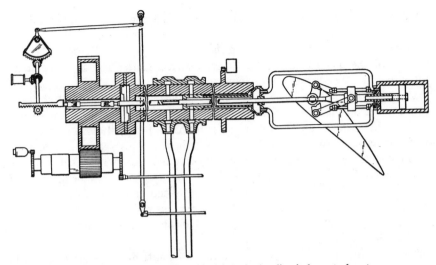

Fig. 70. Schematic representation of the hydraulic pitch-control system.

DESCRIPTION OF THE TEST UNIT

The test unit was a two-bladed wind-turbine, connected through speed-increasing gears and a hydraulic coupling, to a synchronous generator, the whole rotatable on top of a 110-foot tower. Fig. 67 shows the schematic layout of the turbine, with the principal dimensions (Ref. 25-B).

The blades were of constant chord and the cross-section corresponded to the airfoil designated as N.A.C.A. 4418. The blades were supported in the A-frames by shafts, commonly known as the blade shanks, free to rotate in bearings, under the control of the pitch controlling mechanism (Figs. 67 and 1). This mechanism consisted of a hydraulic cylinder connected, through links, cranks, and torque tubes, to the blade shanks (Fig. 70). The A-frames were free to rotate over a limited amount of travel. This movement of the A-frames and the blades, in a plane parallel to the wind direction, is known as coning (Figs. 67 and 1). The limit stops of this motion were spring cushioned and combined with a viscous damping mechanism. The connection between this mechanism and the A-frames was made by means of rocker arms and coning links (Fig. 1). The supporting structure for these damping mechanisms, and also the pitching cylinder, was known as the tailpiece. The tailpiece and A-frames were all connected to the hub-post. The hub-post in turn was connected to the main shaft by means of a taper fit and wedge keys.

The main shaft was a hollow steel forging carried on anti-friction bearings in large cast-steel housings. The center of the shaft contained oil tubes to carry oil in both sides of the pitching cylinder in the tailpiece (Fig. 70), and wires to the instruments which were used to measure coning angle and coning-damping pressures.

There were slip rings on the shaft for the electrical connections and an oil head for the hydraulic connections. The connection between the main shaft and the speed-increaser gear was by means of an Oldham-type coupling. The speed-increaser gear was of the divided-load type, having two jack shafts. The first step-up had spur gears and the second, continuous-tooth herringbone gears. The single high-speed output shaft was connected to the hydraulic coupling through a fast flexible coupling. On this high-speed shaft close to the gear box was placed a gear arranged to be driven either by an electric motor or by hand, for positioning of the blades for maintenance purposes. This mechanism, which was disengageable, also served as a lock to hold the blades in position. The hydraulic coupling was overhung on the generator shaft and coupled directly to it (Fig. 1). The hydraulic coupling was a 48-inch traction-type coupling manufactured by the American Blower Company.

The generator was a General Electric synchronous machine, rated at 1250 kilo-volt-amperes at 2400 volts, operating at 600 revolutions per minute, and with a direct connected exciter.

The entire turbine mechanism was carried on a stiff box-type girder called the pintle girder, which was about 40 feet long by 8 feet wide by 5 feet deep (Fig. 1). A vertical pintle shaft rigidly fixed to this girder was supported by antifriction bearings in the tower cap (Fig. 1). This cap was a weldment attached to the top of the tower, which was of bolted construction.

Attached to this tower cap was a horizontal bull gear 10 feet in diameter (Fig. 1). A mechanism consisting of a hydraulic motor and a speed-reducing gear drove a pinion meshing with this bull gear and rotated the entire assembly aloft above the axis of the tower. This yaw-mechanism was under the control of a damped yaw vane (Fig. 67) which served to keep the turbine always in proper relationship to the direction of the wind.

Also mounted on the pintle girder were the speed-sensitive governor and its auxiliary equipment. This auxiliary equipment consisted of an oil pump mechanically driven by the turbine, an oil pump driven by an electric motor, a sump tank, an accumulator tank, and the necessary piping and controls.

Also mounted aloft were the electrical controls for space and oil heaters, manual controls for the yaw-mechanism, safety devices, and emergency shut-down switches.

The electrical connections were carried down the hollow pintle shaft to a series of slip rings where they were transferred to the stationary structure. The switch-gear was housed in a concrete building some 300 feet from the foot of the tower, and partially protected from flying ice by a shoulder of the summit. The transformers were installed in the open alongside this building where the high-line terminated.

The test unit of the Smith-Putnam Wind-Turbine was designed to be a wholly automatic unattended station. Automatic control operations, shown schematically in Fig. 71, were made dependent on wind speed, as indicated by the action of the propeller under four general conditions. These conditions were:

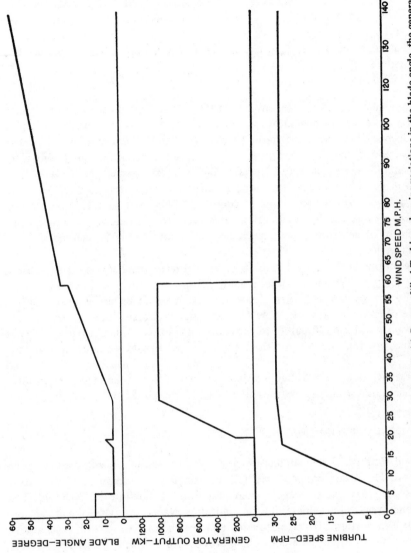

Fig. 71. Schematic control diagram, Smith-Putnam Wind-Turbine, showing variations in the blade angle, the generator output, and the turbine speed, with variations in the wind speed.

1. Wind speed below that necessary to generate power, that is, less than V_{on}.
2. Wind speed above V_{on} and less than that necessary to give rated power (V_{rated}).
3. Wind speed above V_{rated} and less than the maximum allowable operating speed (V_{off}).
4. Wind higher than V_{off}.

In the first condition, the turbine blades were set at a predetermined angle to give not more than 14 revolutions per minute. A speed-sensitive relay closed after a time delay and initiated the starting sequence. The blades were moved to design blade angle and the unit picked up speed. The speed was controlled by the governor at speed-no-load. As the speed increased the governor adjusted the blade angle to keep the speed constant. When the blade angle reached 9 degrees there was sufficient wind to load the generator to about 200 kilowatts. At this point the synchronizing cycle was initiated. Briefly this cycle was as follows: speed matching by control of governor speed adjustment; closing of line breaker; and application of field excitation.

The output in the range V_{on} to V_{rated} was entirely dependent on the wind speed Above V_{rated} the propeller speed, and therefore the output, was held nearly constant by the governor. At first it was thought it would be necessary to take the turbine off the line in winds above 60 miles an hour, because of the greater energy in the gusts and the difficulties of regulation. In this condition the generator was disconnected from the line and the turbine allowed to idle at some predetermined speed (Fig. 72).

The above description is by no means complete. Most of the sequences included time delays, interlocks, and various other devices (Ref. 26-B).

SOME DESIGN PROBLEMS

In completing the design of this test unit, Wilbur was faced with several design problems. His major problem was how to complete the design at all, when the various designing groups were located at California Institute of Technology, in Cleveland, York, Philadelphia, Boston, and other places; and when the design consultants were scattered from coast to coast and partially occupied with other affairs. The onrushing war created a compelling pressure, in the face of which it had been decided to order major forgings in May, 1940, before the design of the structure, or even its stress analysis, had been completed.

This calculated risk turned out badly and contributed to the later failure of one of the blades.

The maximum shank size is determined by the thickness of the blade and the method of connection. At the time of ordering these forgings, the thickness of the blade had been determined only from aerodynamic considerations. The weight of

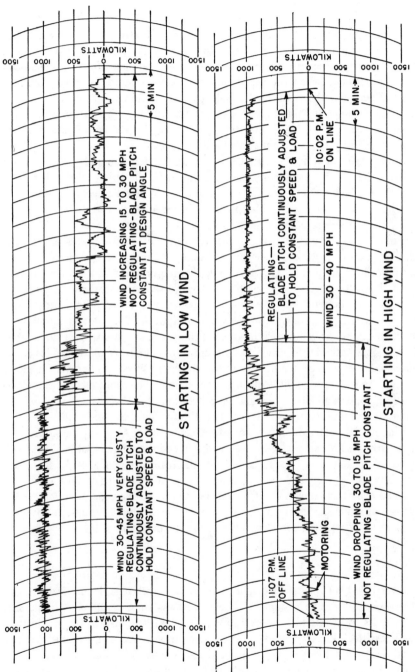

Fig. 72. Operating strip charts of the Smith-Putnam Wind-Turbine showing starting and stopping sequences.

the final blade was as yet unknown. An estimated weight was used, together with approximate aerodynamic loadings, to determine the stresses in the shank and connections. This first analysis indicated that the shanks would be strong enough as then laid out. When the blades were completely detailed, and accurate weight estimates made, and when the aerodynamic loads were more thoroughly studied, it was realized that the shank and shank-spar connection would be the weakest spot in the turbine. By this time it was too late not only to order new forgings, but even to redesign the blades. In order to use larger spars and shanks, it would have been necessary to increase the thickness of the blades.

A similar type of design decision was forced upon Jessop and Wilbur when it was found that 23 5/8-inch bore main bearings were the largest obtainable. This limited them in shaft size to 23 5/8 inches and, in an effort to get the utmost out of the shaft, they took the trouble to roll its surface. In the end it was found that probably there had been ample margin in the stresses of this member.

Having decided upon pitch control in response to speed changes as recorded by the Woodward governor, it became necessary to introduce an element which would permit speed change. As explained, this requirement called for a coupling between the high-speed end of the main gear and the generator; and in 1941 the only hydraulic coupling available was one which had never been continuously operated at a rating equivalent to a generator output of 1250 kilovolt-amperes. In the end, this coupling required some special cooling but otherwise gave no trouble.

The original basic design called for the hinge pins about which the blades coned to be co-axial on the center line of the main shaft. When studies were made to determine the most economical shape of the A-frame and hinge mechanism, it was found that, for ease in fabrication and simplification of the mechanism, it would be desirable to separate the hinges. It was first thought that this separation could be held to about 20 inches on either side of the center line. As finally designed, however, this distance became 32½ inches. While it was thought advisable to keep this distance small, and zero if possible, we did not realize the penalties of increasing it. Actually, the yawing and pitching moments set up by coning about these hinges introduced a certain roughness of operation, so that in the preproduction model, the design of which is described in Chapter 9, the hinge lines have again been made co-axial, at the expense of some complication in the mechanism.

FABRICATION

The tower and blade spars were fabricated in the Ambridge, Pennsylvania, plant of the American Bridge Company. The tower was shipped directly to the site, and the spars to the Budd Company in Philadelphia. The blades, which consisted of a stainless-steel skin supported by ribs built up of stainless-steel structural shapes, were fabricated on these spars. The completed blades were then shipped to the Cleveland Plant of the Wellman Engineering Company, who fabricated the remainder of the structure, and shop-assembled the whole, statically balancing the blades and the rotating system before shipping to Rutland in the Spring of 1941.

ERECTION

The erection of the wind-turbine on Grandpa's Knob presented a number of unusual hauling and lifting problems. In March, 1941, the turbine was shipped by rail from Cleveland, Ohio, to the West Rutland yards of the Vermont Marble Company, where two traveling bridge cranes, of 20-ton and 15-ton capacity, were used to unload the parts. Large trailers then transported the material to the top of the mountain, 10 trips being required to complete the task.

Some loads were heavier than the legal load limit and special permission was obtained from the State Highway Department to move the sections over the State Highway, and two of the bridges on the route had to be reinforced with temporary cribbing. When the blades were moved, crews from the power and telephone companies had to remove and replace low wires across the highways. With the cooperation of these agencies and the State Highway Patrol, the hauling of the turbine from West Rutland to the foot of the mountain took place without mishap, but on the final 2-mile stretch from the foot of the mountain to the top, we were not so fortunate.

The grade on this 2.2-mile road, begun in August, 1940, and completed in six weeks, approached a maximum of 15 percent and in no place was less than 12 percent. We used a half-track in front of the truck and a bulldozer pushing or pulling, as occasion demanded. About 1000 feet below the summit, at a sharp hairpin turn on a steep grade, the pintle girder, weighing about 43 tons, including the parts already assembled on it, broke its lashings and turned over in the ditch by the road, fortunately without serious damage. Three weeks of rigging and blocking were required to get this piece back on the trailer for the remainder of the trip to the top.

The tower foundations had been started in the fall of 1940, as soon as the road was completed. By the early part of December the foundations were in, and the tower was erected during the next two months, this work continuing in temperatures as low as 18 degrees below zero and in winds of 60 miles per hour. Hauling of the wind-turbine parts from West Rutland to the site was done from March 15 to May 1 of the spring of 1941, in a race against spring floods and thawing roads.

The blades were left at the bottom of the hill until all was ready for them at the summit. The pintle girder, with the main shaft assembled on it, was hoisted into place on May 15, 1941. Erection proceeded throughout the summer and the first blade was hoisted into place early in August. After erecting the first blade, it was necessary to rotate the turbine through 180 degrees, lifting the first blade to the vertically upward position, so that the second blade could be put into place (Fig. 73).

Miscellaneous blocking and rigging were then removed and on August 29, 1941, the blades were rotated by wind for the first time.

Fig. 73. Assembling the blades, August, 1940.

SUMMARY

The functional specification of the design was determined in 1937 and the first attempt to discover the best proportions was made in 1939. S. Morgan Smith Company accepted the basic design in 1939, and decided upon the proportions for the test unit, which was engineered under Wilbur's direction, fabricated, erected, and wind-driven for the first time on August 29, 1941, seventeen months after the decision to go ahead, and twenty-three months after the decision to explore the problem of large-scale wind-power.

8
Test and Operation of the Smith-Putnam Wind-Turbine, 1941-1945

Following the adoption of the basic design; its engineering under Wilbur's direction; and its erection on Grandpa's Knob, described in the previous chapter, a test program was carried out, from October 19, 1941, to March 3, 1945 (Ref. 27-B), culminating in the routine operation of the unit as a generating station on the lines of the Central Vermont Public Service Corporation, in the spring of 1945. Operation of the unit ended on March 26, 1945, when a blade failed.

TEST PROGRAM

A test program was laid out to achieve the following aims:

1. Adjust the operation of the test unit until it should run satisfactorily, when it would be turned over to the Central Vermont Public Service Corporation as a generating station.
2. Measure the aerodynmaic and mechanical efficiencies.
3. Measure the stresses, to provide a basis for a production design.

The first aim was achieved, the second approximated, and the third partially approximated (Ref. 22-B).

The risk to observers from flying ice was minimized by nestling the control house into the side of the hill and giving it a heavy concrete roof reinforced with I-beams.

By means of electrical telemetering devices (Ref. 26-B), nineteen indications of pressure, motion, temperature, and electrical output were brought to a single panel in the control house. These were:

1. Coning angle.
2. Wind direction relative to turbine shaft.
3. Up-wind coning damping pressure.
4. Down-wind coning damping pressure.
5. Servomotor pressure to increase blade pitch.
6. Servomotor pressure to decrease blade pitch.

7. Accumulator oil pressure.
8. Yaw motor oil pressure.
9. Blade pitch position.
10. Blade pitch limit.
11. Speed adjustment.
12. Up-wind gear-oil temperature.
13. Down-wind gear-oil temperature.
14. Accumulator-oil temperature.
15. Turbine speed.
16. Wind speed
17. Blade position.
18. Kilowatts.
19. Reactive kilovolt-amperes.

In addition this panel carried a clock, a sweep second-hand, and a date indicator (Fig. 74).

The panel containing these instruments was photographed on 16-mm. film by two cameras operating at speeds of 8 frames per minute and 8 frames per second,

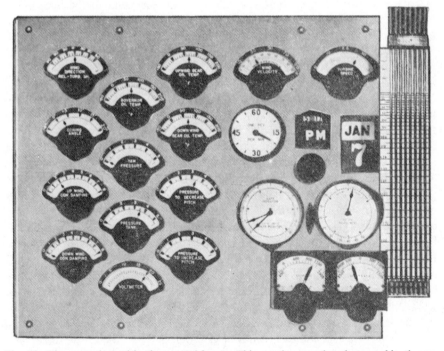

Fig. 74. The control panel in the control house. This panel was under photographic observation by a low-speed camera and a high-speed camera.

respectively. The slower camera was operated continuously when the turbine was in operation. The high-speed camera was run only on occasions when more detailed records were desired.

After processing, the films were viewed in a microfilm news reader or projected on a screen with a 16-mm. projector.

A Foxboro strain-gauge recorder was used to estimate the important stresses. Strain gauges were mounted on the torque tubes for the purpose of investigating flutter in the blades, and were later mounted on the pintle shaft and on a shaft in the yaw mechanism (Ref. 28-B).

From September 29 to October 19, 1941, the unit was gradually brought up to "speed-no-load." At intervals, the blades were checked for flutter by means of the strain gauges on the torque tube and the Foxboro recorder. Dr. J.P. den Hartog supervised this investigation and pronounced the blades free from flutter. After the unit had been brought up to speed, the next step was to adjust the governor so that the speed might be controlled closely. At the same time the electrical circuits were being checked and the individual parts of the switchgear tested. Two weeks were spent in drying out the generator and on October 19 the generator was run on short circuit about one and one-half hours.

Later that same day, at 6:56 P.M., the unit was phased-in to the lines of the Central Vermont Public Service Corporation for the first synchronous generation of power from the wind. The unit was run until 8:35 P.M. carrying loads which varied from 0 to 700 kilowatts. During this time the wind speed varied from 15 to 26 miles per hour from the northeast, and was gusty (Ref. 29-B).

As soon as the turbine had been run it was found that vibration, in both the horizontal and the vertical plane, was greater than was expected. The forces causing motion in the horizontal plane, that is, about a vertical axis, are yawing forces; those causing motion in the vertical plane, that is, about a horizontal axis parallel to the plane of rotation, are pitching forces. Pitching is used here in the nautical sense and has no reference to the pitch of the blades.

The motion produced by the pitching forces was an elastic deformation of the pintle shaft and pintle girder, and sometimes amounted to 3 inches at the up-wind end. The yaw motion was quite different. The turbine was yawable about a vertical axis by means of a hydraulic motor and gearing. This transmission consisted of bevel and spur gears, together with a torque-limiting clutch. It was found that the vibrational forces in yaw, in combination with the component of torque which resulted from the inclination of the axis, produced unbalanced oscillations, the net result of which was to turn the turbine out of yaw to the left when facing up-wind. Wilbur decided, as the first of a possible series of steps to remedy the condition, to insert an irreversible worm gear in the transmission. While this eliminated the progressive creep, the oscillations in yaw were left and these yaw forces were now carried through to the tower, which would absorb the forces to the limit of the torque-limiting clutch in the yaw mechanism. Then the tower would snap back, through as many as several inches at a radius of 6 feet.

The next step was to remove the worm gear and replace it with a three-to-one reduction, at the same time removing the torque-limiting clutch. With the addition of check and relief valves in the hydraulic system, the yaw mechanism now held the turbine in position, although with a good deal of residual oscillation in yaw. A program was instituted to discover and remove the causes of the remaining vibration. This program was twofold. First, the effect of adjustments on the test unit was investigated. Second, a theoretical analysis was made of the factors affecting the oscillating forces on the structure. The theoretical analysis is discussed in Chapter 10. The adjustments were made on the coning-damping system.

The coning motion of the blades was determined from the film records taken under various conditions of operation. It was found that the amplitude of coning was less than that predicted by theory and also that the mean coning angle was less. The coning-damping mechanism was adjusted to give less damping. The amplitude of coning then increased somewhat, but the mean position did not shift. This softening of the coning damping reduced but did not eliminate the vibrational motions in yaw and in pitch. The coning links between the blades and the damping mechanism were then lengthened to allow more negative coning, and the damping was then further softened and readjusted until the vibration was reduced to a minimum, as recorded by the strain gauges. Operation then became satisfactory, although some small amount of oscillation remained (Ref. 30-B).

The determination of the aerodynamic efficiency of the turbine is complicated by the determination of the effective wind speed. Winds with the same hourly average may give widely different values of output, depending on the actual distribution of speed during the hour. Instantaneous readings at one anemometer are not necessarily representative of integrated values over the disc area of a wind-turbine.

A long-term statistical study is necessary to determine how the output varies with variations in wind speed, but the fragmentary data which have been worked up show agreement with von Kármán's predictions based on theory.

A dynamic analysis of the instrumentation showed serious time lag and amplitude distortion in the readings. They had no effect on the adjustment of the test unit, but more accurate values were necessary in the investigation of blade motion, an important part of the design studies of the production unit.

The yawing and pitching forces as determined by strain gauges on the yawing mechanism and on the pintle shaft agreed with the theoretically predicted values.

Although the test program was interrupted many times by mechanical failures, as will be described in the next section, the unit was finally brought to a satisfactory operating condition after about 1100 hours of test running, spread over four and a half war years (Ref. 31-B).

DEBUGGING THE TEST UNIT

The troubles attending the birth of this new design were many and severe. Most of the various oil seals leaked oil out of the hydraulic systems. The packing glands on the pitching cylinder, also those on the oil tubes and on the down-wind end of the main shaft had to be entirely rebuilt. The oil head on the main shaft had to be rebored to take multiple-chevron packing. It was necessary to add oil seals to the shafts in the gear box.

The down-wind generator bearing overheated and had to be returned to Schenectady where it was redesigned by increasing the thrust clearance from 10 to 40 mills and by cutting a new oil groove.

As soon as the unit was started up for the first time, creaking noises were heard in the down-wind end of the turbine. With stethoscopes the noises were traced to the hub-post. A few loose rivets were found. The entire hub-post was then field-welded wherever accessible. It was also found that the hub-post had not been pulled up tight on the tapered shaft, because of an error in measurement during shop assembly. When the hub-post was pulled tight on the shaft, there was no more trouble with these noises.

The test unit was protected by a great many safety devices, most of which consisted of mercury-type switches, which, because of the motion of the structure, would frequently slop over, producing shut-down. With no target indication to tell which relay had caused the shut-down, it sometimes meant a long hunt to find what had happened. All of the mercury switches were replaced by mechanical switches with manually reset target indicators.

At the time when the design of the test unit was frozen, the only hydraulic coupling available for service between the gear and the generator was one whose largest model had never been operated continuously at a rating corresponding to full load in our application. It was soon found that a few hours at full load heated the coupling beyond its allowable limits.

Air-duct cooling was tried and found inadequate even in an ambient of minus 25 degrees Fahrenheit. The coupling was jacketed and sprayed with a solution of ethylene glycol in water. The freezing point of this solution was 55 degrees Fahrenheit below zero. Two radiators were hung below the pintle girder to serve as heat exchangers. The cooling solution was pumped from a sump tank to the spray nozzle, whence it drained by gravity through the radiator to the sump. In the short period during which this device was in use (March 3 to March 26, 1945) it gave satisfactory results. It was required on the test unit, although in a production unit, using an electric coupling, there would be no necessity for such special cooling.

The vertical shaft in the yaw-mechanism, which carried a pinion in its lower end meshing with the bull gear, acted as a shear pin and sheared off several times under the stress of the unexpectedly large yawing moment. A new type of yaw-mechanism was designed, in which the vertical shaft carried on its upper end a worm gear meshing with a worm free to move axially under the restraint of a damping cylinder. Spring loading of the worm kept it centered in its normal position. The final adjustments to coning damping, described elsewhere, reduced the yawing moments substantially. It was felt that this adjustment, in conjunction with a redesign of the A-frame for production, described in Chapter 10, whereby the hinge pins would be co-axial and on the center line of the main shaft, would render this redesigned yaw gear unnecessary.

In May, 1942, after some 360 hours of operation, a routine inspection of the blades showed cracks in the blade skin near the roots of the blades. These cracks were concentrated over the spar, leading to the belief that the loads were being carried to the root by the skin rather than being carried into the spar by the ribs. One of the subtlest design problems is the distribution of loads through complex structures and evidently the assumptions made in the case of the blade had not been entirely valid.

If the loads were coming in in the manner indicated by the cracks, there was only one thing to do and that was to make the section strong enough to take them. To this end a heavy box was built up at the root section, of transverse and longitudinal bulkheads covered with a comparatively heavy skin. Some minor cracks developed after this repair. These further cracks were repaired by arc welding. The effect of this repair on the shank spar connection will be discussed in the section of blade failure.

The really serious interruption to the test program was not apparently a design failure but rather the failure of one of the main bearings, for unknown reasons. In February, 1943, a routine inspection discovered the down-wind main bearing running hot. It was found that the bearing had moved on the shaft and was rubbing on the end-plate. Further inspection showed the inner race cracked through. Replacement of this bearing was a serious jolt to the project. In the first place, it took two years to secure a new bearing. In the second place, installing it meant a major disassembly job. The main shaft was disconnected at the up-wind end and the turbine was allowed to drop down against temporary supports while the bearings were pulled off the exposed shaft. The new double-row spherical roller S.K.F. bearing was pulled back onto the down-wind position on the shaft, following which the up-wind main bearing and Oldham coupling were reassembled, the turbine tipped back into place and the unit placed in operation again on March 3, 1945, twenty-five months after the bearing failure.

Although the test program was interrupted by the necessity of correcting the usual number of minor design defects and although it was further interrupted by the long outage due to the failure of the main bearing, it was possible in the 1100 hours of actual operation under test to smooth out the unit and get it operating satisfactorily as a routine generating station. No evidence was found that the mechanical and aerodynamic efficiencies were not equal to the design estimates.

By 1944 enough experience had been obtained, including strain-gauge recordings, to permit a recalculation of loadings. This was carried out by Wilcox, in collaboration with Holley, under the direction of Wilbur, who concluded that the actual loadings were somewhat higher than those which had been used in the design. His anxiety increased about the blade-root sections, which had originally been very highly stressed. Accordingly, in December, 1944, and just before the unit was to go back on the line as a routine generating station, Wilbur proposed to the S. Morgan Smith Company that the test unit be torn down as soon as it had served its purpose. This decision was accepted by S. Morgan Smith Company.

ROUTINE OPERATION AS A GENERATING STATION

At a planning conference held in Rutland in January, 1945, it was decided that when the turbine was again ready to run, it would be operated as a generating station, with no interruptions for routine testing. The bearing repair was completed on March 3, 1945, and the turbine was then run for the first time since February 21, 1943. Three shifts of operators were supplied by Central Vermont Public Service Corporation, and three shifts of inspectors by the S. Morgan Smith Company as a temporary precaution, until we could be sure that the turbine was functioning properly. The hydraulic coupling cooler and its controls were untried at this time. The new bearing had to be watched carefully during its run-in period, and various other new controls were in use for the first time.

The turbine was operated without incident through an unusually windless three weeks in March; during this time, it generated 61,780 kilowatt-hours in 143 hours and 25 minutes of operation, at an average level of 431 kilowatts.

THE BLADE FAILURE

On March 26 the midnight to 8 A.M. shift came on duty to find only about 5 miles per hour of wind. About 2:30 A.M. the wind freshened, and at 2:50 A.M. there was sufficient wind to start the unit. The unit was phased-in to the line at 2:55 A.M., when it was carrying from 50 to 475 kilowatts of load.

At 3:10 A.M. Harold Perry, the erection foreman, was aloft, standing on the side of the house away from the control panel and separated from it by the 24-inch rotating main shaft. A shock threw him to his knees against the wall. He started for the controls, but was again thrown to his knees. He tried again, and again was thrown down. Collecting himself, he dove over the rotating shaft, reached the controls, and, overriding the automatic controls which were already functioning, he brought the unit to a full stop in about 10 seconds by bringing the remaining blade to full feather.

One of the 8-ton blades had let go when in about the 7 o'clock* position, and had been tossed 750 feet, where it landed on its tip (Fig. 2).

---

*Subsequently computed by von Kármán.

It was estimated that the turbine, with one blade remaining, had made about 3 revolutions at full speed, and about 4 revolutions at diminishing speed.

Visual damage to the standing structure appeared to be confined to the leading edge of the remaining blade, which had coned negatively into one of the stops on a tower leg, but without damaging the tower. Measurements indicated that neither the main shaft nor the pintle shaft was sprung. Nevertheless, the very high unbalanced forces may have caused damage not visible.

Inspection showed that the blade spar had failed along multiple and corroded cracks just outboard of the bulkhead, itself just outboard of the bolted shank-spar connection (Fig. 75). Subsequent inspection showed similar corroded cracks well developed at the same point in the remaining blade spar.

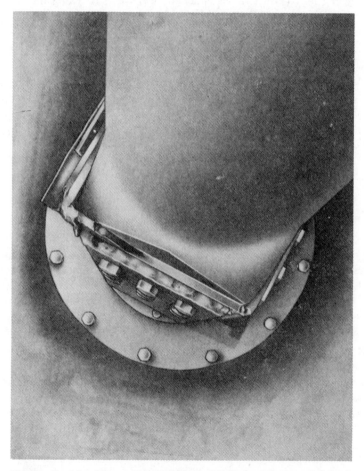

Fig. 75. Looking from outboard along the blade shank in toward the axis of rotation. The broken metallic surfaces show the members that failed.

Without being able to evaluate them, we can list the following causes which directly contributed to the failure of the spar:

A. Stress concentrations produced at the point of failure, because of an abrupt change of section in the box spar.

B. Additional stress concentrations at the point of the abrupt change in section, brought in by the bulkhead located just outboard of the change in section.

C. During the course of some field modifications in July, 1942, both the blade skin over the root area, and the bulkhead located at the point of stress concentration mentioned in A and B, were stiffened, tending to accentuate the stresses at this section.

D. At the same time, acute additional stress concentrations at this point were caused by the notch-effect of an undercut field weld, fastening the bulkhead to the four outside plates of the box spar, on the outboard face of the bulkhead. It was just at this point that failure occurred.

E. Some fatigue due to 1,500,000 revolutions of the turbine, with stress variations (but not complete reversal) in each revolution.

F. An added source of fatigue has been suggested by von Kármán, who points out that during the 2-year outage, the blades, positioned vertically, were locked in rotation and in pitch, although free to yaw. He believes that each blade was subject to continuously varying wind forces, causing it to wave like a fishpole, and setting up stress-reversals in the root sections. This motion was frequently observed during the outage and it was the upper, and more exposed blade, that failed. This stress condition is discussed in Chapters 10 and 12, where it is recognized as being one of the heaviest of the design loading assumptions.

SUMMARY AND CONCLUSIONS

A major structure of novel design had been engineered. In about a thousand hours of operation under test it had been concluded that the basic design was sound. In this rather short time the operation of the test unit became satisfactory as a generating station. It was operated without incident for one month by the Central Vermont Public Service Corporation as a routine generating station. At this point a structural failure occurred in a member which was known to be weak, and which had long since been redesigned for production.

The engineers of the S. Morgan Smith Company feel that they are now in a position to design a large wind-turbine with confidence, and with further improvements which will make the operation smoother, the maintenance simpler and the energy cost less.

9
The Best Size for a Large Wind-Turbine

It seems desirable to describe in an orderly sequence the evolution of our ideas regarding the best size for a large wind-turbine. Accordingly, in this chapter there are summarized four successive studies of the most economical dimensions, made in 1937, 1939, 1943, and 1945.

THE FIRST APPROXIMATION

In 1937 I was the first to carry out, so far as I know, a comprehensive determination of the most economical dimensions of a large wind-turbine (Ref. 10-B). This study was based on the functional specifications described in Chapter 8.

Wilbur had made a preliminary stress analysis and Jackson and Moreland had prepared sketches, but the design was in no sense a detailed one. Rossby furnished an estimate of the vertical distribution of the wind speed above the summit of a ridge in order to arrive at a preliminary notion of the advantages, if any, to be gained from very tall towers. The Bethlehem Steel Corporation furnished weight and cost estimates of a great variety of four-legged towers, with wide variations in the assumed loadings. Only a brief study of the increase of tower cost with increasing height was necessary to show that the increase in available energy with increase in height, as predicted by Rossby, would be insufficient to pay for very tall towers. It was clear that the best tower height over a very wide range of turbine ratings would lie somewhere between 100 feet and 300 feet.

The General Electric Company furnished weight and cost estimates of some seventy-five gears and generators covering a wide range of ratings and speeds. Similarly, the Budd Manufacturing Company and other sources supplied estimating figures of the variations in blade weight and cost with variations in radius and rating.

All costs were quoted in lots of *100 developed units*. These quotations were combined to obtain the installed costs of units rated from 250 kilowatts to 1000 kilowatts, with diameters varying from 140 feet to 320 feet. Tower heights were studied in the range from 50 feet to 300 feet. Outputs were computed from a hypothetical speed-frequency curve, with a mean annual speed of 25 miles an hour, thought to be typical of 3000-foot hilltops in western Massachusetts. Annual charges were set up by Jackson and Moreland at 12½ percent. About one hundred such computations were carried out, from which were selected by inspection the designs in each sub-group that gave the lowest energy cost. These selected points were plotted in Fig. 68.

The curves of Fig. 68 show that the minimum cost is found when a generator with a capacity of 600 kilowatts is driven by a turbine with a diameter of 226 feet on a tower about 150 feet high in a wind regime with a mean annual speed of 25 miles an hour. It was estimated that such a wind-turbine would produce energy at the switchboard at the foot of the tower at a cost of 1.52 mills per kilowatt hour. This unit would cost about $72 per kilowatt installed and would weigh about 350 pounds per kilowatt, including foundation steel.

It is interesting that the energy cost from a 1000-kilowatt unit with a diameter of 300 feet is only 7.2 percent greater than the least cost, found at the rating of 600 kilowatts with a turbine 226 feet in diameter. The extremes of turbine diameter and generator rating which may be had within a variation from the minimum energy cost of 10 percent are found to be 160 to 300 feet and 300 to 1200 kilowatts. This insensitivity of the cost envelope over a rather wide range of diameters and ratings was found to be characteristic of the three later approximations.

Another characteristic of this curve also found in all later ones is that the energy cost increases more quickly when the diameter is reduced below the best value than it does when the diameter is increased above it. (Compare Figs. 69, 76A, and 77.)

The cost of the best 1000-kilowatt unit and the best 500-kilowatt unit is less than that of the best 750-kilowatt unit, indicating some discontinuity in the quotations, probably in the gear or generator price series. Discontinuity in gear cost alone is expected when the number of steps in the gear is changed. Mass production of certain sizes of equipment lowers the cost in steps and introduces discontinuities in the otherwise smooth curves of cost plotted against rating.

This first approximation named 600 kilowatts as the most economical capacity, while later studies showed that 1500 kilowatts was the most economical. At the time this study was made, in 1937, the gear and generator were thought to comprise a larger percentage of the total cost than has since been found to be the case. This error arose because of my lack of detailed knowledge of the rest of the structure and of my belief that it would be lighter and more simple than it proved to be when finally built.

As the cost of the gear and generator decreases relative to the rest of the structure, it is obvious that the most economical generator size will increase, while the most economical diameter will decrease. These relations have been confirmed in later approximations.

THE SECOND APPROXIMATION

In Chapter 1, it was explained that the first step taken by the S. Morgan Smith Company, after having approved my basic design, was to review my estimate of the cost of producing energy from the wind by means of this design (Ref. 24-B).

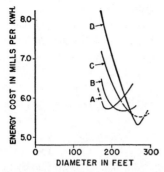

Fig. 76A. Third approximation of the most economical dimensions. Variation in the energy cost with variation in turbine diameter for various generator ratings, but at a constant rotational speed of 1.5 radians per second.

Curve A	1000 kilowatts
Curve B	1500 kilowatts
Curve C	2000 kilowatts
Curve D	2500 kilowatts

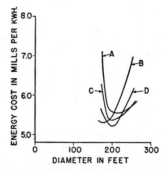

Fig. 76B. Third approximation of the most economical dimensions. Variation in the energy cost with variation in turbine diameter for various generator ratings, but at a constant rotational speed of 2.0 radians per second.

Curve A	1000 kilowatts
Curve B	1500 kilowatts
Curve C	2000 kilowatts
Curve D	2500 kilowatts

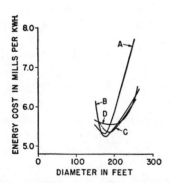

Fig. 76C. Third approximation of the most economical dimensions. Variation in the energy cost with variation in turbine diameter for various generator ratings, but at a constant rotational speed of 2.5 radians per second.

Curve A	1000 kilowatts
Curve B	1500 kilowatts
Curve C	2000 kilowatts
Curve D	2500 kilowatts

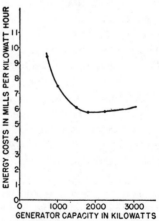

Fig. 77. Fourth approximation of the most economical dimensions. Variation in the energy cost with variation in the capacity of the wind-turbine generator, in a block of 9000 kilowatts of capacity installed on the 4000-foot Lincoln Ridge in Vermont. The plotted values come from line 13 of Table 13.

The general method used in the first approximation was adopted for the second, which was carried out in 1939 by a special computing staff created for the S. Morgan Smith Company at the Budd Manufacturing Company under the general direction of Benjamin Labaree. The wind regime was specified by Petterssen who estimated a mean annual speed of 28.8 miles an hour as being typical of good wind-turbine sites in Vermont. The outputs for a large number of turbines were computed by the special staff, based on a report by von Kármán (Ref. 32-B), who specified the dimensions and characteristics of a series of blades (Fig. 78) designed to give maximum efficiency in a wind speed of 14 miles an hour at a tip-speed of 165 feet per second. Von Kármán had computed the variation in the power output of these blades with varying wind speeds and these computations had been verified by Elliott Reid, who had tested models in the wind-tunnel at Stanford University for the S. Morgan Smith Company (Ref. 33-B). Finally, the Budd Company devoted several man-hours to the preparation of accurate cost estimates of a representative series of these blades.

Again, master computation charts were laid out, but we were able to profit by the work which had been done since 1937, permitting better weight estimates. We investigated diameters between 150 and 300 feet and generator ratings between 750 and 3000 kilowatts, at speeds of 600 and 900 revolutions per minute, and some of the weights were re-estimated by Wilbur.

Tower height and turbine speed were each held constant at values derived from the first approximation. We now know that these two values were each a little low.

A total of 122 computations was carried out. In addition, output only, without cost, was determined for 294 other combinations of diameter, rating, and speed (Ref. 34-B). Jackson and Moreland again set the annual charges at 12½ percent. The total costs were summed, the energy cost computed in mills per kilowatt-hour, and the low cost design of each sub-group was again plotted, yielding the curves of Fig. 69. It is seen that the most economical unit has a diameter lying between 175 feet and 235 feet, based on blade type 2-M (Fig. 78), and a generator capacity lying between 1000 kilowatts and 2000 kilowatts, at a speed of 600 revolutions per minute.

Within these limits the energy cost at the switchboard at the foot of the tower does not vary by more than 2 percent. It seemed likely that the design to be chosen for ultimate production should be selected from the upper or right-hand end of the range, which would mean a turbine with a diameter of at least 225 feet and a capacity of at least 2000 kilowatts. The reason for this is that the second approximation was based on the energy costs of a single unit and did not reflect the effect of such items as access road, connecting transmission line, and erection costs. The effect of charges of this type is to make it desirable to occupy the finite number of sites on a ridge with large rather than small units.

On the other hand, the dimensions of the experimental unit were selected from the lower or left-hand end of the flat envelope, since the S. Morgan Smith Company

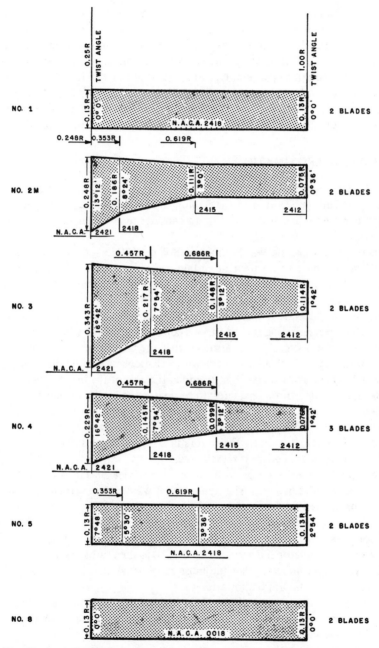

Fig. 78. Second approximation of the most economical dimensions. The series of eight blades studied.

wished to experiment with the smallest unit which would, nevertheless, be dynamically representative of the ultimate production unit. Accordingly, as explained in Chapter 7, it was decided, as a result of the second approximation, that the test unit should be rated at 1250 kilowatts, with a diameter of 175 feet, a generator speed of 600 revolutions per minute, and a hub height of 125 feet.

The installed cost of the most economical unit, as determined in the second approximation, was about $56 per kilowatt, and the weight, including foundation steel, was about 150 pounds per kilowatt, *based on lots of 100 developed units.*

THE THIRD APPROXIMATION

In 1943 a large wind-turbine had been tested in operation for two years. The staff of the Wind-Turbine Division of the S. Morgan Smith Company had accumulated a body of experience and had been able to consider many ideas looking toward the simplification of the test unit.

Accordingly, when the War Production Board in 1943 requested cost estimates for a Victory Model to be used in certain areas where there was a deficiency of power, it was decided to take advantage of the experience gained and to redetermine the most economical dimensions (Ref. 35-B).

Because of the war, it was not feasible to obtain from vendors new smooth curves of cost versus rating for the various components of the wind-turbine. A substitute method was used. It was assumed that the cost of each component of the turbine varied in some regular manner with variations in turbine diameter, turbine speed, and generator capacity. For example, the cost of the blades was assumed to vary with the square of the diameter; to be unaffected by the turbine speed; and to vary with the one-half power of the generator capacity. It was recognized that actually the cost of the blades may vary as the 1.8 power or the 2.2 power of the diameter. However, it was felt that errors in such estimates should tend to cancel out since the number of items is rather large.

The capital charges were again based on the advice of Jackson and Moreland and were set at 12½ percent. The hypothetical wind regime was based on Glastenbury and had a mean annual speed of 21.5 miles an hour.

In Figs. 76A, B, C are plotted the energy costs against diameter for various generator sizes and turbine speeds. In Fig. 79 selected minimum values of cost are plotted against generator capacity, and in Fig. 80 against turbine speed.

The study again indicates that the most economical diameter is about 200 feet and the capacity about 1500 kilowatts.

It is seen that these curves show all the characteristics of the curves of the first and second approximations and substantiate the findings of the second approximation rather closely.

The energy cost at the switchboard at the foot of the tower of the most economical unit was found to be 5.2 miles per kilowatt-hour (Ref. 36-B). This unit

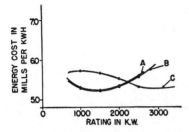

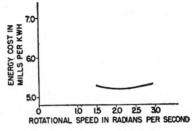

Fig. 79. Third approximation of the most economical dimensions. Variation in the energy cost with turbine rating for various turbine speeds.

Curve A   2.5 Radians per second
Curve B   2.0 Radians per second
Curve C   1.5 Radians per second

Fig. 80. Third approximation of the most economical dimensions. Variation in the energy cost with turbine speed.

was estimated to cost, installed, $149 per kilowatt, and to weigh 450 pounds per kilowatt, including foundation steel.

THE FOURTH APPROXIMATION

In part because of the interest of the War Production Board and in part to make the best of the enforced delay while waiting for the new down-wind main bearing, it was decided in 1943 to move ahead with at least the preliminary stages of the formal design of a preproduction model.

No rigorous attempt had yet been made to determine in detail the most economical blade shape. The first step in the fourth approximation consisted, therefore, of a study of blade shape (Ref. 37-B).

Designs were laid out for sixteen blades, for which outputs and costs were determined for each of two diameters, in each of two wind regimes for each of several generator ratings and each of several turbine speeds.

The ideal blade is a blade in which each elemental section is operating at the maximum theoretical efficiency. This requires that the width of the blade and the blade angle shall vary continuously along the radius, and with a distribution which itself will vary with the wind-speed. Obviously, such a blade is impractical.

The ideal turbine has no profile drag, which is ordinarily ignored in computing the output by classical theory. For our purposes it was necessary to include the profile drag. Accordingly, following a suggestion by von Kármán, an extension of the classical theory was carried out by Wilcox and Holley. By means of this extension they were able to compute the output of an ideal turbine for various values of the lift-drag ratio. This ratio is a measure of the profile drag of a blade section.

The next step in determining the best blade shape was to approximate the ideal blade shape for various values of the tip-speed ratio. This is the ratio between the

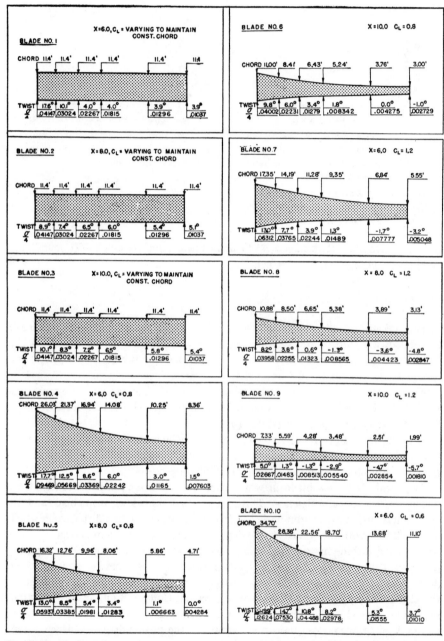

Fig. 81 (Left). Fourth approximation of the most economical dimensions. The sixteen blades which were studied.

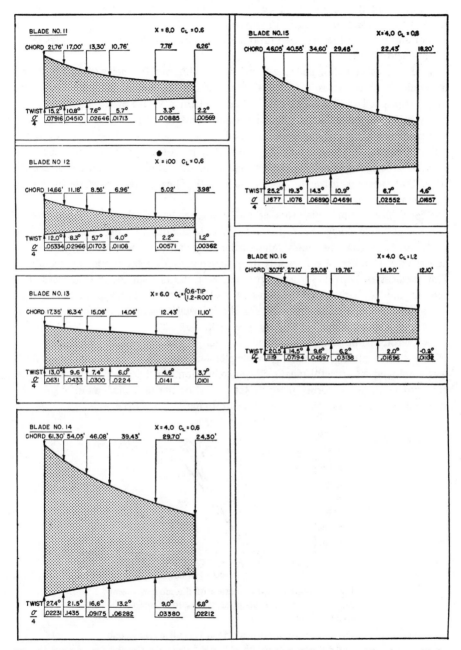

Fig. 81 (Right). Fourth approximation of the most economical dimensions. The sixteen blades which were studied.

linear speed of the tip of the blade, in the plane of rotation, and the speed of the wind at right angles to the plane of rotation.

For any value of the tip-speed ratio there is a corresponding value of the efficiency of an ideal turbine.

Four blades were designed, the respective maximum efficiencies occurring at four different tip-speed ratios, viz., 4.0, 6.0, 8.0, and 10.0. For each of the four values of the tip-speed ratio, Wilcox and Holley used three values of the lift coefficient, viz., 0.60, 0.80, and 1.20, each held constant from root to tip.

In addition, three blades were designed having a constant blade width along the radius (rectangular plan-form). At the suggestion of von Kármán, an additional blade was designed, having a linear variation of lift coefficient from 0.6 at the tip to 1.20 at the root, for a tip-speed ratio of 6.0. These sixteen different blades are shown in Fig. 81.

The choice of the blade cross-section is independent of the other factors determining the blade shape. From the aerodynamic standpoint a high maximum value of the lift coefficient and a low value of the drag coefficient are desirable. Structurally the airfoil must be sufficiently deep to accommodate the supporting members. Flat surfaces or convex surfaces are cheaper to build than concave surfaces. For the test unit we had chosen the N.A.C.A. 4418 section as having good aerodynamic characteristics, sufficient thickness for the internal structure, and no concave surfaces, and it was selected for this study as well.

An aerodynamic analysis was made of each of the blades (Ref. 38-B). In some of our earlier studies, the simple blade-element method of aerodynamic analysis was employed, as a useful means of approximation. In this study, however, we decided to employ the vortex theory (Ref. 11-A) of the propeller, modified for the windmill condition, as being the most accurate method available.

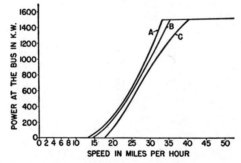

Fig. 82. Fourth approximation of the most economical dimensions. Power-speed curves for three of the sixteen blades of Fig. 81.

Curve A   Output for blade 14 at 1.5 radians per second
Curve B   Output for blade 2 at 2.5 radians per second
Curve C   Output for blade 9 at 4.5 radians per second

TABLE 12. GENERATOR RATINGS USED FOR THE DETERMINATION OF OUTPUT

Diameter	Rating
175	1000
	1250
	1500
200	1200
	1500
	1800

From the aerodynamic output so computed were subtracted the losses caused by hub-windage and tower shadow, to give the gross input to the main shaft. From this input were subtracted the gear losses, the coupling losses, the generator losses, and the power required to operate the servo-mechanisms and the auxiliary equipment, the net balance being the generator output, whose variation with wind speed is plotted, for several of the designs, in Fig. 82.

The variation in annual output of each of the sixteen blades, with variations in the rotational speed, was determined for each of two diameters (175 feet and 200 feet), and for several generator ratings (Table 12). The output of each of these combinations was determined for two different wind regimes, one characteristic of the West Indies, with a mean speed of 18.4 miles per hour, and one characteristic of Glastenbury Mountain (3700 feet) in Vermont, with a mean speed of 21.5 miles per hour. Figs. 83 and 84 show 2 such sets of output curves for the 16 blades investigated. Each figure is for one combination of diameter, wind regime, and rated power. A critical comparison of the 12 sets of curves led to the following conclusions:

1. A single type of blade can be used over a wide range of wind regimes, rated powers, and diameters. The limits of the various ranges were not determined.
2. The best blade (No. 15) gave an output which was 95 percent of the ideal output. We made no investigation of other plan forms and twists which might have shown an improvement over this figure.
3. The largest tapered blade that it seemed reasonable to build (No. 13) gave an output which was 92 percent of the ideal.
4. The best rectangular blade (No. 1) gave an output which was 88 percent of the ideal, or 96 percent of blade No. 13.
5. The gain to be achieved by using a different airfoil section is small since we are limited in our choice by structural considerations.

Production costs were estimated by the Budd Manufacturing Company for the rectangular blade No. 1 and the tapered blade No. 13. The Budd Company costs are given in Fig. 85. It can be seen that for ten units (twenty blades) the tapered blade is twice as expensive as the rectangular blade.

Because the ratio between the cost of the most efficient tapered blade of practical dimensions (No. 13) and the rectangular blade (No. 1) is over 2 to 1, while

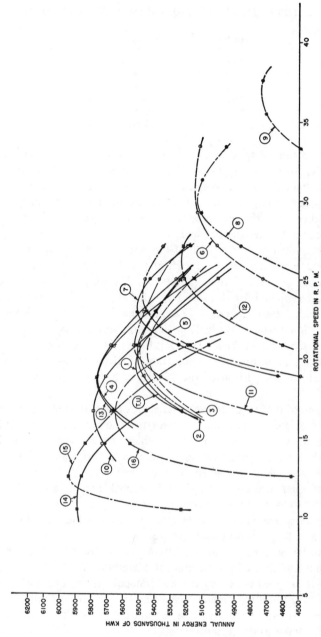

Fig. 83. Fourth approximation of the most economical dimensions. Variation in annual output with variation in the rotational speed of the turbine for each of the sixteen blades of Fig. 81, operating in the wind regime of Mt. Glastenbury, Vermont, the speed frequency distribution of which is described by Curve C of Fig. 36.

Unit rated at 1800 kilowatts; diameter 200 feet

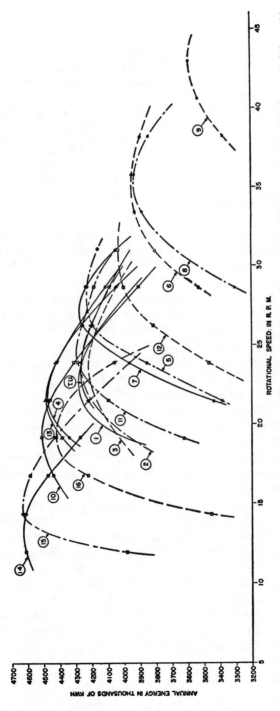

Fig. 84. Fourth approximation of the most economical dimensions. Variation in annual output with variation in the rotational speed of the turbine for each of the sixteen blades of Fig. 81, operating in the wind regime of Mt. Glastenbury, Vermont, the speed frequency distribution of which is described by Curve C of Fig. 36.

Unit rated at 1500 kilowatts; diameter 175 feet.

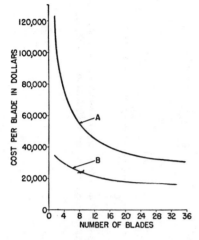

COST PER BLADE IN DOLLARS

120,000
100,000
80,000
60,000
40,000
20,000
0

0 4 8 12 16 20 24 28 32 36
NUMBER OF BLADES

A
B

Fig. 85. Fourth approximation of the most economical dimensions. Variation in 1944 blade costs with quantity, from estimates supplied by the E.G. Budd Company.

Curve A   Represents tapered blade No. 13 of Fig. 81

Curve B   Represents rectangular blade No. 1 of Fig. 81.

the ratio in outputs is not over 1.10 to 1, *it seems unlikely that there is any combination of economically usable wind regime, turbine diameter, and generator rating for which blade No. 1 would not also be the most economical blade for a large wind-turbine.*

Now, the output from a rectangular blade is relatively insensitive to the amount of twist, that is, to the change in blade angle from root to tip. Thus the output from blade No. 1, if untwisted, is 98 percent of the output when twisted.

If the blades are made up in stainless steel, they would be twisted, since in this type of construction twist costs nothing. But, the cost of twisting a mild steel blade is not paid for by the increase in annual output.

The economics of twist are further discussed in Chapter 10.

The maximum efficiency of the most economical blade will, therefore, be 88 percent of the theoretical maximum, if the blade is made up in stainless steel, and 86 percent if made up in mild steel.

The two previous approximations of 1935 and 1939 had indicated that there would be a small but real advantage in increasing the test unit diameter by 25 feet or, possibly, 50 feet. This economic advantage was, however, not a sufficient inducement to depart in the preproduction design from the immediate range of our rather limited experience. Therefore, and also in order to simplify the calculation of what we then expected would be a preliminary cost study, it was arbitrarily decided to hold the diameter to 175 feet in the fourth approximation.

It was felt that if this economic study should prove encouraging, it would then be proper to consider the benefits to be obtained by redesigning the preproduction unit to conform more closely to the most economical dimensions.

A second primary variable, which it was arbitrarily decided to hold constant in the fourth approximation, was tower height. The vertical distribution of speed that we had measured at Grandpa's Knob, Pond, Biddie, Seward, and Mt. Washington, left little doubt that, if a hill has the aerodynamic characteristics enabling it to speed up the wind-flow by 20 percent or more, *the maximum speed will not be found higher than 250 feet above the summit, and the most economical tower height would accordingly lie below 200 feet.* So we decided that the economic gain to be derived from a tower higher than 150 feet, would not, in itself, be sufficient to dominate our conclusions about the economics of large-scale wind-power. Accordingly, for the fourth approximation, we held the tower height constant at 150 feet.

The hypothetical wind regime was based on our accumulated experience in Vermont and represented our expectations on Lincoln Ridge at an elevation of 4000 feet. The assumed mean annual speed was 30 miles an hour.

With the most economical blade identified, the loadings recomputed, and simplifying assumptions made concerning turbine diameter, tower height, and mean wind speed, the fourth and most rigorous approximation of the economically ideal size for a wind turbine was carried out in 1945 (Ref. 18-B). This approximation was based not on a site occupied by a single unit, but on a site occupied by a block of units, and it included such charges as the connecting high-line and access roads.

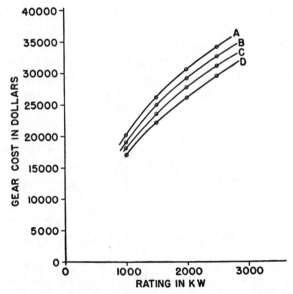

Fig. 86. Fourth approximation of the most economical dimensions. Variation in gear cost with rating, for various rotational speeds.

Curve A   25.0 revolutions per minute
Curve B   27.5 revolutions per minute
Curve C   30.0 revolutions per minute
Curve D   33.0 revolutions per minute

For the fourth approximation entirely new costs were obtained for all parts of the turbine, as will be described in Chapter 11. The items principally affecting the most economical speed and rating are the gear costs, the coupling costs and the costs of the generator and its associated equipment. Gear costs vary with input torque but are not affected by changes in the step-up ratio. Fig. 86 shows variations in the gear cost with rating.

The cost of the coupling and generator is combined and plotted against generator speed in Fig. 87. It is seen that for all generator capacities 900 revolutions per minute is the most economical speed. Since no other items affect the choice of generator speed, we selected 900 revolutions per minute and combined the generator and coupling costs at this speed with the gear costs to determine the most economical turbine speed. This is substantially correct and sufficiently close for our purposes. It is true that the weight and cost of the blades will vary somewhat with turbine speed, as will also the main shaft and certain other parts. However, these variations are minor and we do not think that the uncertainty introduced by them into the determination of the most economical turbine speed exceeds ± 0.5 revolutions per minute.

The variation of output with turbine speed for various generator ratings is shown in Fig. 88. The variations in the cost of the generator, coupling, gear and electrical equipment as the rating and speed are varied, are given in Fig. 89. The energy cost of these components is computed and plotted in Fig. 90. The best turbine speed for various ratings is found by noting the position of the minima of the cost curves. The variation of this best speed as the generator rating varies is plotted in Fig. 91.

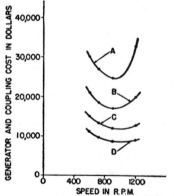

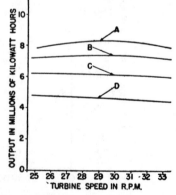

Fig. 87. Fourth approximation of the most economical dimensions. Variation in the cost of the generator and the coupling with variation in speed.

Curve A   2500 kilowatts
Curve B   2000 kilowatts
Curve C   1500 kilowatts
Curve D   1000 kilowatts

Fig. 88. Fourth approximation of the most economical dimensions. Variation of turbine output with turbine speed for various turbine ratings.

Curve A   2500 kilowatts
Curve B   2000 kilowatts
Curve C   1500 kilowatts
Curve D   1000 kilowatts

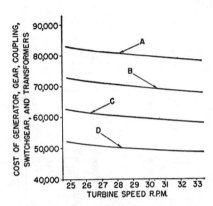

Fig. 89. Fourth approximation of the most economical dimensions. Variation in the cost of the generator, coupling, gear, switchgear and transformers, with turbine speed, for various ratings.

Curve A	2500 kilowatts
Curve B	2000 kilowatts
Curve C	1500 kilowatts
Curve D	1000 kilowatts

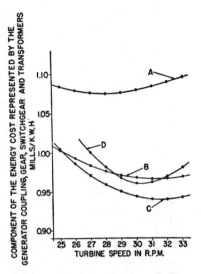

Fig. 90. Fourth approximation of the most economical dimensions. Variation in the component of the energy cost representing the cost of the generator, coupling, gear, switchgear, and transformers, with variation in turbine speed, for various ratings.

Curve A	1000 kilowatts
Curve B	1500 kilowatts
Curve C	2500 kilowatts
Curve D	2000 kilowatts

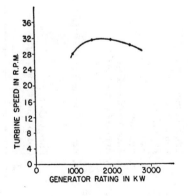

Fig. 91. Fourth approximation of the most economical dimensions. Variation between the best turbine speed and the turbine rating.

TABLE 13. FOURTH APPROXIMATION OF THE BEST SIZE FOR A LARGE WIND-TURBINE

	12 × 750			9 × 1,000			6 × 1,500		
	Dollars per Kilowatt	Dollars per Unit	Dollars per Installation	Dollars per Kilowatt	Dollars per Unit	Dollars per Installation	Dollars per Kilowatt	Dollars per Unit	Dollars per Installation
1. Cost of All Parts Which Vary with Capacity — Includes: Gear, Generator, Coupling, Transformers, and Incremental Costs of Blades, Shafts, Bearings, Pintle Girder and Tower	$ 57.33	$ 43,000	$ 516,000	$ 45.00	$ 45,000	$ 405,000	$ 34.67	$ 52,000	$ 312,000
2. Rest of the Structure	240.00	180,000	2,160,000	180.00	180,000	1,620,000	120.00	180,000	1,080,000
Total F.O.B. Cost	297.33	223,000	2,676,000	225.00	225,000	2,025,000	154.67	232,000	1,392,000
3. Site Selection	1.11	832	10,000	1.11	1,111	10,000	1.11	1,667	10,000
4. Land	0.05	38	450	0.05	50	450	0.05	75	450
5. High-Line	14.65	10,988	131,850	14.65	14,650	131,850	14.65	21,975	313,850
6. Road	3.00	2,250	27,000	2.67	2,670	24,030	2.33	3,495	20,970
7. Erection	26.00	19,500	230,400	24.00	24,000	216,000	22.00	33,000	198,000
8. Total Installed Cost	342.14			267.48			194.81		
9. Annual Charge (10%)	34.21			26.75			19.48		
	Kwh.	Kwh.	Kwh.	Kwh.	Kwh.	Kwh.	Kwh.	Kwh.	Kwh.
10. Annual Output (Lincoln Ridge)	4,500	3,375,000	40,500,000	4,108	4,108,000	36,972,000	3,500	5,250,000	31,500,000
	Mils per Kwh.			Mils per Kwh.			Mils per Kwh.		
11. Energy Cost	7.663			6.511			5.566		
12. Output Correction Factor	1.200			1.145			1.090		
13. Net Energy Cost	9.124			7.455			6.067		

TABLE 13. FOURTH APPROXIMATION OF THE BEST SIZE FOR A LARGE WIND-TURBINE (continued)

	5 × 1,800			4 × 2,250			3 × 3,000		
	Dollars per Kilowatt	Dollars per Unit	Dollars per Installation	Dollars per Kilowatt	Dollars per Unit	Dollars per Installation	Dollars per Kilowatt	Dollars per Unit	Dollars per Installation
1. Cost of All Parts Which Vary with Capacity — Includes: Gear, Generator, Coupling, Transformers, and Incremental Costs of Blades, Shafts, Bearings, Pintle Girder and Tower	$32.58	$58,640	$293,220	$30.67	$69,000	$276,000	$30.38	$91,153	$273,459
2. Rest of the Structure	100.00	180,000	900,000	80.00	180,000	720,000	60.00	180,000	540,000
Total F.O.B. Cost	132.58	238,640	1,193,220	110.67	249,000	996,000	90.38	271,153	813,459
3. Site Selection	1.11	2,000	10,000	1.11	2,500	10,000	1.11	3,333	10,000
4. Land	0.05	90	450	0.05	113	450	0.05	150	450
5. High-Line	14.65	26,370	131,850	14.65	32,963	131,850	14.65	43,950	131,850
6. Road	2.22	3,996	19,980	2.11	4,748	18,990	2.00	6,000	18,000
7. Erection	20.00	36,000	180,000	17.78	40,000	160,000	14.34	43,000	129,000
8. Total Installed Cost	170.61			146.37			122.53		
9. Annual Charge (10%)	17.06			14.64			12.25		
10. Annual Output (Lincoln Ridge)	3,200 Kwh.	5,760,000 Kwh.	28,800,000 Kwh.	2,650 Kwh.	5,962,500 Kwh.	23,850,000 Kwh.	2,100 Kwh.	6,300,000 Kwh.	18,900,000 Kwh.
	Mils per Kwh.			Mils per Kwh.			Mils per Kwh.		
11. Energy Cost	5.335			5.523			5.835		
12. Output Correction Factor	1.072			1.053			1.035		
13. Net Energy Cost	5.712			5.816			6.039		

The value of 9000 kilowatts as the block of capacity to be analyzed was chosen because the Central Vermont Public Service Corporation was not interested in anything over 10,000 kilowatts of capacity and the number 9000 is readily divisible by various standard generator ratings.

Six combinations of generator rating and number of units were considered. The respective costs are tabulated in Table 13 and the energy costs are plotted in Fig. 77. We found the most economical capacity to lie between 1600 and 1800 kilowatts.* For the preproduction unit we selected 1500 kilowatts because this was a standard size. By entering this value in Fig. 91, we find that the corresponding proper turbine speed is 31.5 revolutions per minute.

Thus, the fourth approximation tells us that the preproduction unit will be a turbine with two blades rectangular in plan-form; 175 feet in diameter; operating at 31.5 revolutions per minute; and driving a 1500-kilowatt generator at 900 revolutions per minute.

It will be noted that the general shape of the envelope curve of Fig. 77 is similar to that of the envelopes found in the three previous approximations and, since the dimensions of the most economical unit were found to be so nearly similar, particularly in the second, third, and fourth approximations, the *conclusion is inescapable that, for this general type of large wind-turbine and over a fairly wide range of wind regimes, the most economical dimensions are definitely known, within limits of ± 50 percent.*

Drastic modifications in the design would, presumably, modify the values of some or all of the dimensions of the most economical unit. Thus, if the gear could be eliminated or if the blade cost could be cut in half, it would, doubtless, be found that the most economical ratings and dimensions had increased and this would presumably be the result if some of the modifications and drastic changes, discussed in the next chapter, should prove successful.

In Table 14 the four approximations are summarized and compared with a set of values illustrating the speculations of Chapter 12 on how to reduce the cost of wind power.

The line of data in Table 14 describing "Fourth Approximation 1945" lists the dimensions whose determination is illustrated in Table 13 and Figure 77; the annual charges of 12 percent are those determined as applicable to Central Vermont Public Service Corporation (pp. 197 et seq. and Table 18); the unit cost of $191 per kilowatt and the unit weight of 497 pounds per kilowatt are taken from the detailed cost study of the preproduction unit, described in Chapter 11 and summarized in Table 15; and the net salable unit output of 3320 kilowatt-hours per kilowatt per year is the value determined for the proposed installation on Lincoln Ridge (Table 14).

---

*The outputs used in these computations were not those of Fig. 88, but corrected values thought to be more accurate.

TABLE 14. SUMMARY

The Best Size for a Large Wind-Turbine

	Mean Speed, mph.	Annual Charges, Per Cent	Diameter, Feet	Tower Height, Feet	Generator Rating, Kilowatts	Unit* Cost, $/Kw.	Pound Cost, $/Lb.	Unit* Weight, Lbs./	Net Saleable Unit Output, Kwh./Kw.	Energy* Cost at the Bus, Mils per Kwh.
First Approximation, 1937	25	12½	226	125	600	$72	0.21	350	5900	1.5
Second Approximation, 1940	28.8	12½	200	125	1500	56	0.37	150	4400	1.6
Third Approximation, 1943	21.5	12½	200	(150)**	1500	149	0.33	450	3575	5.2
Fourth Approximation, 1945	26	12	(175)**	(150)**	1600	191	0.39	497	3320	6.9
Based on Chapter XII										
(a) Private Capital	30	12½	235	165	3000	100	0.33	300	4100	3.0
(b) Government Capital	30	8	235	165	3000	100	0.33	300	4100	2.0

*Installed, but exclusive of transformers and connecting transmission line.

**Arbitrarily held to this value.

TABLE 15. COST SUMMARY OF PREPRODUCTION UNIT, 1945

	Dollars	Dollars per Kw.	Weight	Pounds per Kw.	Dollars per Lb.
1. Engineering	$10,500.00	7.00			
2. Manufacturing					
2.1 Standard Items Now in Production					
2.1–1 Generator	8,870.00	5.91	20,250.00	13.50	0.438
2.1–2 Main Gears	20,344.00	13.56	29,000.00	19.33	0.702
2.1–3 Coupling Electric	4,612.50	3.08	9,700.00	6.47	0.476
2.1–4 Governor	2,508.00	1.67	2,000.00	1.33	1.254
2.1–5 Bearings	16,282.00	10.85	11,500.00	7.67	1.416
2.1–6 Switch Gear	5,125.00	3.42	4,000.00	2.67	1.281
2.1–7 Couplings (Flexible)	1,970.00	1.31	5,500.00	3.67	0.358
2.1–8 Elevator	2,665.00	1.78	8,000.00	5.33	0.333
2.1–9 Service Hoist	1,680.00	1.12	5,000.00	3.33	0.336
2.1–10 Miscellaneous Electrical	2,100.00	1.40	6,500.00	4.33	0.323
2.1–11 Tower (Includes Erection)	21,395.00	14.62	180,000.00	120.00	0.122
2.1–12 Paint	691.00	.46	1,500.00	1.00	0.461
2.2 Items Peculiar to the Smith-Putnam Wind-Turbine					
2.2–1 Blades	29,480.00	19.65	67,000.00	44.67	0.440
2.2–2 Hub Assembly	42,935.00	28.67	155,000.00	103.33	0.277
2.2–3 Pintle Assembly	48,600.00	32.40	200,000.00	133.33	0.243
2.2–4 Patterns, Tools, Jigs, and Fixtures	800.00	.53			
Total F.O.B. Cost	221,097.50	147.40	704,950.00	469.96	0.314
Contingency 10%	22,109.75	14.74			
TOTAL	243,207.25	162.14	704,950.00	469.96	0.345
3. Installation					
3.1 Freight**	2,054.00	1.37	437,000.00		0.0047
3.2 Land	0.00	0.00			
3.3 Road	7,460.00	4.97			
3.4 Erection (Includes Foundation Steel)	30,000.00	20.00	40,000.00*	26.67	0.040*
Total Installed Cost	260,611.50	173.74	744,950.00	496.63	0.350
Contingency 10%	26,061.15	17.37			
TOTAL	286,672.65	191.11	744,950.00	496.63	0.385
4. Connection					
4.1 Transformers	3,600.00	2.40	15,000.00	10.00	0.240
4.2 High Line	15,000.00	10.00			
Total Cost of Unit (Connected to Existing System)	279,211.50	186.14	759,950.00	506.63	0.367
Contingency 10%	27,921.15	18.61			
TOTAL	307,132.65	204.75	759,950.00	506.63	0.404

*Foundation steel only.
**Only those items shipped from York, Pennsylvania.

SUMMARY

1. Three determinations of the most economical size for a wind-turbine of my 1937 design were carried out in 1940, 1943, and 1945, respectively. The sum of the evidence indicates that the best dimensions of this design will fall within the following limits, over a fairly wide range of wind regimes, and assuming a mountain-top location:

Tower height	150-175 feet
Generator rating	1500-2500 kilowatts
Turbine diameter	175-225 feet

2. The unit weight of the fourth approximation was 497 pounds per kilowatt. (The test unit weighed 500 pounds per kilowatt.)

3. The mean pound price had nearly doubled in eight years, rising from about $0.21 per pound in 1937 to $0.39 per pound in 1945.

10
The Design of a
Preproduction Model,
1943-1945

INTRODUCTION

When it was found that the third approximation had verified the best size of a large wind-turbine, as found in the two previous approximations, the design of a preproduction model was begun.

The design of any structure, for which the forces associated with the acceleration of gravity are significant, must necessarily proceed by a series of approximations. In the first approximation it is usual to begin by arbitrarily assuming some distribution of the mass. Next, the forces associated with this mass distribution are calculated; then the non-mass forces. Finally, a stress analysis is carried out. A design based on this stress analysis yields a new mass distribution. For this new mass distribution, the above cycle is repeated. Successive cycles are carried through until all members of the structure are designed with consistent strength.

In such structures as bridges or buildings, for which a great deal of previous experience is available, the first estimate of mass distribution usually agrees closely with the final distribution. For these structures, one, or at the most two, cycles will suffice. The wind-turbine differs from these structures in that, first, there is no backlog of experience, and second, the mass forces in the wind-turbine depend largely on accelerations which in turn depend on the mass distribution. For this reason, one must carry through a new dynamic analysis for each new mass distribution, and several successive approximations will be required.

When the test unit was designed, there was not time, as described in Chapter 1, to carry out a sufficient number of such cycles to produce a design which would incorporate consistent strength with maximum economy. Furthermore, the meager knowledge in 1940 of the probable wind conditions to be encountered did not justify a precise analysis and design.

When, in 1944, it was decided to start the design of a preproduction unit, our knowledge had been substantially augmented by actual experience with the test unit and by certain theoretical studies carried out during the test period. It was the intention to carry through as many design cycles as necessary. But the project was abandoned in the early stages of this program, the first cycle of which was not completely carried out.

CHARACTERISTICS OF THE PREPRODUCTION UNIT

The third economic study to determine the best size of a large wind-turbine had fixed some of the characteristics of the preproduction unit, as described in Chapter 9. These were as follows:

1. Diameter—175 feet
2. Turbine Speed—2.71 radians per second, or 26 revolutions per minute
3. Generator Capacity—1500 kilowatts
4. Coupling Type—electric, torque-limiting
5. Blade Plan-form—rectangular
6. Blade Chord—11.4 feet
7. Blade Length—65.6 feet
8. Blade Section—N.A.C.A. 4418
9. Blade Twist—5°, discontinuously, in 4 steps.

During the test program an analysis was carried out to determine the factors affecting the oscillating forces on the structure (Ref. 39-B) described in Chapter 8. The first step in this study was the determination of the blade forces and their effect on the entire pintle structure. The method of determining blade forces was similar to that followed in the redesign of the blades described later in this chapter. The calculation of yawing and pitching moments from the blade forces was straight-forward. Fig. 92 shows schematically the blade forces, the hinge forces and the yawing and pitching moments. In the analysis of the factors affecting the oscillating forces, the wind conditions were limited to certain typical cases.

The magnitude of the following factors was varied one at a time and the effect on each of the others was determined:

a. The angle of tilt (γ).
b. The distance from the pintle axis to the center of rotation (L).
c. The distance from the axis to the hinge-pins (e).
d. The weight of the rotating parts (W).
e. The speed of rotation (ω).
f. The mechanical coning-damping constant (β).
g. The number of blades (B).

In this part of the study only one wind condition was considered. The wind speed selected—30 miles per hour—gives approximately the rated output of the turbine. The distribution of wind speed vertically across the disc—the wind gradient—was assumed to be linear and such as to produce the maximum amplitude of periodic coning which was observed on the test unit.

Figs. 93-96 and 97-98 show the effect of varying the quantities b, c, and f listed above. Based on these results, two units were selected for further study, one with

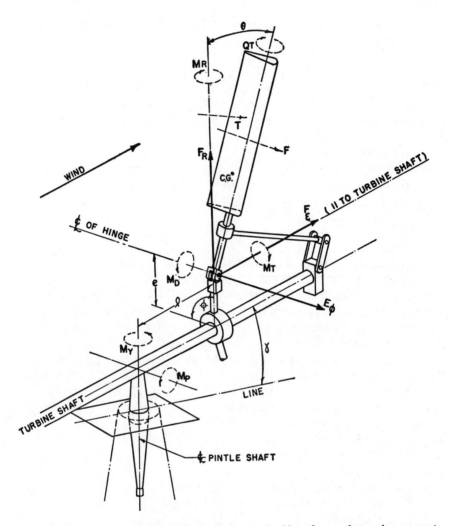

Fig. 92. Schematic representation of the blade forces, the hinge forces, the yawing moments, and the pitching moments.

Note particularly the dimension e, the distance between the hinge line and the turbine shaft.

In the test unit e had the value of 32.5 inches, which was the cause of some roughness in operation. In future units e will equal 0.0 inch.

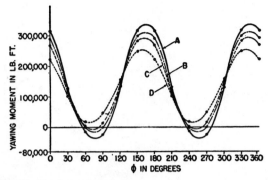

Fig. 93. Yawing moment.

Influence of the length L of Fig. 92.

Curve A	L equals 12.00 feet
Curve B	L equals 8.71 feet
Curve C	L equals 6.00 feet
Curve D	L equals 0.00 feet

Computed for a blade angle of 0 degrees in a wind speed of 30 miles per hour with an assumed vertical speed gradient across the turbine disc such that coning will amount to 5 degrees.

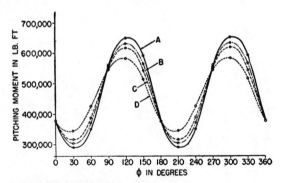

Fig. 94. Pitching moment.

Influence of the length L of Fig. 92.

Curve A	L equals 12.00 feet
Curve B	L equals 8.71 feet
Curve C	L equals 6.00 feet
Curve D	L equals 0.00 feet

Computed for a blade angle of 0 degrees in a wind speed of 30 miles per hour with an assumed vertical speed gradient across the turbine disc such that coning will amount to 5 degrees.

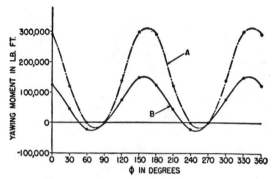

Fig. 95. Yawing moment.

Influence of the length e of Fig. 92.

        Curve A   e equals 2.71 feet
        Curve B   e equals 0.00 feet

Computed for a blade angle of 0 degrees in a wind speed of
30 miles per hour with an assumed vertical speed gradient
across the turbine disc such that coning will amount to 5
degrees.

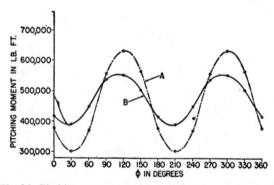

Fig. 96. Pitching moment.

Influence of the length e of Fig. 92.

        Curve A   e equals 2.71 feet
        Curve B   e equals 0.00 feet

Computed for a blade angle of 0 degrees in a wind speed of
30 miles per hour with an assumed vertical speed gradient
across the turbine disc such that coning will amount to 5
degrees.

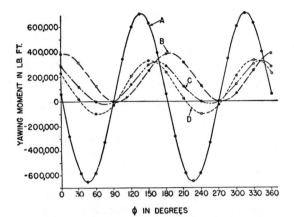

Fig. 97. Influence on the yawing moment of the value of the mechanical coning damping constant, β.

Curve A	$\beta = \infty$
Curve B	$\beta = 0.0$
Curve C	$\beta = 0.5$
Curve D	$\beta = 1.0$

These curves are computed for a blade angle of 0 degrees and a wind speed of 30 miles per hour with an assumed vertical wind speed gradient across the disc such that the coning angle will amount to 5 degrees.

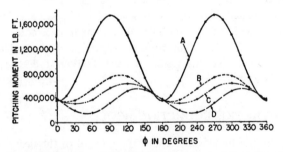

Fig. 98. Influence on the pitching moment of the value of the mechanical coning damping constant, β.

Curve A	$\beta = \infty$
Curve B	$\beta = 1.0$
Curve C	$\beta = 0.5$
Curve D	$\beta = 0.0$

These curves are computed for a blade angle of 0 degrees and a wind speed of 30 miles per hour with an assumed vertical wind speed gradient across the disc such that the coning angle will amount to 5 degrees.

two blades, and the other with three. The three-bladed turbine was considered at this time because it was thought by some that it would be superior to a two-bladed unit, independently of the values selected for the variables listed above.

In comparing these two units, three conditions of wind, blade angle, and vertical gradient of wind speed were assumed. These three conditions were:

1. Wind 30 miles per hour; design blade angle; gradient to give maximum coning.
2. Wind 30 miles per hour; design blade angle; gradient to give average coning.
3. Wind 60 miles per hour; design blade angle +24 degrees; gradient to give maximum coning.

The comparison between two- and three-bladed wind-turbines can best be shown by Figs. 99 to 104 on which are plotted the variations, respectively, of the yawing and pitching moments with variation in the blade position. It was found that this three-bladed unit, when exposed to a linear wind gradient, was not the best from the standpoint of yawing and pitching moments. The best three-bladed unit for a linear wind gradient would have fixed blades, i.e., no coning, and there would be no periodic yawing or pitching moment associated with such a unit. However, for a non-linear wind condition, such a unit would have large periodic yawing and pitching moments. Since the wind gradient usually is non-linear, it was concluded that the three-bladed wind-turbine holds no dynamic advantage over a well-designed two-bladed unit which, as was indicated previously, was more economical.

The greatest single improvement in a two-bladed unit, as compared with the design of the test unit, results from reducing the hinge distance e to zero. If circumstances had permitted, further studies would have been made, based on the assumption of zero hinge distance, in order to determine more accurately the effect of such things as tilt angle, blade weight, rotational speed, and coning-damping.

However, we believe that the results of such further studies would not have changed the conclusions.

LOADINGS FOR THE PREPRODUCTION MODEL

Having fixed all of the dimensions under the control of the designer (other than member sizes), the next step was to compute loadings for an assumed mass distribution, beginning with the blades (Ref. 40-B). Since the dimensions of the preproduction unit were similar to those of the test unit, the mass distribution of the rotating parts was assumed to be the same as that of the test unit.

The distribution of mass along the blade axis from the root to the tip was assumed to be that of the test unit. The possibility of ice-formation on the blade was allowed for by adding a constant mass per foot of blade length. The maximum ice load was that specified by von Kármán for the test unit—132 pounds per linear foot. (For an intermediate condition 25 percent of this ice loading was used.)

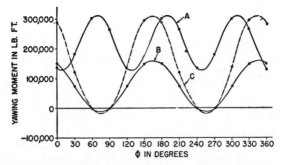

Fig. 99. The influence of the number of blades on the yaw-ing moment. The conditions are: wind, 30 miles per hour; blades set at design blade angle; wind speed gradient across the disc is taken to be that which will produce maximum coning.

> A. A three-bladed unit
> B. The two-bladed optimum unit
> C. The two-bladed test unit

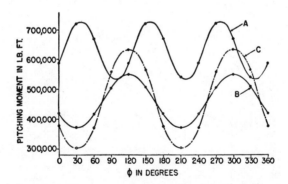

Fig. 100. The influence of the number of blades on the pitching moment. The conditions are: wind, 30 miles per hour; blades set at design blade angle; wind speed gradient across the disc is taken to be that which will produce maxi-mum coning.

> A. A three-bladed unit
> B. The two-bladed optimum unit
> C. The two-bladed test unit

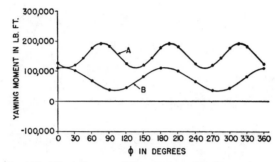

Fig. 101. The influence of the number of blades on the yawing moment. The conditions are: wind, 30 miles per hour;
blades set at design blade angle; wind speed gradient across
the disc is taken to be that which will produce average coning.

A. A three-bladed unit
B. The two-bladed optimum unit

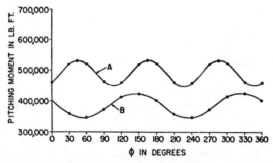

Fig. 102. The influence of the number of blades on the pitching moment. The conditions are: wind, 30 miles per hour;
blades set at design blade angle; wind speed gradient across
the disc is taken to be that which will produce average coning.

A. A three-bladed unit
B. The two-bladed optimum unit

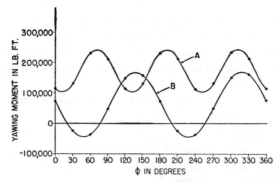

Fig. 103. Influence of the number of blades on the yawing moment. The conditions are: wind, 60 miles per hour; blades set at design blade angle plus 24 degrees; wind speed gradient across the disc is taken to be that which will produce maximum coning.

      A. A three-bladed unit
      B. The two-bladed optimum unit

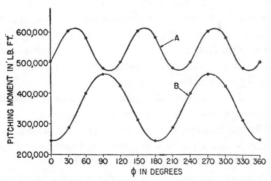

Fig. 104. Influence of the number of blades on the pitching moment. The conditions are: wind 60 miles per hour; blades set at design blade angle plus 24 degrees; wind speed gradient across the disc is taken to be that which will produce maximum coning.

      A. A three-bladed unit
      B. The two-bladed optimum unit

The conditions for which loadings should be computed are not clear-cut in the case of wind-turbines, because we lack knowledge of wind behavior. The conditions investigated were set up as a result of the accumulated knowledge gained in operating the test unit, and in carrying out the special wind-research programs (Ref. 41-B).

These conditions fell into two general classes, operating and non-operating, which were further sub-divided:

I. Operating conditions
   A. Fatigue loading conditions
      1. Constant speed condition
      2. Varying speed condition
   B. Maximum loading conditions
II. Non-operating conditions
   A. Idling conditions
   B. Locked conditions

The forces on the blades—both mass and aerodynamic—can be expressed in terms of certain known coefficients and the unknown motions of the system.

Thus, it is possible to write differential equations of the motion of the system. The solution of these equations yields the unknown motion components, that is, displacements, velocities, and accelerations. These quantities, when multiplied by the proper coefficients, become the values of the aerodynamic and mass forces.

In addition to gravity forces, mass forces associated with the following accelerations were computed:

1. The rotational centrifugal acceleration, which acts radially inward along a line perpendicular to the plane passing through the axis of turbine rotation.
2. The rotational tangential acceleration, which acts along a line perpendicular to the plane containing the mass and the axis of turbine rotation.
3. The centrifugal acceleration of coning, which acts along a line perpendicular to and passing through the coning axis.
4. The tangential acceleration of coning, which acts along a line perpendicular to the plane containing the mass and the coning axis.
5. The coriolis acceleration, which acts along a line parallel to the coning axis.
6. The pitching acceleration, which acts along a line perpendicular to the plane containing the mass and the pitching axis of the blade.

The forces associated with the above accelerations are given by the products of the respective masses and their accelerations. Fig. 105 shows these forces in relation to the geometry of the turbine.

The aerodynamic forces are obtained in the form of lift, drag, and moment components. These are computed at a sufficient number of stations along the blade to permit smooth curves to be drawn. Values at other stations were obtained from the curves by interpolation.

All the elemental forces are resolved into three components acting respectively: parallel to the blade axis; perpendicular to the blade axis and perpendicular to the coning axis; and perpendicular to the blade axis and parallel to the coning axis.

The forces were integrated to give the following quantities at each cross section of the blade:

A. Total force, parallel to the blade axis.
B. Shear and bending moment in the coning plane.

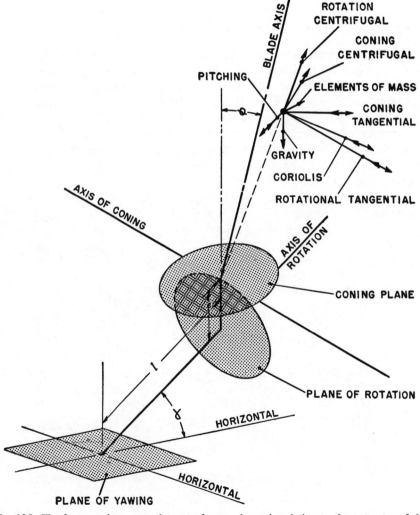

Fig. 105. The forces acting on an element of mass, shown in relation to the geometry of the turbine.

C. Shear and bending moment in the plane perpendicular to the coning plane.

D. Torque about the blade axis.

The four quantities provided the data for blade design.

The integrated forces at the root of the blade plus the integrated forces on the other parts were used for the design of the remainder of the structure. In so far as loadings for the rotating parts of the structure were concerned, the first cycle was completed. Loads for the remainder of the structure were obtained in a more approximate fashion. This was because information was urgently needed for rough layouts and for cost estimates.

PRELIMINARY DESIGN STUDIES AND LAYOUTS

Several types of blade were considered for the production unit. The Budd Company, builders of the original blades, designed new blades of two different types. One was very similar to the original design, but with a modified and strengthened root section. The other was a flexible-skin blade in which all loads were carried by the spar. The skin of this blade was in sections approximately 6 feet long with no provision to transfer load from one section to the next. Both of these blades would have stainless-steel skin and ribs on an alloy steel spar. The spar would have a bolted, flanged connection to the shank. In these blades twist can be built in at no extra cost.

A stressed-skin blade which could be fabricated in short sections and field-assembled was designed by S.D. Dornbirer, Chief Construction Engineer, Wind Turbine Division of the S. Morgan Smith Company (Fig. 106). This design was

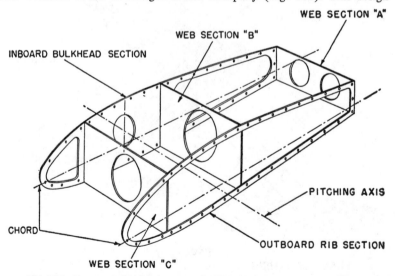

Fig. 106. Frame section of the stressed skin blade proposed by Dornbirer.

originally an annealed stainless-steel skin on a carbon steel foundation. Annealed stainless is very little stronger than some of the carbon steels. For a weight penalty of 10 percent, we could reduce the cost of material 25 percent. The 100 percent carbon steel construction would make the entire fabrication and assembly job much easier.

A mock-up of several sections was made in the S. Morgan Smith shops to aid in completing the design. Fig. 107 shows a view of this mock-up. If this type of blade were used, it would be built with no twist, since there is a distinct gain in fabrication by so doing. Sufficient investigation of this blade was made to indicate that it held promise.

Plywood blades for a 200-foot diameter wind-turbine were designed by the United States Plywood Corporation (Ref. 42-B). These blades were designed prior to the determination of the loadings to which the stainless and mild steel blades were designed. These blades are not, therefore, directly comparable to the steel blades. However, certain general comparisons between plywood blades and steel blades can be drawn. The initial cost of plywood blades is probably less in both material and labor than that of metal blades. The weight is about the same. The efficiency should be slightly better, at least for the first few years.

Maintenance costs over the 20-year life would be much higher for the plywood blades. In fact, we do not know how the plywood blades would behave over such a long period when exposed to high winds, ice, oil, dirt, and other hazards. With the present knowledge of wood glues and surface protection, it is probably a little optimistic to say that plywood blades will stay in operation for 20 years.

Blades fabricated of an aluminum alloy were briefly discussed with several groups interested in that material. Preliminary studies were started with the Aluminum Laboratories Ltd. of Montreal, Canada, but were suspended when S. Morgan Smith abandoned the project.

Preliminary layouts indicated the following design features for the preproduction model, as shown in Fig. 108.

1. A redesigned A-frame with a tubular member and side tie rods. The name A-frame is no longer applicable.
2. The coning hinges are co-axial about the center line, thereby reducing e to zero (Chapter 8).

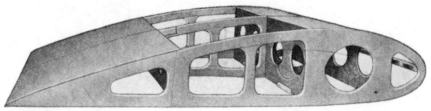

Fig. 107. Mock-up of the section of the frame of the stressed skin blade proposed by Dornbirer.

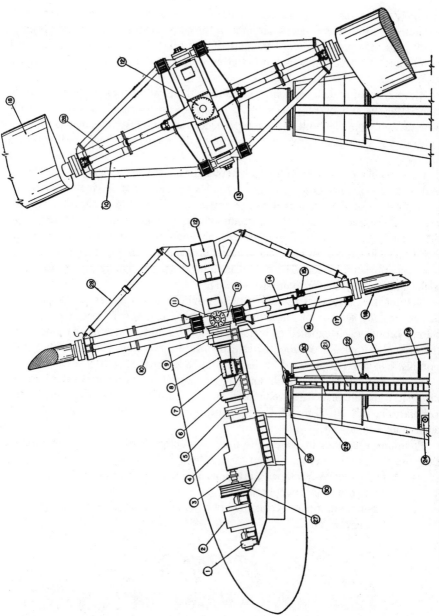

Fig. 108. Schematic layout aloft of the proposed preproduction unit of the Smith-Putnam Wind-Turbine. Rated at 1500 kilowatts; diameter 175 feet.

3. A redesigned pitching mechanism eliminates universal joints.

4. Oleo strut type of coning-damping mechanism.

5. A entirely welded hub-post bolted to a flange on the main shaft.

6. All welded pintle girder.

7. Hollow tubular pintle shaft.

8. Streamlined house and spinner.

9. All-welded tower.

The mechanical features of the turbine have been changed but little. Most bearings are anti-friction-ball, straight roller, or spherical roller.

The gear box incorporates a three-speed step-up rather than two as in the test unit. Minor simplifications are incorporated in this layout as a result of experience with the test unit, primarily to add to the ease of erection and maintenance.

The control problem is simplified by the substitution of an electric coupling for the hydraulic coupling. The generator would be phased on the line manually and left on until an emergency required taking it off. When the wind was too low to generate power, the coupling would be de-energized and the generator would run as a synchronous condenser under automatic voltage control. Several devices have been suggested to energize and de-energize the coupling at that wind speed when the output becomes zero. With the full complement of safety devices, the turbine would operate as an automatic substation.

SUMMARY

The objectives of the 1945 redesign of the Smith-Putnam Wind-Turbine were to provide: a structurally safe unit; a smooth operating unit; an economically competitive unit.

The first two objectives have probably been met, but the 1945 design, which weighed as much per kilowatt as the 1940 test unit, failed, as will be described in Chapter 11, to reach the third objective. Chapter 12 describes speculations on how to reach this economic objective.

11
Economics of Large-Scale Wind Power

INTRODUCTION

The preproduction unit, whose design is described in Chapter 10, and the model of which is shown in Fig. 109, is a simplified and cleaned-up version of the test unit of 1940. It does not incorporate the modifications, still less the radical suggestions, put forward in Chapter 12.

This was because the engineers of the S. Morgan Smith Company had acquired six years of familiarity with the test unit. They felt confident that they could now go into production with a design of this general character and size. Furthermore, this familiarity provided the only basis for cost studies. To attempt to estimate closely the production costs of a design whose stresses had not been analyzed would obviously be a waste of time.

However, a comparison of Fig. 1 with Fig. 108 will show that, although the preproduction design was fundamentally identical with that of the test unit, it had been sufficiently modified in detail to require new shop drawings, which it was not feasible to prepare in the summer of 1945. Without shop drawings it was impossible to establish the production costs with finality. The methods by which we approximated the cost estimates are described in detail with respect to each item.

INSTALLED COSTS

This 1945 cost study (Ref. 43-B) is based on a hypothetical production run of twenty 1500-kilowatt units, of which six would be installed on Lincoln Ridge in Vermont, while the other fourteen would be installed at unknown sites, the access costs of which were arbitrarily assumed.

Engineering Costs

The man-hours spent in designing the test unit in 1939-1941, and in carrying out preliminary engineering studies of the preproduction unit in 1944 and 1945, formed the basis on which were estimated the costs of completely engineering a preproduction unit. At a conference in York, Pennsylvania, attended by the Chief Engineer of the S. Morgan Smith Company and most of the members of the Wind-Turbine

Fig. 109. Scale model of the proposed preproduction unit of the Smith-Putnam Wind-Turbine. Rated at 1500 kilowatts; diameter 175 feet.

TABLE 16. BREAKDOWN OF ENGINEERING COST

			Per Month	Per Year
1 Chief Engineer	at		$800.00	$ 9,600.00
1 Assistant Chief Engineer	at		500.00	6,000.00
3 Principal Engineers	at		400.00	14,400.00
1 Construction Engineer	at		400.00	4,800.00
4 Designers	at		300.00	14,400.00
6 Draftsmen				
(maybe 12 for 6 months)	at		200.00	14,400.00
2 Stenographers	at		150.00	3,600.00
		Total Labor		$ 67,200.00
		Overhead (30%)		20,160.00
				$ 87,360.00
		Travel		7,200.00
		Consulting		15,000.00
		Furnishings		7,000.00
				$116,560.00

Total man hours excluding stenographers 32,000.

Add $5000 per unit after the first unit for design and production changes and to provide for engineering during erection and testing.

For the purpose of this estimate, we used the following:

	Total	Per Unit
1 Unit	$115,000	$115,000
5 Units	135,000	27,000
6 Units	140,000	23,333
20 Units	210,000	10,500

Division, an engineering schedule was set up, on the assumption that the design of the preproduction unit could be completed in twelve months. The engineering cost of such a design was estimated at $115,000, as itemized in Table 16. An additional sum of $5000 per unit was allocated to cover the cost of design and production changes, and field engineering. For the twenty units, then, the total engineering cost would be $210,000, or $10,500 per unit.

Manufacturing Costs

It is convenient to distinguish between the costs of standard items now in production and the costs of those items peculiar to the Smith-Putnam Wind-Turbine.

To allocate handling and other charges, the standard production items were further classified into three groups: Group one—items shipped by the manufacturer directly to the turbine site; Group two—items shipped to the S. Morgan Smith Company at York, Pennsylvania, for shop assembly; Group three—items shipped to York for machining and fitting before shop assembly.

The handling charges on these three groups were set by the S. Morgan Smith Company at 2½ percent, 5 percent, and 10 percent of the net purchase price, respectively.

ITEMS IN STANDARD PRODUCTION

Synchronous Generator

As a result of the fourth approximation of the most economical design, described in Chapter 9, the synchronous generator selected for the preproduction unit had been rated at 1500 kilowatts, 2300 volts, with a speed of 900 revolutions per minute. Quotations were received from the Electric Machinery Manufacturing Company, the General Electric Company, the Allis-Chalmers Manufacturing Company, and the Westinghouse Electric and Manufacturing Company.

Main Gears

In 1943, when the third study was under way to determine the most economical generator speed and rating, the Falk Company of Milwaukee, Wisconsin, had sent their Chief Engineer, Walter Schmitter, to Rutland to explore the problem. Until the S. Morgan Smith Company abandoned the project on December 31, 1945, the Falk Company gave it close cooperation and made many design and cost studies to assist in the determination of the most economical capacity and speed of the generator and gear. When the results of the fourth determination in 1945, described in Chapter 9, indicated a 1-to-30 gear to drive the generator, the Falk Company designed a series of gear boxes for this duty. At one end of this series was an ultralight gear assembly of a type used in destroyers, and with a high pound price. At the other end of the series was a heavy assembly with a low pound price. Considering the effect of weight on the rest of the turbine structure, an intermediate design was selected.

Estimates of the gear assembly were also supplied by the General Electric Company and the Westinghouse Electric and Manufacturing Company.

The Electric Coupling

The problem of providing a coupling between the generator and the gear box, to serve as a torque-limiting device and also to provide a speed change to which the Woodward Governor could respond, was explored by the Dynamatic Corporation of Kenosha, Wisconsin, and the Electric Machinery Manufacturing Company of Minneapolis, Minnesota.

Governor

The governor specifications were explored with the Woodward Governor Company of Rockford, Illinois. Experience with the test unit indicated that the production governor could be far simpler than the one used on the test unit.

Bearings

There were many anti-friction bearings in the Smith-Putnam Wind-Turbine, the largest being 48 inches in diameter, at the head of the pintle shaft (Fig. 108). Among the most important bearings were the blade-shank bearings, the main-shaft bearings, and the pintle-shaft bearings. These bearings were either special, or at least nonproduction, items. Most of the other bearing requirements could be met by standard catalogue items.

The S. K. F. Industries and the Bantam Bearing Division of the Torrington Company cooperated closely with us in studying our bearing problems. Where necessary, new bearings were designed. After giving consideration to many alternative schemes, an assembly was selected which appeared to be the most economical.

Switchgear

A functional specification, far simpler than in the case of the test unit, was prepared by the customer, the Central Vermont Public Service Corporation, whose engineers specified the switchgear necessary to discharge the required functions. Cost estimates were obtained from the Allis-Chalmers Company, the Westinghouse Electric and Manufacturing Company, the Electric Machinery Manufacturing Company, the General Electric Company, and the G & N Engineering Company.

Flexible Couplings

The cost of the flexible couplings for the low-speed shaft was obtained from the Falk Company and for the high-speed shaft from the Farrell Birmingham Company.

Elevator

S. Morgan Smith Company engineers estimated the cost of the production elevator, based on that of the test unit design.

Service Hoist

Experience on the test unit had indicated that, while a service hoist was essential, it need not be elaborate. The cost of such a hoist was estimated.

Miscellaneous Electrical Equipment

The cost of miscellaneous electric equipment including such items as slip rings, lighting circuits and fixtures, control circuits, and safety circuits, was estimated by S. Morgan Smith Company engineers.

The Tower

Quotations were obtained for two different types of tower. The American Bridge Company of Pittsburgh, Pennsylvania, who had designed and built the four-legged tower for the test unit, supplied quotations for a similar four-legged preproduction tower.

The Chicago Bridge and Iron Company, of Chicago, Illinois, quoted costs of a shell-type tower, looking like a truncated cone, and built up of sheets of steel plate. Quotations were submitted for each of four design conditions. Two of the design conditions varied with the foundations to be encountered, and two with the buckling stresses to be allowed. It was not feasible to include in this study a determination of the actual foundation conditions on the summit of Lincoln Ridge, so we assumed what we thought would be the worst condition, that is, a crumbling schist, like that found on the summit of Grandpa's Knob. Also, we used the lower of the two buckling stresses, with the result that the estimated cost of the shell-type tower, which we used for comparison, was the highest of the four estimates supplied by the Chicago Bridge and Iron Company.

There was little to choose between this estimate for the shell tower and the estimate for the four-legged tower.

Paint

We estimated the area to be painted at 30,000 square feet, including blades. E.I. DuPont DeNemours estimated $2.25 per 100 square feet for a four-coat spray application.

ITEMS PECULIAR TO THE SMITH-PUTNAM WIND-TURBINE

In this category are the blades, the hub assembly, and the pintle assembly. It must be emphasized that detailed shop drawings of the production design were not available. To make cost estimates of these components of the wind-turbine, it was necessary first to estimate the over-all weight of each component, and second, to apply to this weight a unit pound price based on experience with similar structures in the S. Morgan Smith Company shops.

The Blades

Studies had indicated that mild-steel blades, with surfaces either painted or galvanized, or treated in some other way, to resist corrosion, would be somewhat more economical than stainless-steel blades. Various types of structure had been roughed out, from the nearly monocoque and heavily skin-stressed, to a spar and rib structure covered with a non-stressed skin. An intermediate type had been selected, which could be manufactured to advantage in the shops of the S. Morgan Smith Company and which looked as though it would reduce the cost of field erection. Based upon the new blade loadings described in Chapter 10, the approximate thicknesses of the various members were determined. From this information the blade weight was estimated at 67,000 pounds for two blades.

The pound price of this blade structure was independently estimated by J. Oerter and W.H. Hollingshead of the S. Morgan Smith Company Estimating Department. The two estimates were in close agreement, and averaged $0.484 per pound F.O.B. York, Pennsylvania, in lots of one pair, and $0.440 per pound in lots of 20 pairs, equivalent to a cost of $29,480 for each pair, in lots of 20 pairs.

Hub Assembly

This includes all special items between, but not including, the blades and the main shaft. The over-all weight of these items was estimated after considering two somewhat independent approaches:

A. The weight of all corresponding parts on the test unit was found to be 175,900 pounds, and, of this total, about 115,800 pounds consisted of parts which might reasonably be expected to go down in weight through redesign and the use of welding instead of riveting. Weight reduction from this cause was estimated at 17.5 percent. Thus, the estimated weight of the preproduction hub assembly would be 155,600 pounds.

B. The weight of these parts estimated from Dornbirer's 1945 sketches for the preproduction unit was 136,620 pounds. Since these sketches were based on inadequate analysis, it was felt that the weights should be increased by an arbitrary factor of 10 percent, yielding a weight estimate of 150,300 pounds, broken down approximately as follows:

Plate steel . . . . 50 percent
Cast steel . . . . 25 percent
Forged steel . . . . 15 percent
Miscellaneous . . . . 10 percent

The weight estimate adopted by the engineers was 155,000 pounds.

The pound price was again estimated independently by J. Oerter and W.H. Hollingshead. They compared the known costs of similar structures built by S. Morgan Smith Company, in which the proportions of plate steel, castings, and forgings were approximately the same as in the foregoing tabulation. Again, the two estimates were very close and averaged $0.310 per pound for single units and $0.277 per pound in lots of 20, giving a total cost for the hub assembly of $42,935.

Pintle Assembly

This assembly includes all special items from, and including, the main shaft to, and including, the tower cap. It also includes the yaw-mechanism, although part of this may be located below the tower cap.

The weight of the pintle assembly was estimated by considering the evidence found in two more or less independent approaches:

A. The weight of all corresponding parts on the test unit was found to be 200,000 pounds.

B. Dornbirer's 1945 sketches for the preproduction unit showed weights itemized as follows:

> Plate steel . . . . 175,000 pounds
> Castings. . . . . 20,000 pounds
> Forgings . . . . 30,000 pounds

The engineers felt that the weight of the plate steel items should come down as a result of a more careful analysis and design. This item was, therefore, arbitrarily reduced to 150,000 pounds, giving a total weight of 200,000 pounds.

For estimating purposes the engineers accepted 200,000 pounds as the weight of the pintle assembly, itemized as follows:

> Plate steel . . . . 73 percent
> Forged steel . . . . 15 percent
> Cast steel. . . . . 10 percent
> Miscellaneous . . . . 2 percent

The pound price was determined in the same manner as for the hub assembly, and averaged $0.260 per pound for a single unit and $0.243 per pound in lots of 20 units, making a total cost for the pintle assembly of $48,600 for each of 20 units.

Patterns, Tools, Jigs, and Fixtures

The cost of patterns, tools, jigs, and fixtures was estimated by the respective departments of the S. Morgan Smith Company.

INSTALLATION COSTS

Freight

The Rutland Railroad estimated that the average freight rate from York, Pennsylvania, to Rutland, Vermont, would be $0.47 per 100 pounds.

The shipping weight of the unit was taken to be the weight of the items manufactured at York, Pennsylvania, plus the weight of the items shipped to the site via York, giving a total shipping weight of 437,000 pounds.

Although this freight rate applied only to the first 6 units, we arbitrarily applied it also to the remaining 14 units.

Land

On the recommendation of the Central Vermont Public Service Corporation this item was estimated to cost nothing.

Road

The cost of an access road is obviously variable. On paper we laid out a road which would serve 6 sites on Lincoln Ridge. We reconnoitered the ridge on foot and explored the problem with road-building contractors, whose lowest estimate was $44,760 for the 6 units, or $7460 per unit. We arbitrarily assumed that this cost would not be exceeded in the case of the remaining 14 units.

Erection

Erection costs were estimated by S.D. Dornbirer, who had erected the test unit. He made a complete breakdown of the erection procedure, which he had greatly simplified. He estimated the erection cost for one unit on Lincoln Ridge as $57,000, and for 6 units as $192,000, or $32,000 per unit. It was assumed that some of the remaining 14 sites would be more accessible and subject to less rugged weather; and, accordingly, an average erection cost of $30,000 per unit was assumed for the 20 units.

Connection Cost—Transformers

A study showed that for a block of 6 wind-turbines, the most economical arrangement was to use a single transformer bank and run lines at generator voltage to each turbine. The cost of such a transformer bank, with its associated equipment, lightning arresters, disconnects, etc., is approximately $22,000 or $3600 per unit.

High-Line

The cost of the high-tension transmission line is an item which obviously will vary greatly with circumstances. At Grandpa's Knob it was necessary to build only 2.8 miles of transmission line. At Lincoln Ridge, on the other hand, 24 miles of transmission line would be needed. The Central Vermont Public Service Corporation estimated the cost of a 44-kv line at about $5000 per mile, making a total cost for Lincoln Ridge of $120,000 or $20,000 per unit. It was felt that Lincoln Ridge is probably as far from an existing power line as one would find in New England, or in many other places; and for this reason it was arbitrarily assumed that $15,000 would be the average cost of the transmission line for each of the 20 units.

TOTAL INSTALLED COST

The total installed cost of the wind-turbine is summarized in Table 15. It is seen that the cost at the switchboard is $191.11 per kilowatt, and the cost at the point of connection to the existing high-line, including contingency, is $204.75 per kilowatt. The installed weight, including foundation steel, but excluding transformers and connecting line, is 497 pounds per kilowatt, and the average unit cost is $0.39 per pound.

ANNUAL CHARGES

The annual charges are determined by multiplying the total investment by a percentage determined by a consideration of the following factors:
   a. Interest rate on bonds.
   b. Dividend rate on stocks.
   c. Property tax rate.
   d. Income tax rate.
   e. Depreciation rate.
   f. Operating costs.
   g. Maintenance costs.
The proportionality between the amount of the investment put into bonds and that into stocks has a large effect on the total rate since current dividend rates are so much higher than current rates. All of the above listed factors will vary with different utilities.

In general, the annual charge which a private company would make for a wind-turbine installation would be in the range from 12 percent to 15 percent. Similar charges by a government agency would be in the range from 6 percent to 10 percent. An example of the computation for a specific utility will be given in the section on the cost of wind-power to the Central Vermont Public Service Corporation.

ANNUAL OUTPUT

It is convenient to measure the annual output in terms of the number of kilowatt-hours put out in one year per kilowatt of rated generator capacity. Table 1 of Chapter 1 summarizes in these units the outputs to be expected on the windy islands of the world, using a scale which runs from "over 6000" kilowatt-hours per kilowatt per year to "less than 2000." At Grandpa's Knob we realized about 1200 kilowatt-hours per kilowatt per year. In 1945 we estimated that a battery of six 1500-kilowatt wind-turbines on Lincoln Ridge would generate an average of 3500 kilowatt-hours per kilowatt per year. Since then, wind-speed measurements on the Horn of Mount Washington, where the deformation of the balsams is similar to that on the southern end of Lincoln Ridge, have confirmed our belief that the average output on Lincoln Ridge will be about 3500, while, in the case of a unit erected on Mount Abraham, the southernmost end of the ridge, the output would reach 4500 kilowatt-hours per kilowatt per year, or more.

ENERGY COSTS

The cost of energy is computed by dividing the annual charges in dollars per kilowatt by the annual output in kilowatt-hours. The result, in dollars per kilowatt-hour, is converted to mills per kilowatt-hour by multiplying by 1000.

Fig. 110 is a nomogram showing the relationship between installed costs, annual charges, annual output, and the resultant energy cost at the switchboard at the foot of the tower.

Thus, if private capital installs a wind-power plant for $100 per kilowatt with annual charges of 12 percent in a wind regime such that the output is 4000 kilowatt-hours per kilowatt per year, the energy cost will be 3 mills per kilowatt-hour. If, on the other hand, the installation had been made by a Government agency, with annual charges at 6 percent, the energy cost would have been 1.5 mills per kilowatt-hour.

COMPONENTS OF THE WORTH OF WIND-POWER

The worth of wind-power to any system has four components: capacity value, reactive value, value because of predictability, and energy value. A study of the worth of these components was carried out by Jackson and Moreland, whose findings are summarized in the following sections.

Capacity Value

A wind-power installation is not an isolated source of firm power. In order that it may create some capacity value, there must be other generating equipment available on the same system. Thus, if there is an excess of water-wheel capacity accompanied by adequate pondage on a certain system, the addition of a block of wind-power would create some additional firm capacity. Alternatively, if a hydro-electric system was already in balance with respect to its water-wheel capacity, it would be necessary to add some water-wheel capacity in order to firm up whatever wind capacity it was proposed to add.

The probable capacity value of wind-power as found on Lincoln Ridge would vary from $5 per kilowatt on a system with short hydro-storage periods (and high daily load-factors during the months of low wind-power) to $40 per kilowatt on

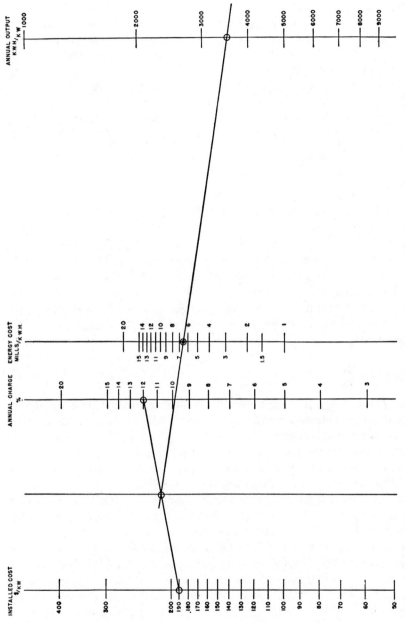

Fig. 110. Monogram relating the unit cost of wind generated energy to the unit installed costs of a wind power plant, the annual charges, and the annual unit output.

systems with long hydro-storage periods (realizing thus the average wind-power) and modest annual load factors.

Of course the usable wind-energy may at times exceed the system load, while simultaneously the hydro system may be incapable of storing more water. In these circumstances, not all the wind-energy can be used to back off fuel-generated energy. Such excess wind-energy has no value.

No credit is given to wind-power for any capacity value arising from the diversity in output of many wind-power plants spread over a vast area. The worth of such capacity value would be exceeded by the transmission costs incurred in getting the wind-energy from the spots where the wind was blowing to the spots where it was not.

Reactive Value

In certain types of wind-power plants the synchronous generator may be left on the line as a synchronous condenser when there is not enough wind for power generation. When acting as a condenser it will provide reactive kilowatt-amperes for the system. The worth of this will vary from $0.00 per kilowatt, when the system is adequately supplied with reactive power or when the wind-turbine is so located that the reactive power cannot be used, to $8.50 per kilowatt when the system is in need of all the reactive power that can be provided from idling wind-turbines and when these turbines are so located that all the wattage is useful for system purposes.

Predictability Value

In Chapter 5 it was reported that it would be possible to predict 12 hours in advance that one of three conditions would prevail at a wind-power plant during a 6-hour period. The three conditions were:
1. No output—no winds of more than 17.5 miles per hour.
2. Some output—winds of more than 17.5 miles per hour.
3. Full output—no winds of less than 34 miles per hour.

It was reported that the results of several hundred such predictions, analyzed qualitatively, show that the correlation factors were high enough to be useful, averaging about 0.80 for the 12-hour prediction.

Under certain circumstances this should make it possible for a power dispatcher to reduce some of his "floating" and "banked" standby fuel generating capacity. The economic value of these predictions would tend to offset any increase in the cost of dispatching which might result from the fluctuating nature of wind-energy. An analysis of individual systems might show that in certain cases this degree of predictability would endow wind-power with a small additional capacity value.

Energy Value

The worth of wind-energy is measured by the extent to which a block of wind-power reduces the cost of energy which must otherwise be generated on the interconnected system. In a steam or a Diesel power plant the cost of generating energy is chiefly the cost of fuel, with some small additional charges for lubricants, supplies, and maintenance resulting from the actual output of energy from the plant, as distinguished from similar expenses which are fixed and continue whether the plant is operating or not.

Such energy costs in 1945 typically ranged from 2.5 mills per kilowatt-hour at a large, modern steam station, to 6 mills per kilowatt-hour at a small, older station, and 10 mills per kilowatt-hour in the case of a small isolated Diesel plant.

Table 17 shows typical 1945 costs of fuel generation on systems capable of absorbing blocks of wind-power varying in size from 100,000 kilowatts to 100 kilowatts.

Evaluation of Wind-Power

The components of the worth of wind-power vary in value depending on whether the wind-power installation stands alone or is backed up by hydro, or by fuel, or by combined hydro and fuel (Ref. 44).

Wind-Power Alone

Where wind-power is backed up by other power sources, its applicability is as broad as the applicability of electric power generally. Where back-up from other sources is not feasible, the applicability of wind-power is limited to those uses which can practicably or economically employ intermittent and variable power.

Practicability requires that operation of the process or service can be reduced in rate, or interrupted, without damage to equipment or disorganization of the process and without undue inconvenience or economic loss. Processes requiring much time and expense for start-up after an interruption, or involving loss of expensive material due to interruption, are not suitable. Processes requiring heavy capital investments or heavy fixed operating costs in relation to the value of power would probably be uneconomic since variable and intermittent power reduces the use-factor of the investment and organization.

The use of intermittent power requires either availability of operators to start up equipment which may have been shut down and to regulate its operating rate in accordance with the variable power supply; or it requires investment in automatic controls to perform the same functions. Expense for maintaining operators in idleness, to start up processes which have been shut down, may be mitigated in some instances if the duration of the idle period and its termination may be predicted

TABLE 17. VALUE OF WIND-ENERGY UNDER VARIOUS ASSUMPTIONS

Size of Block of Wind-Power	Kwh. of Wind per Year		Typical 1945 Costs of Fuel-Generated Energy		Capitalized Value per Kw. of Wind Generator at				
	Available in Wind	Assumed Usable in Load	Mils/Kwh.	$/yr.	15%	12%	10%	8%	6%
Very large blocks, say, 100,000 kw.	2500	2200	2.5	$ 5.50	36.67	45.83	55.00	68.75	91.67
	3000	2700		6.75	45.00	56.25	67.50	84.38	112.50
	3500	3200		8.00	53.33	66.67	80.00	100.00	133.33
	4000	3700		9.25	61.67	77.08	92.50	115.63	154.17
	4500	4200		10.50	70.00	87.50	105.00	131.25	175.00
	5000	4700		11.75	78.33	97.92	117.50	146.88	195.83
Large blocks, say, 50,000 kw.	2500	2300	3.0	6.90	46.00	57.50	69.00	86.25	115.00
	3000	2800		8.40	56.00	70.00	84.00	105.00	140.00
	3500	3300		9.90	66.00	82.50	99.00	123.75	165.00
	4000	3800		11.40	76.00	95.00	114.00	142.50	190.00
	4500	4300		12.90	86.00	107.50	129.00	161.25	215.00
	5000	4800		14.40	96.00	120.00	144.00	180.00	240.00
Medium blocks, say, 10,000 kw.	2500	2400	3.5	8.40	56.00	70.00	84.00	105.00	140.00
	3000	2900		10.15	67.67	84.58	101.50	126.88	167.17
	3500	3400		11.90	79.33	99.17	119.00	148.75	198.33
	4000	3900		13.65	91.00	113.75	136.50	170.63	227.50
	4500	4400		15.40	102.67	128.33	154.00	192.50	256.67
	5000	4900		17.15	114.33	142.91	171.50	214.38	285.83
Small blocks, say, 1000 kw.	2500	2500	5.0	12.50	83.33	104.17	125.00	156.25	208.33
	3000	3000		15.00	100.00	125.00	150.00	187.50	250.00
	3500	3500		17.50	116.67	145.83	175.00	218.75	291.67
	4000	4000		20.00	133.33	166.67	200.00	250.00	333.33
	4500	4500		22.50	150.00	187.50	225.00	281.25	375.00
	5000	5000		25.00	166.67	208.33	250.00	312.50	416.67
Very small blocks, say, 100 kw.	2500	2500	8.0	20.00	133.33	166.67	200.00	250.00	333.33
	3000	3000		24.00	160.00	200.00	240.00	300.00	400.00
	3500	3500		28.00	186.67	233.33	280.00	350.00	466.67
	4000	4000		32.00	213.33	266.67	320.00	400.00	533.33
	4500	4500		36.00	240.00	300.00	360.00	450.00	600.00
	5000	5000		40.00	266.67	333.33	400.00	500.00	666.67

with some degree of reliability, thus making it possible to achieve a degree of coincidence between the normal time-off for operation—or assignment to other duties—and the period of deficient wind.

Listed below are some types of power application which, under favorable conditions, might profitably employ variable and intermittent power.

1. Pumping
   a. Pumping water to reservoirs for industrial and municipal supply purposes.
   b. Pumping water for irrigation and land reclamation.
   c. Pumping water into salt deposits for brine production or pumping up brine solutions.
2. Electrolytic deposition of metals
   a. Refining electrolytic copper from blister.
   b. Manufacture of electrolytic lead.
   c. Manufacture of electrolytic powdered iron for powdered iron metallurgy, etc.
3. Inorganic electrolyses: Electrolysis of water for production of hydrogen and oxygen (which could be compressed into cylinders by use of the same power source).
4. Mechanical power uses
   a. Manufacture of ice in standard equipment for isolated communities.
   b. Manufacture of distilled water by the vapor compression process for isolated communities having no regular fresh water supply, such as islands.
   c. Compression of gases for storage, either compressed or liquefied, as at the Cleveland liquefied natural gas storage plant.
5. Untended airway beacons, as in the Arctic or the great deserts.

It is obviously impossible to generalize about the worth of wind-power in these applications. The worth would vary with each case, and could be determined only by a special study.

Wind-Power with a Hydro System

In conjunction with a simple hydro system without fuel auxiliary, and on which there exists a water-wheel capacity for supplying secondary unfirmed power, the addition of wind, where adequate pondage is available, would firm such surplus capacity to approximately the extent of the average rate of wind generation during the low water season.

If no such surplus exists the same additional firm power can be created by the simple addition of water-wheel generators in the amount of the average wind generation.

There are regions where the costs of hydro development are such that auxiliary fuel plants are not economical. In some of these regions, all the power necessary in the foreseeable future can be generated by hydro developments under these same economic conditions. In this case, it is not likely that wind-power can be justified.

The use of wind-power with pumped storage in combination with an existing hydro system will parallel the foregoing case in all respects except when the pumped storage can be so located that its water can be discharged through an existing hydro plant. In such a case, if excess wheel capacity is available in the existing plant, the benefit will obviously be the cost of this wheel capacity plus the costs for additional pumping and storage facilities due to the increased demands on the storage occasioned by the wind.

Wind-Power with a Fuel System

When used in conjunction with fuel-generated power, with no hydro capacity, where the capacity of the steam or Diesel bears a normal relation to the system load, it seems reasonable to say that a wind-power plant can create no capacity value, for it will not reduce the size of the steam plant required to carry the load when there is no wind.

However, the wind-energy will displace the fuel-generated energy; and the worth of the wind-energy may be measured by the incremental cost of the fuel-generated energy so displaced. If, for example, the wind-power installation generates 3000 kilowatt-hours per year per kilowatt of wind-generated capacity, and if the incremental cost of the fuel-generated energy is 3 mills per kilowatt-hour, then the wind-energy would be worth $9 per kilowatt per year. It is understood that the rate of wind generation is here assumed to be never in excess of the difference between the system load and the minimum practicable rate of generation from the fuel plant.

Table 17 shows typical costs of generating energy in steam and Diesel power plants of various sizes and at various fuel prices. It will be seen that the cost of fuel-generated energy goes down as the scale of the operation increases, thus reducing the value of wind-energy as the block of wind-power increases in size.

Wind-Power with a Combined Fuel and Hydro System

The energy value of wind-power to a combined water and fuel system is the same as on a simple fuel system, provided the fuel plants are similar in both cases.

Wind-power, however, added to such a system, will frequently also have a capacity value when backed up by the necessary new or existing water-wheel capacity and pondage. Where physical conditions are favorable, some of the wind-energy may be used to reduce the draw-down from the storage reservoirs while the wind is blowing, using this increment of stored water in the surplus hydro-generating capacity during hours when the wind is deficient, thereby increasing the firm capacity of the existing surplus water-wheels; or even making it economic to add further water-wheel capacity. As a result, the requirement for fuel capacity is decreased and the investment in additional fuel plants is avoided.

The addition of wind-power to a combined stored water and fuel system can create capacity value equal to the average wind-power generation during the period of draw-down at the storage reservoirs. The worth of this capacity value will be the saving in investment cost in providing this amount of generating capacity less the cost of providing water-wheel capacity in an equal amount. It is normal for hydro systems to have water-wheel capacity in excess of the firm hydro value and on such systems the immediate saving by the installation of wind-power is the gross value of the kilowatts created without such a deduction for installing an equal amount of water-wheel capacity. It should be recognized that this gross value, in contrast with the net value, will persist only so long as the water-wheel capacity would have been in excess of the firm capacity on the system load. For, with a growing system load, as the peak increases, the need for water-wheel capacity per unit of flow in the river increases.

Of course, as in the case of the use of wind-power on a simple fuel system, if the load at any time drops to a value as low as the minimum practicable steam plus hydro-generation, then the energy value of the wind-power during such hours of operation would be lost.

THE WORTH OF WIND-POWER TO THE CENTRAL VERMONT PUBLIC SERVICE CORPORATION

In October, 1945, the worth to the Central Vermont Public Service Corporation of a block of 9000 kilowatts of wind-energy installed on Lincoln Ridge was determined by Jackson and Moreland (Ref. 45-B). This study was made in collaboration with the Utility Company, but the conclusions have not been reviewed or approved by them. However, Jackson and Moreland believe that they have made the best estimate of the worth of wind-power to this Utility Company that could be made in the time available.

The analysis was confined to the main Central Vermont Public Service Corporation system comprising plants located on Otter Creek and its tributaries. The plants in this group are interconnected electrically and there are connections with Bellows Falls Hydro-Electric Corporation for purchased power which supplies about half of the system's total requirement. Isolated properties of Central Vermont Public Service Corporation were excluded.

Some eighteen hydro stations varying in capacity from 100 to 3400 kilowatts comprise this system, whose full load generating capacity totals some 19,000 kilowatts. For purposes of this study the plants were considered in three groups. The first group, totaling 11,200 kilowatts, has large storage reservoirs more than capable of handling the variations in stream-flow which occur within a period of a year, and with some capacity, although limited, to handle differences in annual flow as between one year and the next. The second group comprises two plants totaling 3900 kilowatts, and has a small reservoir capacity considered to be adequate to handle variations in load shape for a period of a week or two but with insufficient capacity to handle seasonal fluctuations in flow. The third group comprises a considerable number of small run-of-river plants, also totaling 3900 kilowatts, but with no storage and with ponds capable of maintaining rated output in dry weather for very short periods—in some cases not at all.

The system has no steam plants of its own, but the interconnection with Bellows Falls Hydro-Electric Corporation, which in turn interconnects with large steam plants, reflects the costs of steam-generated power. The contract provides a demand charge of $1.25 for each kilowatt of monthly demand and an energy charge of $0.009 per kilowatt-hour for the first 250 hours use of the demand, and $0.005 on additional energy with a discount of $0.001 on all kilowatt-hours. There is a coal charge which, at the time of the study, just about offset the discount. The demand charge is based upon the demand in certain peak hours of the heavy load season, as defined in the contract, with provisions to permit exceeding this demand in off-peak hours and seasons up to certain limits. For the purposes of this study the contract terms are the equivalent in cost to $0.005 per kilowatt-hour of energy and $27 per year per kilowatt of demand, as defined in the contract. To broaden the analysis, a second set of calculations was made assuming a situation in which the energy charge would be $0.0045 and the demand charge $15 per year per kilowatt.

FIXED CHARGES AND OPERATING COSTS

For the block of capacity on Lincoln Ridge, Jackson and Moreland, in conference with Central Vermont Public Service Corporation, agreed upon the fixed charges and operating costs in Table 18. The 3.36 percent for depreciation of the wind-

TABLE 18. ITEMIZATION OF ANNUAL CHARGES

	Wind-Power Development	Transmission Lines	Hydro-Plant Additions
Interest 40% bonds at 3.25%	1.30%	1.30%	1.30%
Dividends 60% stock at 6.50%	3.90	3.90	3.90
Property Tax	2.00	2.00	2.00
Federal Income Tax at 25%	1.30	1.30	1.30
Depreciation	3.36	3.36	1.78
	11.86	11.86	10.28
Operation and Maintenance	$ 2.50/kw-yr	1.00	1.00
		12.86	11.28

power development and the transmission lines corresponds to twenty years of life, with reinvestment in a sinking fund at 4 percent.

Similarly, the 1.78 percent for the depreciation of the hydro plant corresponds to a life of thirty years.

In the absence of actual wind speed measurements on the summit of Lincoln Ridge, it was necessary to assume a wind regime by interpolating in the meteorological and ecological evidence found in New England on higher and lower summits. It was recognized that the outputs of the six units on the ridge would vary considerably among themselves, as indicated in Table 9, but it was assumed that the average output of the block would be not less than 3500 kilowatt-hours per year per kilowatt of wind-generated capacity, of which there would be a net delivery to the Central Vermont Public Service Corporation system of 3320 kilowatt-hours after allowing for transformation and transmission losses.

The monthly distribution of this wind-energy was estimated by S. Morgan Smith Company to be as follows:

Month	Percent
January	11.3
February	10.5
March	11.0
April	7.5
May	6.8
June	6.3
July	5.0
August	5.8
September	5.7
October	8.5
November	9.6
December	12.0
Total	100.0

The energy value of the wind-power is measured by the worth of the resultant decrease in the quantity of purchased energy. The capacity value of the wind-power is measured by the resultant decrease in purchased demand, which depends upon the existence of hydro capacity to meet such demand when the wind does not blow. The estimates indicated that about 3600 kilowatts of additional hydro capacity would be required on the Central Vermont Public Service Corporation System to firm the 9000 kilowatts of wind. This is about 0.4 kilowatt of hydro per kilowatt of wind.

Against this background of physical plant, contractual obligations, and assumed wind speeds, Jackson and Moreland studied the rated and actual capacities of the hydraulic turbines on the system, the predictions of load growth, and the estimates of operating and maintenance costs of wind-power, as supplied by the Central Vermont Public Service Corporation. To arrive at a simplified dispatch, tentative agreements were reached with the Utility engineers as regards typical load shapes; reservoir operation and grouping of water flows into seasons; and grouping of stations into storage, semi-storage and strictly run-of-river. Agreement was reached as to the locations of the future development of the 3600 kilowatts of hydroelectric power required to firm the 9000 kilowatts of wind-power.

Based on this information, the most economical dispatch of the generated capacity, with and without wind, was worked out. This was held to be a sound approach since it was felt that corrections for departure in practice from these idealized operations would tend to cancel out. Finally, Jackson and Moreland estimated that the average cost of adding the additional 3600 kilowatts of hydro capacity would be $130 per kilowatt.

The economics of adding wind-power to the system of the Central Vermont Public Service Corporation is summarized in Tables 19, 20, 21, and 22, the first two showing the value of wind-power based only on savings in energy, and the last two including savings in the demand charges. Table 23 is a comparative summary of the foregoing tables. The last line represents the worth of the wind-power development and transmission line, including an additional credit for reactive kilowatts of $4 per kilowatt of installed capacity of wind-turbines.

On Fig. 111 are plotted the variations in the worth of wind-power to Central Vermont Public Service Corporation as the annual output varies from 2000 to 6000 kilowatt-hours per kilowatt. The high, low, and most likely values are plotted.

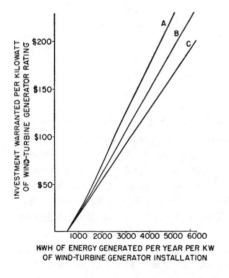

KWH OF ENERGY GENERATED PER YEAR PER KW
OF WIND-TURBINE GENERATOR INSTALLATION

Fig. 111. Variation in the worth of wind-power to the Central Vermont Public Service Corporation with variations in annual output.

Curve A   Charges for power purchased from other systems:
Energy Charge—$0.0050 per kilowatt-hour
Demand Charge—$27.00 per kilowatt-year

Curve B   Most likely value

Curve C   Charges for power purchased from other systems:
Energy Charge—$0.0047 per kilowatt-hour
Demand Charge—$15.00 per kilowatt-year

TABLE 19. COMPARATIVE FINANCIAL STATEMENTS FOR 9000 KILOWATT WIND-POWER DEVELOPMENT

(Assuming Purchased Energy at $0.0050/kwh.: No State Tax on Wind-Energy; No Associated Hydro Plant for Reducing Purchased Demand)

BALANCE SHEET

ASSETS	Without Wind	With Wind	Net	LIABILITIES	Net
Hydro Plant Additions	——	$ ——	$ ——	Bonds (40%)	$ 424,300
Wind-Power Development	——	950,700	950,700	Stock (60%)	636,400
Transmission Line	——	110,000	110,000		
	——	$1,060,700	$1,060,700		$1,060,700

INCOME STATEMENT

Revenue:	Without Wind	With Wind	Net
Purchased Energy Savings	——	$149,400	$149,400
Purchased Demand Savings	——	——	——
		$149,400	$149,400

Annual Costs:	Hydro	Wind	Trans.	With Wind	Net
Operation and Maintenance	——	$22,500	$1,100		
Property Tax	——	19,014	2,200		
Depreciation	——	31,944	3,696		
	——	$73,458	$6,996	80,454	80,454
				$ 69,946	$ 68,946

Net Income:		
Bond Interest at 3.25%		13,789
Taxable Income		$ 55,157
Federal Income Tax (25%)		13,789
Available for Dividends at 6.5%		$ 41,368

TABLE 20. COMPARATIVE FINANCIAL STATEMENTS FOR 9000 KILOWATT WIND-POWER DEVELOPMENT

(Assuming Purchased Energy at $0.0045/kwh.; No State Tax on Wind-Energy;
No Associated Hydro Plant for Reducing Purchased Demand)

BALANCE SHEET

ASSETS	Without Wind	With Wind	Net	LIABILITIES	Net
Hydro Plant Additions				Bonds (40%)	$373,900
Wind-Power Development		$824,700	$824,700	Stocks (60%)	560,800
Transmission Line		110,000	110,000		$934,700
		$934,700	$934,700		

INCOME STATEMENT

	Without Wind	With Wind	Net
Revenue:			
Purchased Energy Savings		$134,460	$134,460
Purchased Demand Savings			

Annual Costs:	Hydro	Wind	Trans.		
Operation and Maintenance		$22,500	$1,100		
Property Tax		16,494	2,200		
Depreciation		27,710	3,696		
		$66,704	$6,996	73,700	$134,460

Net Income:		
		$ 66,760
Bond Interest at 3.25%		12,151
Taxable Income		$ 48,609
Federal Income Tax (25%)		12,152
Available for Dividends at 6.5%		$ 36,457

TABLE 21. COMPARATIVE FINANCIAL STATEMENTS FOR 9000 KILOWATT WIND-POWER DEVELOPMENT

(Assuming Purchased Energy at $0.0050/kwh.; Demand at $27/kw.; No State Tax on Wind-Energy)

BALANCE SHEET

Assets	Without Wind	With Wind	Net
Hydro Plant Additions	$358,800	$ 468,000	$ 109,200
Wind-Power Development		1,182,400	1,122,400
Transmission Line		110,000	110,000
	$358,800	$1,760,400	$1,401,600

Liabilities	With Wind	Net
Bonds (40%)	$149,400	$ 560,700
Stock (60%)	97,200	840,900
	$246,600	$1,401,600

INCOME STATEMENT

Revenue:	Without Wind	With Wind	Net
Purchased Energy Savings	———		$189,200
Purchased Demand Savings	$57,400		
	$57,400	$246,600	

Annual Costs:	Without Wind	With Wind			Net
	Hydro	Hydro	Wind	Trans.	
Operation and Maintenance	$3,588	$4,680	$22,500	$1,100	
Property Tax	7,176	9,360	23,648	2,200	
Depreciation	6,389	8,330	39,729	3,696	
	17,151	115,243			98,092

Net Income:	Without Wind	With Wind	Net
	$40,249	$131,357	$ 91,108
Bond Interest at 3.25%			18,225
Taxable Income			$72,885
Federal Income Tax (25%)			18,221
Available for Dividends at 6.5%			$54,664

TABLE 23. SUMMARY COMPARISON OF WIND-POWER VALUES ON CENTRAL VERMONT PUBLIC SERVICE CORPORATION SYSTEM

COMPETITIVE CONDITIONS

Purchased Power Contract:

	Total Value		Energy Value Only		Total Value		Energy Value Only	
Net energy charge (per kwh.)	$.0050				$.0045			
Cost of "demand" (per yr. per kw.)	$27				$15			

INVESTMENTS WARRANTED BY VALUE OF WIND

	Total Value $	$/kw. of wind	Energy Value Only $	$/kw. of wind	Total Value $	$/kw. of wind	Energy Value Only $	$/kw. of wind
Total investment warranted	1,401,600		1,060,700		1,412,900		934,700	
Less associated extra hydro plant	109,200				468,000			
Warranted for wind power including transmission line	1,292,400	143.60	1,060,700	117.85	944,900	104.99	934,700	103.86
Add $4/kw. for reactive kva. value		4.00		4.00		4.00		4.00
Total warranted for wind-power and transmission line including reactive kva. value		147.60		121.85		108.99		107.86

TABLE 22. COMPARATIVE FINANCIAL STATEMENTS FOR 9000 KILOWATT WIND-POWER DEVELOPMENT

(Assuming Purchased Energy at $0.0045/kwh.; Demand at $15/yr./kw.; No State Tax on Wind-Energy)

BALANCE SHEET

Assets	Without Wind	With Wind	Net	Liabilities		Net
Hydro Plant Additions	—	$468,000	$468,000	Bonds (40%)		$ 565,200
Wind-Power Development	—	834,900	834,900	Stock (60%)		847,700
Transmission Line	—	110,000	110,000			
		$1,412,900	$1,412,900			$1,412,900

INCOME STATEMENT

Revenue:	Without Wind	With Wind			Net
Purchased Energy Savings	—	$134,460			
Purchased Demand Savings	—	54,000			
					$188,460

Annual Costs:	Hydro	Wind	Trans.		Net
Operation and Maintenance	$ 4,680	$22,500	$1,100		
Property Tax	9,360	16,698	2,200		
Depreciation	8,330	28,053	3,696		
	$22,370	$67,251	$6,996		96,617
					$ 91,843

Net Income:					
Bond Interest at 3.25%					18,369
Taxable Income					$ 73,474
Federal Income Tax (25%)					18,368
Available for Dividends at 6.5%					$ 55,106

WORTH VERSUS COST

Based on fixed and operating costs, as tabulated in the previous section, the total warranted investment by the Central Vermont Public Service Corporation for a 9000-kilowatt wind-power development, including 22 miles of transmission line, apparently lies between a maximum of $148 and a minimum of $108 per kilowatt of rated wind-generated capacity, with $125 as a likely value.

The higher figure also presupposes that Central Vermont Public Service Corporation, after further study, would find it practicable to embark upon 8600 kilowatts of additions to their hydro plants for the purpose of backing up the wind.

However, the cost of this 9000-kilowatt wind-power development, as determined by S. Morgan Smith Company in October, 1945, was about $205 a kilowatt. The only valid conclusion was that such a wind-power plant was not economically justified. Means of reducing the installed cost are explored in the next chapter. As fuel prices continue to rise, it seems likely that the gap could be bridged and that wind-turbines could be developed into additional sources of revenue for this and other utility systems.

SUMMARY

The costs of large-scale wind-power installations were determined, in 1945, based on a single hypothetical production run of 20 units rather similar in design to the test unit, to be about $190 a kilowatt installed at the switchboard, exclusive of transformers and connecting transmission line, and about $205 a kilowatt at the point of connection with the existing high-line.

The worth of wind-power was evaluated by Jackson and Moreland, who concluded that Central Vermont Public Service Corporation, for example, could afford to pay about $125 a kilowatt for a block of 9000 kilowatts.

Means of bridging this $80 gap are explored in the next chapter.

12
Ways to Reduce the Cost of Wind-Power

The cost, in October, 1945, of installing a block of 9000 kilowatts of wind capacity on Lincoln Ridge in Vermont, is reported at $191.11 a kilowatt exclusive of transformers and connecting high-line. This block of capacity would consist of six 1500-kilowatt units, weighing 497 pounds per kilowatt including the tower and foundation steel.

The possibilities for reducing this cost fall into six categories:
1. Cost reductions in the 1945 design by means of competitive bidding.
2. Cost reductions by means of refinements in the 1945 design.
3. Cost reductions by means of major modifications of the 1945 design.
4. Cost reductions by means of radical departures from the 1945 design.
5. Cost reductions resulting from the quantity production necessary to support a national wind-power program.
6. Cost reductions inherent in technological development.

The validity of estimating cost reductions under these six headings, item by item, and then adding them up to arrive at a possible total cost reduction, is open to a good deal of question. Some of my associates think it inevitable that the sum total of the savings described in this chapter will be realized and exceeded. Other associates doubt this. I shall itemize the possibilities, and leave it to the reader to make his own evaluation.

1. COST REDUCTIONS IN THE 1945 DESIGN BY MEANS OF COMPETITIVE BIDDING

As explained in Chapter 10, a large wind-turbine is made up partly of items which are in standard production and partly of items peculiar to the wind-turbine.

Reference is made in Chapter 11 to quotations received from various suppliers from items in standard production. In no way should these figures be considered as competitive bids. In the first place, our specifications were not, in general, rigid enough to permit rigorous competitive bidding, with the result that each manufacturer modified the specifications somewhat according to his best judgment of our requirements. In the second place, in any such preliminary proposal engineering, there is inevitably present more or less of a cushion to provide a factor of safety against the uncertainties in the specifications and in general to provide for contingency. Thirdly, the estimates were requested in lots of 1, 5, 10 and 20 units, but it was not possible to specify the production schedule. Lastly, it was understood by the vendors that the quotations were for information only and were not final competitive bids.

In our judgment, true competitive bidding based on shop drawings detailed for a definite production schedule would have shown a cost reduction of not less than 5 percent of the total cost of these standard production items.

As regards the second class of material entering into the Smith-Putnam Wind-Turbine, that is, items peculiar to the turbine, we lack even competitive estimates. The costs are based only on estimates prepared by the S. Morgan Smith Company. There is no reason to question the efficiency of the production engineering or the accuracy of the cost accounting of the S. Morgan Smith Company, both of which are thought to compare favorably with the practices of other manufacturers of comparable equipment. However, this is not to say that the items peculiar to the wind-turbine cannot be manufactured more cheaply than estimated by the S. Morgan Smith Company. The farming out of sub-assemblies to small shops frequently results in lower pound prices, particularly if a shop is found where the addition of a little load is welcomed as a means of justifying a second or a third shift; or of stabilizing employment; or of distributing overhead.

Our consensus is that the reduction in cost of the parts peculiar to the turbine by such manufacturing methods would be not less than 5 percent.

2. COST REDUCTION BY MEANS OF REFINEMENTS IN THE 1945 DESIGN

In Chapter 9 the recomputation of the loadings was discussed. These are the loadings used as the basis for the weight estimates of the preproduction design, the cost of which is analyzed in Chapter 10. It will be recalled that the 1940 test unit weighed 500 pounds per kilowatt, and the 1945 preproduction design weighed 497 pounds per kilowatt, despite much simplification and cleaning up.

There are two ways in which the weight of the unit, and also its cost, may be reduced through a modification in the loadings. The first may be called "Refinement by Successive Approximations" and the second "Refinement by Modifying the Functional Requirements."

Refinement by Successive Approximations

It has been pointed out in Chapter 10 that the determination of the weight of a rotating structure is made by a process of successive approximations. Since gravity or other acceleration forces are at work, the weight of the rotating members contributes to the total stresses. One assumes a distribution of weight, and runs through the calculations. If there is found to be an excess of weight, some may be removed in the 2nd approximation, bearing in mind that its removal will reduce some of the forces, and thus justify the removal of a little more.

It had been intended to carry through enough approximations of the preproduction unit to assure a cleaned-up model, designed to consistent strength. This has not been done. If done, it seems likely that there would be a weight reduction of not less than 7.5 percent.

Refinements in the Loading Assumptions, by Modifying the Functional Requirements

A review of the assumptions underlying certain of the loading conditions has revealed that acceptable alterations in operating and maintenance procedures would result in a reduction in the assumed maximum loading.

Maximum Aerodynamic Loading. For example, the most severe case of aerodynamic loading discussed in Chapter 10 occurs when the blades are positioned vertically, and locked in both rotation and pitch, in the maximum wind. If this condition could be eliminated by a change in operating practice, which would leave the blades free to move in rotation and in pitch, then the maximum blade loading would be reduced by about 10 percent. If a major repair does necessitate locking the blades vertically in periods of high wind, special supports and bracing can be rigged into the structure. There are other similar examples.

It is estimated that by these means the design stresses used in Chapter 10 could be reduced by 10 percent, with some resultant reduction in weight.

Reduction in Maximum Torque by Means of the Electric Coupling. A further small but real reduction in loading comes from a different source. When a synchronous generator is driven through an hydraulic coupling, the maximum torque that can be delivered by the turbine is limited only by the pull-out torque of the generator, that is, by the torque necessary to jump the generator out of phase with the transmission system with which it is interconnected. The pull-out torque of a generator is within the control of the designer. In practice on the Smith-Putnam Wind-Turbine, whenever the input torque reached 2.5 times the rated capacity of the generator, the generator was tripped off the line by a safety relay.

However, for reasons explained in Chapter 10, it is felt that any future large wind-turbine should not use an hydraulic coupling, but rather an electric coupling. The maximum torque transmitted by the electric coupling is in the control of the designer. If the specified maximum torque is exceeded, the generator is not affected, the excess torque merely causing an increased slip in the coupling. Thus, it is possible with an electric coupling to specify that the maximum input torque will not exceed 1.10 times the full-load torque. Fig. 112 shows the torque-speed curves for a Smith-Putnam Wind-Turbine with an hydraulic and electric coupling, respectively.

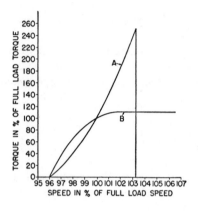

Fig. 112. Torque-speed curves, of the Smith-Putnam Wind-Turbine.

Curve A Using an hydraulic coupling

Curve B Using an electric coupling

The reduction in maximum input torque from 2.5 to 1.10 times the full-load torque reduces the maximum stresses in the structure by an amount which it is not easy to evaluate. Our best estimate is that this stress reduction would be reflected in a reduction in the total weight of the unit of about 2.5 percent.

Refinements in Loading Assumptions on Minor Members

Finally, it seems likely that a complete redesign campaign would secure as one of its dividends similar refinements of the loadings on various of the minor parts of the structure. It is too much to expect that a complete illumination of this problem would decrease *all* the loadings, but it seems equally improbable that the net result would not be an over-all reduction in weight.

Our consensus is that the total cost of the unit might be reduced 15 percent by refinements in the loading assumptions.

3. COST REDUCTIONS BY MEANS OF MAJOR MODIFICATIONS OF THE 1945 DESIGN

Flaps

One of the most attractice modifications would seem to be the use of controllable flaps, instead of controllable pitch, for the regulation of power input.

The rotational speed and the output of the test unit were controlled by changing the pitch angle of the blades. Fig. 66 shows schematic layouts of several types of flap to accomplish the same end. If flaps proved satisfactory aerodynamically, their use would result in very substantial structural advantages.

For example, the connection of the blade to the A-frame becomes not only a simple problem, since there is no need for relative motion between the two parts, but, also, a region of high stress concentration in moving parts is replaced by a simple structure with much lower stresses.

Among the expensive items eliminated by this arrangement are the shank forging and its costly supporting bearings. It is true that the operating mechanism becomes more complex, but the individual parts become much smaller so that the mechanism to operate the flaps should cost less and require less servo power than the mechanism to operate pitch control.

Von Kármán raised the question of flaps in 1939, but no definitive study has been made to my knowledge to determine whether control of the turbine under all conditions would be satisfactory with flaps. Today, we are in about the same position as regards flap control as we were in regarding pitch control in 1939.

We think the cost of the unit would be reduced about $25 per kilowatt by the use of flaps.

Elimination of Coning Damping

Another modification which should simplify the unit somewhat is the elimination of the coning-damping mechanism. The blades would then cone freely. Provision for coning-damping was built into the test unit and the initial runs were made with the coning heavily dampened. Operation was rough. The coning damping was relaxed and the operation became progressively smoother; when the coning damping was virtually eliminated, the operation was at its smoothest. We lack test data adequate to settle the point whether there should be a small amount of coning damping or whether the blades could be designed to cone freely without paying any important penalty in the operation of the unit. If the latter should prove to be the case, there would be a saving of about $0.50 per kilowatt.

Elimination of Yaw-Damping

Another modification in the direction of simplification would be the elimination of the yaw-damping mechanism. This device had been introduced into the redesigned test unit as a modification calculated to smooth out the operation of the unit in yaw. However, the engineers felt that, although this device was included in the cost schedules of Chapter 11, it was perhaps not absolutely essential, even on the test unit as designed. And the reduction of e to zero (page 159), and the decrease in the overhang of the plane of rotation, both incorporated in Dornbirer's plans for the preproduction unit, would go so far toward eliminating the residual roughness in the operation of the test unit as to make it possible to eliminate the yaw-damping mechanism and to substitute for it a direct connected yaw-mechanism similar to that installed on the test unit. The substitution of this simpler mechanism should reduce the cost of the unit about $1 per kilowatt.

Skin-Stressed House

This is another modification which is probably in the direction of cleaner structural design and which might result in a small reduction in cost. In the design of Chapter 10, a stiff, box-type pintle girder supported all the machinery aloft, which was protected from the weather by wrapping a light sheet steel house around the whole thing. In a production design one might make the house and pintle girder into one unit. The house would be at least partially skin-stressed and, therefore, something more than self-supporting; such a structure would probably show an appreciable saving in weight but might cost more per pound, especially in small production runs. Arbitrarily, therefore, we estimate that by combining these two structures we would save $0.50 per kilowatt.

Conical Sheet Tower

A further modification was partially explored in 1945. I refer to a conical tower constructed of sheet steel and looking like an ice-cream cone upside down. The greatest advantage of this type of tower compared with the four-legged tower was found to occur where the foundation rock was sound and the tower could be secured by anchor bolts directly imbedded in the rock, eliminating the necessity for an expensive foundation excavation, of the sort we had to make at Grandpa's Knob. Costs of this type of tower were obtained from the Chicago Bridge and Iron Company, based on a somewhat incomplete stress analysis, whose uncertainties were guarded against by using plate thicknesses which were probably a little heavier than would be selected after a rigorous design. Ignoring this factor, however, it was found that in lots of 20 and where the foundation conditions were good, the use of the conical tower design would result in a saving of only about $0.15 per kilowatt; where the foundation conditions required excavation there would be no saving.

Conforming to the Most Economical Dimensions

We have now made four determinations of the most economical dimensions of a large wind-turbine, with increasing confidence in the results. The dimensions used in the cost analysis of Chapter 10 were substantially those of the test unit which, it will be remembered, had been deliberately selected as being within range of the most economical dimensions, but on the small side.

We think that by conforming to the most economical dimensions, as determined by a fifth approximation, there would be realized a reduction in unit cost of about $1 per kilowatt.

Recapitulation of Possible Cost Reductions by Means of Major Modifications in the 1945 Design:

Modification	Cost Reduction
1. Flap control.........................	$25.00/kw.
2. Elimination of coning-damping	0.50/kw.
3. Elimination of yaw-damping	1.00/kw.
4. Combined house and pintle girder.........	0.50/kw.
5. Shell-type tower	—
6. Optimum dimensions..................	1.00/kw.
Total saving	$28.00/kw.

Recapitulation of the Total Cost Reductions Which Might Be Realized by Competitive Engineering and Bidding, Refinements in the Loading Assumptions of the 1945 Design and Major Modifications of the 1945 Design:

Estimated installed cost at the switchboard on Lincoln Ridge, Vermont, of six 1500-kilowatt units, diameter 175 feet, turbine speed 31.5 revolutions per minute, generator speed 900 revolutions per minute, tower height 150 feet—August, 1945......... $166.77/kw. Cost reductions by means of:

a. Competitive bidding in the 1945 design, 5%.....	$ 8.34
b. Refinements in the 1945 design, 15%	25.02
c. Major modification in the 1945 design.........	28.00
Total cost reduction.......	$61.36
Net cost..............	$105.41/kw.

4. COST REDUCTION BY MEANS OF RADICAL DEPARTURES FROM THE 1945 DESIGN

Elimination of Gear and Coupling

The gear is the bottleneck in the design of a large wind-turbine. Many schemes have been proposed looking to its elimination. One of these, made by Honnef of Berlin, is described in Chapter 6, together with reasons why that particular suggestion was considered impractical.

It might be possible to connect a 2000-kilowatt generator directly to a large wind-turbine with a rotational speed of about 30 revolutions per minute. The construction of such a generator is practical if the members are sufficiently rigid to retain the air-gap within the proper limits. The elimination of the gear and other equipment would reduce the cost by about $21 per kilowatt. Development studies would be required to determine if the increase in cost of the special low-speed generator would be less than this.

Combination of Generator and Electric Coupling in One Unit

A less drastic suggestion is the combination of a conventional generator and an electric coupling into one unit. Such a composite design has been developed on an experimental basis. Whether it could be made satisfactory in wind-turbine service and what the cost per kilowatt would be in limited mass production can only be determined by further study.

Elimination of Coning

A structural suggestion frequently repeated is that the blades should be fixed at the hub with no freedom to cone. It is possible that definitive cost studies of designs with and without coning would show that a design with three fixed blades would prove economically superior to one with three blades free to cone. Our dynamic studies have indicated that for the two-bladed design coning is better, and it is questionable whether there would be any economic gain in going to three fixed blades.

There are still other possibilities, too speculative to be mentioned here, and it is futile even to guess at the economic gains, if any, to be achieved by this sort of thinking.

5. COST REDUCTIONS RESULTING FROM QUANTITY PRODUCTION NECESSARY TO SUPPORT A NATIONAL WIND-POWER PROGRAM

The costs of Chapter 11 were based on a total production of twenty 1500-kilowatt units, and the pound price quoted by S. Morgan Smith Company for the 30,000-pound mild-steel blades, for example, was $0.44 a pound. The over-all manufacturing pound price for the unit was $0.39 a pound.

Some comments are necessary to complete the assessment of such cost estimates.

The design had not been analyzed for production. The costs had been estimated in lots of 1 and 20. The percentage reduction in the unit cost of a lot of twenty blades, hub assemblies, and pintle assemblies was 9 percent, 10 percent, and 7 percent, respectively, as compared with the estimated costs of lots of 1. What this reduction amounts to when elaborate jigging and fixtures can be used to advantage is seen in Fig. 85, which shows that the unit cost of twenty pairs of stainless-steel blades is estimated by the Budd Company to be only about 50 percent of the cost of one pair.

If a national program were to get under way, in Scotland, New Zealand, or the United States, the price of $0.44 per pound for the blades would tend to shift toward the price of $0.06 per pound at which 45,000-pound mild-steel box cars and gondolas were being sold in August, 1945. It is true that the car builders were able to offer such a price because their shops were tooled for a production of tens of thousands of freight cars a year, and it is equally true that even national wind-power programs would hardly consume more than a few thousand 2000-kilowatt or 3000-kilowatt wind-turbines over a period of years. Still, this would be something better than a total production run of 20 units, and the over-all competitive pound price of a design engineered for production in lots of 100 per year would be something less than $0.39, by perhaps 15 percent, bringing the over-all pound price to $0.33 (1945 prices).

6. COST REDUCTIONS INHERENT IN TECHNOLOGICAL PROGRESS

It is a truism that the history of many technical developments is one of relatively decreasing unit costs. Today we can buy for $1500 a better car than we could get for $5000 in 1915. So it is also with household refrigerators, radios, frozen foods, and many products where competition has resulted in aggressive engineering.

It is very unlikely that the speculations of this chapter have exhausted the possibilities which lie in the future of the large-scale wind-turbine. After all, our experience is limited to 1100 hours of operation of a single test unit, not designed with an eye to low production costs.

It seems inevitable that the hundred and first production unit, and even the eleventh, would contain many refinements contributing to lowered unit costs not foreseen in this chapter.

SUMMARY

In Chapter 11 it was found that a preproduction unit closely following the test unit in design would cost $191.11 a kilowatt installed on Lincoln Ridge in Vermont, exclusive of transformers and connecting high-line. Six means of reducing this cost are considered. Estimates of cost reduction by each of these means are offered. Some of these estimates rest on good evidence; some are highly speculative. If all of the suggestions could be realized and if no unfavorable factors remain to be discovered, it is possible that 2000-kilowatt or 3000-kilowatt wind-turbines, produced in quantities sufficient to support a national wind-power program, could be installed for about $100 a kilowatt, exclusive of transformers and connecting high-line (1945 prices).

13
The Future of Wind-Power

J.B.S. Haldane, in *The Last Judgement,* records verbatim a Children's Hour broadcast from Venus in the year 40 million. In describing the end of human life on Earth, the commentator remarks: "It was characteristic of the dwellers on Earth that they never looked ahead more than a million years, and the amount of energy available was ridiculously squandered."

Oil. It is true that our reserves of oil contain no more heat than we receive from the Sun in the course of a day or so, and that we are burning them up with the abandon of spendthrifts. The United States, once the greatest producer, now imports oil.

Coal. It is also true that one of the world's great industrial countries, Great Britain, has run short of coal. The mines are old, the working faces are far from the pit-heads, and the miners are telling their sons to quit mining. In the winter of 1946-1947 Britain, who exported as much as 40,000,000 tons of coal annually before the war, exported virtually none. Indeed, she did not have enough for herself.

And in this country it seems likely that coal prices will go higher.

Nuclear energy has been brilliantly harnessed. Whether the Atomic Energy Commission will feel free in the years just ahead to release uranium for general use in power piles remains to be seen. Without much to go on, it has been estimated that a 100,000-kilowatt nuclear power plant would cost about $270 per kilowatt and would break even with $9 coal. And it has been pointed out that the general application of nuclear fuels would probably be confined to the very large central stations. Oppenheimer, testifying at Lake Success as a guest witness before the Atomic Energy Commission of the United Nations in June, 1947, estimated that the application of atomic fuels to large central stations lay twenty to thirty years in the future.

On July 24, 1947, the *New York Times* quoted the U.S. Atomic Energy Commission as saying, in its first semiannual report: "But a number of basic advances in physics, chemistry and metallurgy will be required before power is produced at satisfactory efficiency and cost. The technical problems to be overcome are many, but we confidently expect them to be solved."

In any event, such substitution of nuclear energy for coal as the fuel in central stations would not appear to have more than incidental bearing on the future of wind-power.

214

The Sun. The power available in the direct radiation from the sun is colossal, but only in deserts in low latitudes is it possible to extract some of this power economically.

The Tides. Widespread extraction of power from the tides remains an economic fantasy.

Wind-power. It is hardly possible to make a tally of the world's available wind-power as one does of water-power. We know the world stream-flow in cubic feet per second fairly accurately and annual rainfall gives us a useful cross check. We know the average height through which this mass of water falls in getting to the sea or to the land-locked lakes. The product, in foot-pounds per second or horse-power, is a matter of arithmetic. Thus, the world's potential water-power amounts to about 5×10^8 kilowatts. Of this, about 5×10^7, or 10 percent has been developed.

The case of wind is different. True, we can weigh the atmospheric envelope and assume a mean speed for it. Brunt (Ref. 12-A), who has done this, estimates the total power in the atmosphere to be 3×10^{17} kilowatts. But not all of this power is available to man—only that portion in the lowest stratum of the atmosphere.

Willett,* after making certain assumptions, concludes that the wind-power available to wind-turbines amounts to about 2×10^{10} kilowatts. Only national wind-power surveys can determine what percentage of this 20 billion kilowatts of wind-power is to be found on the type of site described in Chapter 4, and near load centers.

What is the potential role for some of this 20 billions of kilowatts of wind-power, especially in conjunction with water and fuels, including nuclear fuels?

THE EXTENT OF THE MARKET

In Vermont, we have found a good wind-power site—Lincoln Ridge—with a capacity for some 50,000 kilowatts. Like a good site for a large dam, a good site for a large block of wind-power is a topographic rarity. To provoke discussion, I will guess that somewhere on earth there are 49 more sites like Lincoln Ridge, close to heavy load centers. Most of these sites would be found in such windy, industrialized regions as Scotland, northern Ireland, Iceland, Newfoundland, the Maritime Provinces of Canada, New England, other parts of the United States, southern Chile, New Zealand, Tasmania, and possibly high in the Italian Appennines, and in Scandinavia and other regions itemized in Tables 1 and 2 in Chapter 2. In the aggregate these 50 sites would amount to a potential market for large wind-turbines of 2,500,000 kilowatts.

---

*Private communication.

If governments should sponsor wind-power projects, using a standard design of about 2500 kilowatts in order to obtain the benefits of mass production and standard maintenance, then conceivably 100 such units, aggregating about 250,000 kilowatts, might be installed annually for a number of years.

At the other extreme lies the wind-power set, rated at 1 kilowatt or less, used for farm lighting and radio battery charging in districts without rural electrification. About 10,000 are being sold each year by United States manufacturers. If it is assumed that this market will not be saturated for twenty years, it will have amounted to 200,000 kilowatts by 1967.

Between these two extremes much interest has been shown in wind-power plants of a few hundred kilowatts or less.

In the period 1940-1946, the S. Morgan Smith Company, without having advertised the experiment on Grandpa's Knob, received hundreds of inquiries from all quarters of the globe for wind-power plants rated at about 100 kilowatts, for isolated use; or use in conjunction with hydro alone; or with fuel generation alone; or on small combined water and fuel systems. Although one or two experimental plants of about this size have been built, as described in Chapter 6, such a plant has never been designed for production and the market for it has never been estimated.

THE WORTH OF WIND-POWER INSTALLATIONS

Table 22 (Chapter 11) shows how the worth of wind-power varies with the size of the installation; with the windiness of the site; with the amount of the annual charges; and with the credit for capacity value. The 1-kilowatt set for farm lighting, when properly installed on a tower of suitable height, commands a market price of $300 to $600 a kilowatt. The 100-kilowatt set is worth from about $100 a kilowatt based on energy value alone to over $600 a kilowatt where a credit for capacity value can be earned. The 3000-kilowatt set similarly is worth from about $50 to over $300 a kilowatt.

THE INSTALLED COSTS OF WIND-POWER INSTALLATIONS

The family of non-linear curves of Fig. 113 has been prepared to suggest how the 1945 installed cost per kilowatt would have varied with variation in the capacity in kilowatts, and with variation in the manufacturing rate. This family of curves has been developed around only two rather vague fixes—the known unit cost of the 1-kilowatt set in lots of 10,000 per year, and the estimated unit cost of the 1500-kilowatt set in lots of 6 per year—the whole embellished by the speculations of Chapter 12 concerning the least cost of a 2000- or a 3000-kilowatt unit, in lots of 100 per year. It must be realized, therefore, that the curves have no serious quantitative meaning.

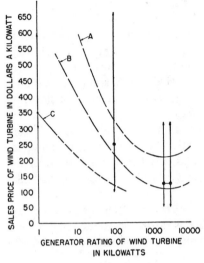

Fig. 113. First approximation of the variation in the 1945 worth of a wind-turbine with variation in the rating, for three assumed annual manufacturing rates.

Curve A	6 units per year
Curve B	100 units per year
Curve C	10,000 units per year

Vertical arrows show ranges of worth taken from Table 17. Double circles indicate most probable values.

For comparison, the range of the worth of each of the three sizes, taken from Table 17, is also shown in Fig. 113 by means of three vertical arrows. The bull's-eyes indicate representative values.

Assuming that the 1-kilowatt set was being sold at a profit in 1945, then, if the speculations of Chapter 12 have validity, it would have been possible in 1945 to manufacture and install the 2000- or 3000-kilowatt (and perhaps the 100-kilowatt) unit at a profit, provided that the market could have absorbed 100 or more units a year.

In the absence of both a national wind-power survey and cost studies based on production designs, it does not seem profitable to push these computations further. The crystal ball is not yet in focus. The conclusions to be drawn from the blurred image so dimly seen follow.

CONCLUSIONS REGARDING THE FUTURE OF WIND-POWER

1. Nuclear energy is another source of heat, but its economics are not yet accurately known, and in the foreseeable future nuclear energy will have no effect on the market for large-scale wind-power.

2. In the foreseeable future, nuclear energy will not displace water-power.

3. Neither solar nor tidal-power will be harnessed on a large scale in the near future.

4. As long as water-power remains economically justified, special partnerships between wind and water will be justified.

5. Coal prices will continue to increase.

6. The market for wind-power plants may fall into four groups, characterized by the size of the unit:

A. The largest unit would be rated at 2000 or 3000 kilowatts, for addition to existing power systems, principally in conventional support of water and steam. This application is limited to those selected sites near heavy load centers which occur most frequently in windy regions between latitudes 30 and 60 degrees, North and South. The market may range in size from 1,000,000 kilowatts to 10,000,000 kilowatts.

B. The medium unit would be rated at 100 to 500 kilowatts, for use in conjunction with small hydroelectric installations, or Diesel sets, in windy, isolated communities, such as the Shetlands, the Orkneys and some of the islands in the trade winds. The market may range in size from 250,000 kilowatts to 2,500,000 kilowatts.

C. A small unit of about 10 kilowatts would have a special limited use in charging batteries for untended airway beacons, as in the Arctic, and perhaps some desert regions. The market may range from 1000 kilowatts to 10,000 kilowatts.

D. The smallest unit of 1 kilowatt or less, is for farm lighting. The market may range in size from 250,000 kilowatts to 2,500,000 kilowatts.

7. The range in the worth, measured in dollars per kilowatt, at the 1-, the 100-, and the 3000-kilowatt sizes, respectively, is discussed in Chapter 11, tabulated in Table 16, and compared with very uncertain estimates of manufacturing costs in Fig. 113, which indicates that production runs of about 100 units a year would permit selling at a profit in the 100- and the 3000-kilowatt markets.

8. Possible applications for intermittent wind-energy, in isolated packages, are listed in Chapter 11.

9. The first step toward harnessing wind-power in blocks larger than a few kilowatts should be a national wind-power survey, the specifications for which are discussed in Chapter 5.

10. Grandpa's Knob has demonstrated that the technical problems of the 1250-kilowatt wind-turbine are understood and have been solved. To solve the economic problems of putting this or a larger wind-turbine into low-cost production probably requires Government aid.

BIBLIOGRAPHY A–GENERAL REFERENCES*

1. Baker, O.E.
   Atlas of American Agriculture, Government Printing Office–1936
2. Sherlock, R.H. and Stout, M.B.
   Relation Between Wind Velocity and Height During a Winter Storm
3. Wells, B.W. and Shunk, I.V.
   Salt Spray–An Important Factor in Coastal Ecology, *Bulletin* Torrey Botanical Club
   65, 485-492–October 1938
4. Willhofft, F.O.
   Industrial Applications of the Flettner Rotor, *Mech. Eng.,* Vol. 49–No. 3
5. *The Electrician,* Electricity from the Wind–November 24, 1933
   Electrical World, Wind Rotor Experiments "Decidedly Satisfactory"–October 29, 1933
6. Savonius, S.J.
   The S Rotor and Its Applications, *Mech. Eng.*–May 1931
   Klemin, Alexander
   The Savonius Wind Rotor, *Mech. Eng.*–November 1925
7. *Scientific American,* Aerodynamic Windmills–June 1929
8. Darrieus
   Les Moteurs a Vent. Les Colines Electriques, *La Nature*–December 15, 1929
9. Sectorov, W.R.
   Report on the Operating Characteristics of the Initial 100 KW Aero-electric Unit at
   Balaklava (Translated), *Elektrichestvo*–No. 2 of 1933
10. *The Electrician,* Electricity from Wind Power–December 8, 1933
11. Glauert, H.
    Aiscrew Theory–Vol. 4, Section L of *Aerodynamic Theory,* edited by Durand
12. Brunt, D.
    Physical and Dynamical Meteorology, Cambridge University Press–1934

BIBLIOGRAPHY B

A partially annotated bibliography of upwards of a thousand references dealing with the prior art, in English, French, German, Italian, Spanish, and Russian, is available at the York, Pennsylvania, offices of the S. Morgan Smith Company.

The following references are to be found only in the S. Morgan Smith Company files:

1. Petterssen, Dr. Sverre
   Wind Regimes of the World, Preliminary Report–April 15, 1940
   Second Report–August 9, 1940
   Appendix 1 to Reports–December 10, 1940

---

*Reference numbers followed by an A in the Text, such as 1-A, are general references, whereas those followed by a B are to be found in the S. Morgan Smith Company files only.

2. Petterssen, Dr. Sverre
   Preliminary Report on the Energy of the Winds in the New England Area—April 1940
   Report on Additional Research in Connection with the Energy of the Winds in the New England Area—April 1940
   Reports on Wind Observations in Vermont—July 1–October 23, 1940
   Appendix 1 to Report on Wind Observations in Vermont—January 20, 1941
   Summary Report on Site Investigations—June 10, 1940
   Appendix 2 to Report on Wind Observations in Vermont—March 14, 1941
   Site Specifications for Test Site and Memoranda on Site Specifications for the Akron Symposium—July 29, 1940
3. von Kármán, Dr. Th.
   Preliminary Report on the Aerodynamic Characteristics of the Smith-Putnam Wind-Turbine—January 5, 1940
   Second Report on the Aerodynamics of the Smith-Putnam Wind-Turbine—1940
4. Wilcox, Carl J.
   Memorandum on the Computation of Mean Annual Weighted Density—April 1941
5. Wilcox, Carl J.
   Computation of Mean Annual Weighted Density at Sea Level, Blue Hill, Mt. Abraham, Lincoln Mt., Mt. Washington and 10,000 Ft.—September 1945
6. Brooks, Dr. Charles F.
   Construction of Maps of Maximum Icing and Maximum Wind for the U.S., Southern Canada and the South Coast of Alaska—January 6, 1941 (Bound with Appendix 2 to Report 3 by S. Petterssen)
7. Griggs, Dr. Robert F.
   Reports on Field Trips to Central Vermont—May 25, 29, 1940, and June 30, 1940
8. Troller, Dr. Theodore
   S. Morgan Smith Wind-Turbine Site Tests, Pond Mt.—April 1940
   Further Report on Pond Mt.—June 1940
   East and Glastenbury Mts.—April 1940
   Mt. Washington—April 1940
   Summary Report—June 1940
9. Wilcox, Carl J.
   Anomaly Study—April 1941
   Petterssen, Dr. Sverre
   Appendix 1 to Report on Wind Observations in Vermont—January 20, 1941
10. Putnam, P.C.
    Predictable Wind Power—June 14, 1939
11. Lange, Dr. Karl O.
    Anemometry Suggestions and Recommendation—March 1940
12. Harvard University Aerodynamics Laboratory
    Report No. 102—March 15, 1940—William Bollay
    Report No. 103—November 1940—A.E. Puckett
    Report No. 104—February 25, 1941—A.E. Puckett
    Report No. 105—no date—A.E. Puckett
    Report No. 106—April 28, 1941—A.E. Puckett
13. Wilcox, Carl J.
    Handbook of Aerology, Vol. 3, Site Factor and Variation with Height Computations Test Site
    Memorandum on the Computation of Variation with Height—February 21, 1941
    Mt. Washington with Correction Factors for Anemometer Locations—September 1945

Original Data—Comparison of Old and New Masts
Miscellaneous Correspondence—Original Data for Report
Original Data for Grandpa's Knob by Months, Southwest Winds Only
Report on Vertical Velocity Gradients on Grandpa's Knob—December 1941

14. Griggs, Dr. Robert F.
Tree Reactions to Wind—February 14, 1940
Report on Field Trips to Central Vermont—May 1940
Report on Examination of Mount Ellen Ridge—March 1, 1945
Report on Field Trip—May 26 through June 4, 1945

15. Wilcox, Carl J.
Report on Vertical Velocity Gradients on Grandpa's Knob—December 1941

16. Wilcox, Carl J.
Computations on Variation of Velocity with Height, Vol. 3, Handbook of Aerology—
1941

17. Wilcox, Carl J.
Mt. Washington Gradient with Correction Factors for Anemometer Locations—September 1945

18. Wilcox, Carl J. and Dornbirer, S.D.
Large-Scale Wind Power Analysis—October 1945

19. Wilcox, Carl J.
Static and Dynamic Characteristics of the Smith-Putnam Wind-Turbine—1940
Memorandum on the Computation of Outputs—February 1941

20. Wilcox, Carl J.
Output Summaries—1940-1945, Vol. 4, Handbook of Aerology
Anomaly Study—April 1941-December 1946

21. Wilcox, Carl J.
Memorandum on Diurnal Variation of Output—February 18, 1944

22. Voaden, Grant H.
Field Test Report No. 1—Preliminary Report of Field Tests on Wind-Turbine Speed
Regulation—November 14, 1941
Smith-Putnam Wind-Turbine Field Test Report No. 2—December 10, 1941
Smith-Putnam Wind-Turbine Field Test Report No. 3—February 2, 1942
Smith-Putnam Wind-Turbine Field Test Report No. 4—March 18, 1942
Smith-Putnam Wind-Turbine Field Test Report No. 5—March 26, 1942
Smith-Putnam Wind-Turbine Field Test Report No. 6—April 4, 1942
Smith-Putnam Wind-Turbine Field Test Report No. 7—May 20, 1942
Smith-Putnam Wind-Turbine Field Test Report No. 8—September 21, 1942
Smith-Putnam Wind-Turbine Field Test Report No. 9—January 14, 1943
Smith-Putnam Wind-Turbine Field Test Report No. 10—April 1, 1943

23. Haynes, S.S.
Comparison of Forecasts—April 29, 1942
Reviewed and brought up to date by G.H. Cheney—February 1944

24. Putnam, P.C.
Report to Board of Directors, S. Morgan Smith Company—March 1940

25. Wilbur, Dr. J.B.
General Specifications for Smith-Putnam Wind-Turbine—April 8, 1941

26. Voaden, G.H.
Book of Instructions for Smith-Putnam Wind-Turbine—2 Vols.

27. Wilbur, Dr. J.B.
Test Program Smith-Putnam Wind-Turbine Test Unit—May 10, 1941

28. Holley, M.J.
    Operating Instructions for Foxboro Recorder—June 11, 1941
29. Wilcox, Carl J.
    Evaluation of the First Run of the Turbine—October 19, 1941
30. Wilcox, Carl J. and Holley, M.J.
    Memo on Coning Prepared for von Kármán—July 1942
31. Test Engineers and Operators
    Log Sheets Daily—September 4, 1941-March 26, 1945
32. von Kármán, Dr. Th.
    Report on Aerodynamic Data for Choice of the Optimum Design—February 5, 1940
33. Reid, Elliot G.
    Model Tests Smith-Putnam Wind-Turbine—February 3, 1941
34. Wilcox, Carl J.
    a. Outputs and Frequency Distribution Curves, New England Sites, Vol. 7, Handbook
       of Aerology—Eastern U.S.
    b. Outputs and Frequency Distribution Curves, New England Sites, Vol. 12, Handbook
       of Aerology
    c. Outputs and Frequency Distribution Curves for Oceanic Islands and the Maritime
       Littorals, Vol. 13, Handbook of Aerology
35. Wilcox, Carl J.
    Cost Study of Induction Generation—1943
    Synchronous Generation—Optimum Study—1943
    Summary of Energy Costs Induction and Synchronous Generation—1943
36. Wilcox, Carl J.
    Report on Energy Costs—May 9, 1945
37. Wilcox, Carl J.
    Report on Choice of Blade Shape—1943
38. Rutland Office
    Basic Aerodynamic Calculations
39. Wilcox, Carl J. and Holley, M.J.
    Report on Yawing and Pitching Moments Transmitted to the Pintle Axis of the Smith-
    Putnam Wind-Turbine—April 1943
40. Wilcox, Carl J. and Holley, M.J.
    Determination of Design Loadings for Blades of Smith-Putnam Wind-Turbine—May 14,
    1945
    Summary
    Reports on each of 10 Loading Conditions
41. Wilbur, Dr. J.B.; Wilcox, Carl J.; and Holley, M.J.
    Tentative Loading Specifications for the Blades of the Preproduction Model—Decem-
    ber 22, 1943
42. U.S. Plywood Corporation
    Report on Preliminary Design of Plywood Blades for the Smith-Putnam Wind-Turbine—
    June 1, 1944
43. Wilcox, Carl J.; Holley, M.J.; and Dornbirer, S.D.
    Report on Cost Study of October 1945
44. Atkinson, K. (Jackson & Moreland)
    Preliminary Memorandum on Value of Wind Power—April 27, 1945
45. Atkinson, K. (Jackson & Moreland)
    Recommended Draft for "The Worth of Wind Power to Central Vermont Public Ser-
    vice Corporation" (included in "Large-Scale Wind Power"—an analysis by C.J. Wilcox
    and S.D. Dornbirer, October 26, 1943)

Part 2

1
The Wind Resource and
Its Potential for
Electric Generation

THE POWER IN THE WIND

It is useful to know the upper limit for the power in the wind. Recently, Gustavson (1979) estimated that the radiant energy from the sun is transferred to the winds of the earth at a rate which maintains the wind power capacity at approximately 3.6 X 10^{15} watts. This power level corresponds to an annual energy of 107,000 quads.* He employed rough working assumptions to obtain an estimate for the amount of energy which could be extracted by wind machines and obtained a global extraction limit of 4,000 quads per year. The extraction limit for the conterminous United States was estimated to be 60 quads per year. The total U.S. energy use for 1975 was 75 quads.

Three national wind energy assessments have been performed to evaluate the geographical distribution of the power in the wind over the United States. The studies were performed by Reed (1975) of Sandia Laboratories, the General Electric Company (Garate, 1977, and the Lockheed-California Company (Coty, 1977). Reed's study of wind power climatology was completed before the other studies began and acted as a data base. The General Electric and Lockheed-California wind energy assessments were components of mission analysis studies which were conducted for the Federal Wind Energy Program. These studies provided the first overview estimates of the national wind energy potential in terms of the national wind resource, the economic impact and benefit of wind systems, potential wind energy applications and markets, and the performance and requirements of potential wind power users. The discussion of the results from the mission analysis studies here will be limited to wind energy assessments and scenarios for the production of electrical energy using wind power technology.

---

*One watt-hour equals 3.412 Btu and one quad equals 10^{15} Btu.

MATHEMATICAL DESCRIPTION OF WIND POWER

The kinetic energy (E) of a mass (m) of air which moves through an area (A) at normal incidence during a period of time (t) may be expressed as

$$E = (\frac{1}{2}) m V^2. \tag{1-1}$$

The mass is given by the product of the air density (ρ) and the volume of the air which passes through the surface during the period (AVt). The energy E may thus be expressed as

$$E = (\frac{1}{2}) \rho A V^3 t. \tag{1-2}$$

The power of the wind $(P = E/t)$ is given by

$$P = (\frac{1}{2}) \rho A V^3 \tag{1-3}$$

and the power density of the wind $(P_d = P/A)$ is given by

$$P_d = (\frac{1}{2}) \rho V^3. \tag{1-4}$$

If ρ and V are expressed in kilograms per cubic meter (kg/m^3) and meters per second (m/s), respectively, the power density is obtained in watts per square meter (W/m^2).

The mean wind power density for a given period $(\langle P_d \rangle)$ is usually calculated using the expression[*]

$$\langle P_d \rangle = \frac{1}{2} \langle \rho \rangle \langle V^3 \rangle \tag{1-5}$$

where $\langle \rho \rangle$ and $\langle V^3 \rangle$ are the mean air density and the mean of the wind speed cubed for the given period, respectively. In the wind energy assessment studies, mean wind power densities were evaluated using frequency distributions prepared by the National Climatic Center (NCC). Values were calculated using the expression

$$\langle P_d \rangle = (\frac{1}{2}) \langle \rho \rangle \sum_{i=1}^{n} f_i (V_i^m)^3 \tag{1-6}$$

where f_i is the fraction of wind speed measurements in class i, V_i^m is the median wind speed for class i, n is the number of wind speed classes, and $\langle \rho \rangle$ is the mean

---

[*]If n measurements of the air density and wind speed are made at regular intervals during a given period, the mean wind power density may be evaluated using the expression

$$\langle P_d \rangle = (1/2) \sum_{i=1}^{n} \rho_i V_i^3 = (1/2) \langle \rho V^3 \rangle.$$

If an expression of the form $\langle P_d \rangle = (1/2) \langle \rho \rangle \langle V^3 \rangle$ is employed to evaluate the mean power density, the appropriate mean density $\langle \rho \rangle$ for the calculation is the weighted mean $\sum_{i=1}^{n} \rho_i V_i^3 / \sum_{i=1}^{n} V_i^3$ rather than the period mean $\sum_{i=1}^{n} \rho_i / n$.

density for the period. A wind speed class refers to a range of wind speeds and the classes span the full range of observed wind speeds. The accuracy of the wind power density depends upon the number of classes and the number and type of measurement samples.

NATIONAL WIND ENERGY ASSESSMENTS

Reed (1975) employed data from 758 North American weather stations to evaluate seasonal and annual mean wind power densities. The results were presented in the form of maps which show isodyn controus, i.e., the locus of points of equal wind power density. The results are shown in Figure 1-1. Reed's map shows high wind power densities in the Texas Panhandle and Western Great Plains regions, and off the New England and Pacific Northwest coasts.

  The General Electric researchers developed composite maps which show the distribution of annual available wind energy at surface (approximately 10 m) and 50-m levels. To obtain the distributions at 50-m elevation above ground level, the General Electric researchers developed an extrapolation procedure based on surface data and 150-m rawinsonde data from the 65 stations in the upper-air reporting network. Since few wind measurements have been made at mountaintop locations, Crutcher's (1959) upper-air climatological charts at 850-millibar (mb) and 700-mb pressures were used to estimate the wind energy in mountainous regions. The 850- and 700-mb levels correspond to mean elevations of approximately 5000 ft (1520 m)

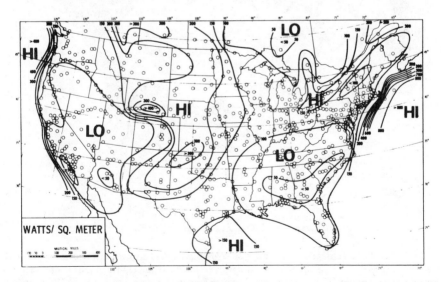

Fig. 1-1. Reed's map of annual mean available wind power for the conterminous United States. The isodyne values are in W/m^2. The results are based on surface wind data from the weather stations indicated by circles. From Reed (1975).

and   10,000 ft   (3050 m),   respectively.   Over high-terrain areas, values of
the free-air wind energy based on the extrapolation of Crutcher's results to 50-m
elevation above terrain level were used, with no correction for terrain effects on the
wind flow, to estimate the wind energy distributions.

The General Electric researchers presented results in the form of maps which
show wind isopleth contours, i.e., the locus of points of equal annual mean wind
energy density. Maps were also developed which show low, moderate, and high
wind regimes based on the contour values of wind energy density. Maps of both
types were prepared for both the surface and 50-m levels. The relationship between
the annual mean wind energy and power densities ($\langle E_d \rangle$ and $\langle P_d \rangle$, respectively) is
given by the expression

$$\langle E_d \rangle = 8760 \, \langle P_d \rangle.$$

(1-7)

If $\langle P_d \rangle$ is expressed in W/m^2, $\langle E_d \rangle$ is obtained in watt-hours per square meter per
year ($Wh/m^2/y$). Figure 1-2 shows the General Electric Company map of favorable

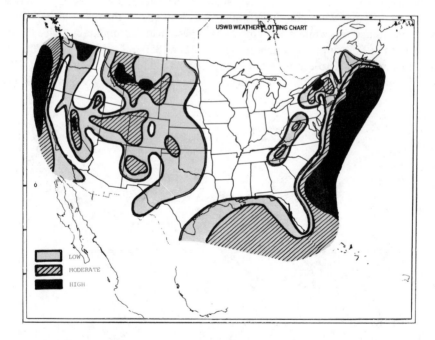

Fig. 1-2 General Electric Company map of favorable wind energy regimes based on the geographic distribution of annual mean wind energy density for the conterminous United States.
Regions corresponding to low, moderate, and high energy wind regimes are shown. The regions
are enclosed by isopleths which give the locus of points of equal annual mean wind energy density ($\langle P_d \rangle$ x 8760 in $mWh/m^2/Y$) according to the classes: Low, 2 to 4 $MWh/m^2/y$, ▢; Moderate, 4 to 7 $MWh/m^2/y$, ▨; and High, > 7 $MWh/m^2/y$, ▮. From Garate (1977).

wind energy regimes. The low, moderate, and high wind regimes span the ranges 2 to 4, 4 to 7, and > 7 megawatt-hours per square meter per year ($MWh/m^2/y$), respectively. Areas with an annual mean energy density less than 2 $MWh/m^2/y$ were judged unfavorable for economical operation of large wind turbines. Columns one and two of Table 1-1 show the relationship between values of annual mean wind power and energy densities. Column three gives a range of the annual mean wind speed for each value of the annual mean wind power density. Each range is based on the Rayleigh wind speed frequency distribution and a ± 10 percent range in the mean air density about the value for the standard atmosphere (1.225 kg/m^3).

The distribution of wind energy shown on the maps prepared by the General Electric Company is different from the distribution predicted by Reed's analysis. The difference is due to Reed's reliance on surface wind data and the General Electric Company's improved estimates of wind energy in high terrain areas. The General Electric researchers concluded that (1) the highest amounts of annual mean wind energy in the conterminous U.S. are likely to be found in the mountainous areas of northern New England, northern New York State, and the Pacific Northwest Region, (2) many significant maxima of wind energy will be found over high terrain areas in both the western and eastern mountains; however, greater wind energy is found at a height of 1520 m (5000 ft) over the northeastern U.S. than at 3050 m (10,000 ft) above the western U.S., (3) moderate amounts of wind energy are to be found over the Great Plains Region, and (4) minima of wind energy are believed to exist across the southern tier of states from the Mississippi Valley to

Table 1-1. Relationship between annual mean wind power and energy densities, and annual mean wind speeds.

$\langle P_d \rangle$ W/m²	$\langle E_d \rangle$ MWh/m²/y	Annual Mean Wind Speed Range[a] m/s	(mph)
100	0.88	4.27–4.56	(9.55–10.2)
200	1.75	5.38–5.75	(12.0–12.9)
300	2.63	6.15–6.58	(13.8–14.7)
400	3.50	6.77–7.24	(15.2–16.2)
500	4.38	7.30–7.80	(16.3–17.5)
600	5.26	7.75–8.29	(17.3–18.5)
800	7.01	8.53–9.13	(19.1–20.4)
1000	8.76	9.19–9.83	(20.6–22.0)

[a]This column provides a range for the annual mean wind speed for each annual mean wind power density which is based on the Rayleigh wind speed frequency distribution and a ± 10 percent variation in air density from the value for the standard atmosphere, $\rho_0 = 1.225$ kg/m^3. The Rayleigh distribution provides the relationship: $\langle V^3 \rangle = (6/\pi) \langle V \rangle^3$. Since $\langle P \rangle = (1/2) \langle \rho \rangle \langle V^3 \rangle$, it follows that $\langle V \rangle = (\pi \langle P \rangle/3 \langle \rho \rangle)^{1/3}$. The values of $\langle V \rangle$ in column three were obtained by using the values $0.9\rho_0$ and $1.1 \rho_0$ for $\langle \rho \rangle$.

Florida, in the major lowland areas of the west, and between the Appalachians and the Atlantic Coast.

The Lockheed-California researchers employed a data base which was similar to that employed by the General Electric Company. Less emphasis was placed on the use of rawinsonde data. The Lockheed-California researchers assumed that the mean wind speed at a mountain summit is approximately one-half of the mean free-air wind speed at the summit height and one-eighth of the free-air wind power. Maps were prepared which show station values of the annual mean wind power densities for 10-, 50-, and 100-m levels. Isodyns or isopleths were not drawn in order to avoid misrepresentation caused by both subjective and objective errors.

The results from the three wind energy assessment studies were reviewed by Elliott (1977, 1978) of Battelle Memorial Institute's Pacific Northwest Laboratory (PNL). Pacific Northwest Laboratory is responsible for the Wind Characteristics Program Element of the Federal Wind Energy Program. Elliot found that, although the results from these studies had many common features, there were significant inconsistencies involving both the estimated magnitude and geographical distribution of wind power. For some areas, the estimates for the annual mean wind power density differed by more than a factor of two. Elliott was able to account for the inconsistencies in terms of differences in assumptions, data sets, and analytical methods. He selectively employed information from the assessments and other sources in conjunction with refined techniques to prepare a synthesis of the national wind energy assessments. The results of the refined analysis of the distribution of the annual mean wind power density at 50-m elevation above exposed areas is shown in Figure 1-3. The term "exposed areas" refers to locations where the wind is unobstructed by nearby terrain features, e.g., hilltop locations over regions of gently rolling and hilly terrain and capes and open shoreline sites along coastal regions. The estimates were obtained to represent lower limits for exposed sites and may be low by 50 to 100 percent or more for isolated areas.

The national synthesis estimate for the annual mean wind power density for exposed areas in northern New York, Vermont, and New Hampshire is 500 W/m². Since this estimate is based on upper-air statistical data, it is informative to compare this result with more accurate estimates based on wind speed measurements at mountaintop and mountain ridge locations which were made during the Smith-Putnam project. Table 1-2 gives estimates for annual mean wind energy and wind power densities for Smith-Putnam stations based on measured mean wind speeds and the method of quantitative ecology. The estimates were calculated by using the Rayleigh wind speed frequency distribution. The estimates of the annual mean wind power density based on measured mean wind speeds vary from 70 to 9100 W/m².

During the Smith-Putnam project, Putnam originated a wind prospecting technique which he named quantitative ecology. He recruited Robert F. Griggs, an

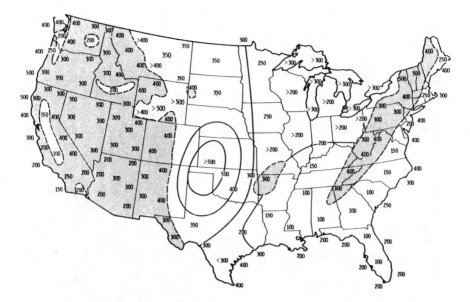

Fig. 1-3. Annual mean wind power density in W/m² estimated at 50-m elevation above exposed areas. Over mountainous regions, which are shown as shaded areas, the estimates are lower limits expected for exposed mountain tops and ridges. From Elliott (1978).

ecologist, and together they studied the relationship between measured wind speeds and certain types of wind-deformed vegetation. The mean wind speeds determined using this method are cited with an asterisk in Table 1-2. The estimates of the annual mean wind power density based on the application of this method range from 950 to 4900 W/m². The wide variation of wind power density shown in Table 1-2 shows that although the large-scale wind energy assessments can provide a basis for preliminary scoping studies, more refined assessments are required on a regional scale.

Pacific Northwest Laboratory has developed a regional wind energy assessment program. The Northwest region was used for a prototype study (Wendell, 1979). In the refined analysis, the data coverage for the region was increased from the 80 stations used in the national assessment to 462 stations. The final product of the study was a wind power atlas for the region. The atlas includes an analysis of the wind power density on a seasonal and annual average basis for each state in the region. The rest of the U.S. was divided into eleven regions for regional analyses. The refined analyses for all of the regions were completed in 1981. The distribution of the annual mean wind power density for the Northeast Region is shown in Figure 1-4 (Pickering, 1980). Despite the increase in resolution provided in the regional analysis, the wind power maps can only provide preliminary information for wind turbine siting studies.

Table 1-2. Estimates of annual mean wind energy and power densities for Smith-Putnam survey stations.

Elevation at Hub Height (140 feet above ground level) m (ft)	Station	$\langle V_{140}\rangle$ Annual Mean Wind Speed at Hub Height m/s	(mph)	Annual Mean Atmospheric Density (kg/m³)	$\langle E_d\rangle$ Annual Mean Wind Energy Density at Hub Height (MWh/m²/y)	$\langle P_d\rangle$ Annual Mean Wind Power Density at Hub Height (W/m²)
73 (240)	Crown Point	4.60	(10.3±15%)	1.30	.6 – 1.6	70 – 180
457 (1501)	Pond	5.72	(12.8±18%)	1.22	1.0 – 3.1	120 – 360
759 (2490)	Chittenden	7.51	(16.8±1.8%)	1.17	3.9 – 4.4	450 – 500
655 (2150)	Biddie Proper	7.11	(15.9±1.6%)	1.20	3.4 – 3.8	390 – 430
677 (2220)	Seward	7.96	(17.8±1.5%)	1.20	4.8 – 5.3	550 – 600
649 (2130)	Grandpa's Knob	7.45	(16.66±0.0%)	1.20	4.15	474
629 (2065)	Biddie I	7.51	(16.8±1.5%)	1.21	4.1 – 4.5	470 – 510
823 (2700)	Herrick	7.78	(17.4±0.9%)	1.15	4.4 – 4.7	500 – 530
914 (3900)	Glastenbury	9.39	(21.0±5.0%)	1.10	6.5 – 8.8	750 – 1000
1252 (4107)	Pico	10.1	(22.6±1.5%)	1.10	9.1 – 9.9	1030 – 1130
1250 (4100)	Lincoln Ridge	12.1	(27±20%)*	1.10	8.3 – 28	950 – 3210
1278 (4192)	Abraham	13.9	(31±20%)*	1.10	13 – 43	1440 – 4900
1959 (6428)	Washington	19.7	(44.0±7%)	1.02	52 – 80	6000 – 9100

The data in columns one through three for all but the Lincoln Ridge and Abraham stations appears in Putnam's Table VII. The values of the annual mean wind speed for the Lincoln Ridge and Abraham stations are Grigg's estimates based on quantitative ecology. The value for the Lincoln Ridge corresponds to the mean for five stations. These estimates were obtained from a report: "The Future of Large Scale Wind Power" by C.J. Wilcox and S.D. Dornbirer which was prepared for the S. Morgan Smith Company in 1945. The uncertainty of ±20 for these entries corresponds to Putnam's estimate of the accuracy of this method. The annual mean atmospheric densities of column four were obtained from Putnam's Figure 19. The ranges in annual mean wind energy and power densities given in columns five and six are based on the Rayleigh wind speed frequency distribution and the uncertainties in the annual mean wind speeds of column three. The annual mean wind power and energy densities were calculated using the equations $(3/\pi)\langle\rho\rangle\langle V_{140}\rangle^3$ and $8.76 \times 10^{-3}\langle P_d\rangle$, respectively, and the annual mean wind speeds corresponding to the limits of the uncertainty range.

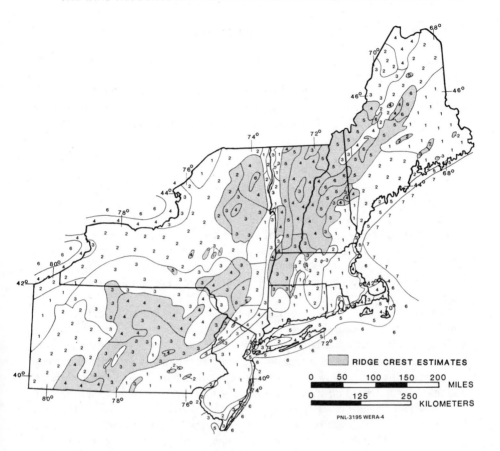

Classes of Wind Power Density at 10 m and 50 m[a]

Wind Power Class	10 m (33 ft)		50 m (164 ft)	
	Wind Power Density, watts/m²	Speed,[b] m/s (mph)	Wind Power Density, watts/m²	Speed,[b] m/s (mph)
	0	0	0	0
1	100	4.4 (9.8)	200	5.6 (12.5)
2	150	5.1 (11.5)	300	6.4 (14.3)
3	200	5.6 (12.5)	400	7.0 (15.7)
4	250	6.0 (13.4)	500	7.5 (16.8)
5	300	6.4 (14.3)	600	8.0 (17.9)
6	400	7.0 (15.7)	800	8.8 (19.7)
7	1000	9.4 (21.1)	2000	11.9 (26.6)

[a] Vertical extrapolation of wind speed based on the 1/7 power law.
[b] Mean wind speed is based on Rayleigh speed distribution of
equivalent mean wind power density. Wind speed is for standard sea-
level conditions. To maintain the same power density, speed increases
5%/5000 ft (3%/1000 m) of elevation.

Fig. 1-4. Northeast annual mean wind power density at 10m and 50m elevations above ground level (Pickering et al., 1980). The insert identifies the wind power classes shown.

PRELIMINARY ASSESSMENT OF WIND GENERATION CAPACITY

The mission analysis studies indicated that the most advantageous application for wind energy is the generation of electricity by utilities as a supplement to their existing generating facilities. Wind turbines can be connected directly to the network and can provide from 5 to 15 percent of the network power without inducing instabilities. The amount of penetration depends on the composition of the generating mix which feeds the network.

General Electric Study

The General Electric researchers used a simple approach to estimate the wind generation capacity of the conterminous U.S. The capacity depends upon the distribution of wind energy, the amount of land in the favorable wind regimes which is suitable for the installation of wind turbines, and the size, efficiency and packing density of wind turbines in each wind regime. The contribution from offshore areas was not included in the estimate. The procedure used to obtain the saturation estimate is outlined in Table 1-3.

Figure 1-2 shows the areas identified in the General Electric study as having low, moderate, and high wind regimes. The amount of land available for wind turbine installations depends upon land use and the suitability of the terrain for the construction and installation of generating systems. In estimating the amount of available land in the favorable wind regimes, institutional areas such as national parks, national monuments, Indian reservations, and military areas were excluded from consideration. Land use categories were developed and land use factors (i.e., the fraction of land which can be used for wind turbines) were estimated for each category. Topographical factors were also developed and applied to the land areas to account for terrain suitability. The results are summarized in Part A of Table 1-3. Only 13.5 percent of the total land area within the favorable wind regimes was judged available and suitable for wind turbine installations. The amounts of available land in the low, moderate, and high wind regimes correspond to 74, 25, and 1 percent, respectively.

The electrical outputs used for wind turbines installed in the low, moderate, and high wind regimes were based on three modified versions of the General Electric, base-line 1.5-MW wind turbine. The design of each machine was optimized to produce energy at a minimum cost at a site where the mean wind speed matches the design wind speed. The design and operating characteristics for these machines are given in Part B of Table 1-3. The 5.4-, 6.7-, and 8.0-m/s wind turbine designs were selected to characterize performance in the low, moderate, and high wind regimes. The rated wind speed is the wind speed at which the wind turbine produces its rated power. The capacity factor is the ratio of the energy produced during a given period to the energy which would be produced if the machine operated at its rated output (here 1.5 MW) during the entire period. The design wind speed has a significant

Table 1-3. General Electric estimate of the saturation limit for wind generation.

A. Land area available for wind turbines (thousands of miles^2)

		Not Available			Available		
Regime	Regime Area	Institutional Lands	Other Land Uses	Poor Topography	Area	Fraction of Regime (%)	Distribution of Area (%)
Low	1104	201.5	325.5	420.4	156.8	14.2	74
Moderate	415	85.6	79.3	196.2	53.9	13.0	25
High	55	2.2	10.8	39.0	3.0	5.5	1

B. Design characteristics of three 1.5MW wind turbines

Design wind speed, m/s (mph) at 10 m	8.0 (18)	6.7 (15)	5.4 (12)
Rotor diameter, m (ft)	58 (190)	67 (219)	85 (278)
Rotor speed, rpm	40	31.5	20.6
Rated wind speed, m/s (mph) at 10 m	9.92 (22.2)	9.07 (20.3)	7.51 (16.8)
Design annual energy output, MWh	6620	5860	5290
Design capacity factor	0.50	0.45	0.40

C. Operational characteristics of 1.5-MW wind turbines in the favorable wind regimes for use in saturation limit calculation

Wind Regime	Low	Moderate	High
Annual energy output, MWh	2920	4330	5010
Capacity factor	0.22	0.33	0.38

D. Wind turbine packing densities

Wind Regime	Packing Density (units/mile^2)
Low	1.301
Moderate	1.851
High	3.01

E. Saturation limits for wind turbines

Regime	Available Area (thousands of miles^2)	Number of Installed Units	Installed Electrical Capacity (GW)	Annual Energy Output (billions of kWh/y)	Efficiency Factor (millions of kWh/mile^2)
Low	156.8	204,800	307.0	592	4
Moderate	53.9	99,800	150.0	433	8
High	3.0	8,960	13.4	45	15
Total	214.0	313,500	470.0	1070	5

From Garate (1977).

impact on the rotor diameter and annual energy output. The conservative values of the annual outputs used to characterize the operation of wind turbines in the low, moderate, and high wind regimes are given in part C of Table 3-1.

The number of wind turbines per unit area is referred to as the packing density. The packing density is limited by the minimum distance which yields no power reduction due to aerodynamic interference between machines. The General Electric researchers used a minimum distance of 15 rotor diameters, and the packing densities were calculated for a hexagonal machine arrangement, i.e., the machines were located at the centers of the hexagons. The packing densities for the low, moderate, and high wind regimes are given in Part D of Table 1-3.

The calculation of the maximum number of 1.5-MW wind turbines (saturation limit) which can be installed in the low, moderate, and high wind regimes is illustrated in Part E of Table 1-3. The saturation values of the installed wind turbine capacity and annual electrical output are also shown. The amounts of available wind energy in the low, moderate, and high wind regimes correspond to 55, 41, and 4 percent, respectively. The last column of Part E gives an efficiency factor which is defined as the ratio of electrical output to the available land area. This factor provides a useful measure of the effectiveness of wind turbine installations. It accounts for the efficiency of the wind turbine as well as land use and topographical limitations. Although the contribution to the total energy at saturation is small, the economic viability of installations in high wind regime areas where the electrical output amounts to approximately 15,000 MWh per square mile will ensure that these areas are developed first.

The General Electric study indicates that 313,500 wind turbines of 1.5 MW rated capacity, which amounts to 470 GW of installed capacity, would produce 1.070 trillion kWh of electrical energy. General Electric used estimates of 1715 GW and 7.90 trillion kWh for the year 2000's installed capacity and electrical demand, respectively. The saturation estimates correspond to 27.4 percent of the installed capacity and 13.6 percent of the demand.

Estimates of the future demand for electrical energy vary widely. Different assumptions have produced different projections which are usually presented in terms of low and high estimates. Bodansky (1980) has summarized results from recent studies. Estimates for the year 2000's demand range from the low estimate of 2.41 trillion kWh obtained in a study by the National Academy of Sciences to the high estimate of 9.13 trillion kWh obtained in a study by the Electric Power Research Institute.

Lockheed-California Study

The Lockheed-California researchers identified five high potential applications for wind energy: utilities, industry, farms, residences, and pipelines. The utility market was estimated to be eight times greater than the next largest, industry. For each application, a forecast was developed to predict the maximum potential market in 1995. The potential was limited by the economic trade-off between wind generation and generation from conventional systems. The analysis was performed for each of the nine Federal Power Commission Regional Electric Reliability Councils which comprise the conterminous U.S. A detailed description of the wind energy resource, energy demand, conventional versus wind generation costs, and available land for each region were used in the analysis. The result of the study was an estimate of the wind generation that can be obtained at a cost less than or equal to the cost for generation from conventional systems. The required investment and the quantity of fossil fuel saved were also estimated.

The results of the study of utility applications will be reviewed here. The analysis presumes that utilities in each region will use wind generation if it costs less than the fuel consumed by conventional generating systems. The economic analysis of wind generation was based on a horizontal-axis, two-bladed wind turbine generator with a diameter of 107 m (350 ft) and a rated power of 2 to 5 MW. The machine was designed to produce energy at a minimum cost at sites with mean wind speeds of 6 to 8 m/s (13.4 to 17.9 mph). An economic analysis yielded energy costs in the range 1.7 to 2.8 ¢/kWh (1975 dollars).

Five utility applications of wind turbine generators were considered: integration with hydroelectric facilities, operation with pumped hydro facilities, supplemental power without storage, operation with short-term storage using batteries, and operation by nongenerating utilities. The first four applications are given in order of increasing cost of energy production. The analysis of these applications is summarized in Table 1-4 for the case of high demand and high fuel cost. High demand and high fuel cost correspond to an electrical demand in 1995 of 6.4 trillion kWh and a rise in fuel prices at an annual rate of 10 percent. Some of the assumptions used to obtain the results in this table and additional findings are described briefly below.

Integration with Hydroelectric Facilities. In this application, wind turbines are connected to a network which is also served by hydroelectric power stations. When the wind turbines produce power, the drawdown rate is reduced by an equivalent power level. This water is used to produce peak power above the level which is produced without the conserved water. Additional hydroelectric turbines are installed to accommodate the higher drawdown rates and these units are credited to the peaking capacity of the utility. The wind turbines can thus be credited with the capacity of the new hydroelectric units.

Hydroelectric plants can accommodate changes in the load or wind power generation quickly. For example, in the Bonneville Power Authority network, power variations of more than 10 percent can be made in less than 15 seconds. The increase in generating capacity offsets the requirement to build equivalent fossil steam plants and the saving may be used for part of the wind turbine investment costs. It is assumed that the hydroelectric capacity factors are reduced to 0.25 by additional generating capacity while the annual flow of water is preserved.

The analysis indicated that if wind turbines were integrated with U.S. hydro-electric facilities, 62,500 MW of wind power capacity would be used to produce 292.5 billion kWh per year. The water conserved could provide approximately 133,500 MW of peaking power in addition to the original 119,800 MW of hydro-electric capacity. If the wind generation replaced oil fueled steam plants, 517 million barrels of oil per year would be saved. The wind turbine investment would be $23.4 billion (1975 dollars). The payback period was estimated to be four years.

Operation with Pumped Storage Facilities. Utilities in the conterminous U.S. plan to increase their pumped hydropower capacity for peaking from 8,500 to 24,600 MW by 1995. In this application, wind generation is used to pump up the hydrosystem, except during peaking hours. During peaking hours, the wind generation is used directly and water in the high reservoir is conserved. It is assumed that wind gener-ation will save base plant fossil fuel and that nuclear base plants will not be used for pumping because their capacity is less than the regional demand for base-load power. It is also assumed that the capacity factor for pumped hydrosystems is 0.25.

The analysis indicated that if, instead of fossil steam plants, wind turbines are used in 1995 to pump water at night, 54 billion kWh of hydroelectric energy would be produced by the pumping unit's generating systems. To provide this energy, approximately 81 billion kWh of wind generation is required. The installed capacity of the wind turbines is 17,700 MW and their cost is $11 billion (1975 dollars). The annual saving in fossil fuel amounts to 143 million equivalent barrels of oil. The estimated payback period is seven years.

Supplemental Power without Storage. Without storage, a rapid increase or decrease in wind generation will disrupt the stability of the network due to the limited inertia and spinning reserves of the primary generating equipment. Surges or outages which amount to a significant fraction of the demand may cause voltage or phase instability, trip the overload breakers that protect generators and transformers, and produce a blackout. Wind generation without storage must be limited to 5 to 15 percent of the network power. For this application, an 8 percent change in the system-generated power is used to limit the rating of the wind turbines which feed the network. Up to 8 percent of the average power produced by fossil fuel gener-ating plants can be replaced by wind generation if the cost of wind energy equals the cost of fuel saved (break-even cost) when the wind blows. No capacity credit is given for the wind turbines.

The results for all nine regions indicate that if the annual fuel cost escalation rate is 7 percent, all regions will receive the full benefit from this application before 1995. A total of 8300 units will be required at a cost of $16.5 billion (1975 dollars). These units would produce 126 billion kWh per year, an amount which corresponds to 2 percent of the estimated national demand for electrical energy in 1995 and a savings in fuel which is equivalent to 223 million barrels of oil.

Operation with Short-Term Storage. Stability limitations may be overcome by adding storage capability to the network. This capability provides time for the primary generating plants to match the demand when the wind power falls off. The penetration of wind turbines may be increased to the point where the backup thermal systems can be throttled down to their lowest practicable level during good wind conditions. Some of the assumptions used in the analysis were: (1) the battery storage units have an efficiency of 70 percent, a life of 10 years, and an investment cost of $15/kWh of storage capacity plus $48/kW power rating, (2) each unit can produce the wind turbine rated power for 30 minutes, (3) when the wind power falls off, battery storage supplies the power until the thermal plant picks up the lead, (4) when the wind power returns, wind generation recharges the storage before it is used to meet demand, (5) when wind generation is meeting demand, thermal steam plant output is reduced (minimum capacity factor = 0.2), (6) the thermal plants may be brought up to power in less than 30 minutes, (7) nuclear plants are excluded because their power cannot be recycled rapidly, and (8) one-third of the hydroelectric generation is excluded because it is assumed to be committed to the base-load.

The analysis indicated that 13 percent of the 1995 utility energy demand could be supplied by wind generation. The estimated cost of the wind turbine and battery storage units is $124 billion (1975 dollars); 56,500 wind turbines with an installed capacity of 187,000 MW would produce an annual output of 835 billion kWh. The wind turbine installations would affect 54,000 km^2 (20,900 mile^2) of land; however, they would occupy only 3500 km^2 (1350 mile^2) of land—approximately 0.12 percent of the low land value-low use land in the conterminous U.S. The study indicated that all regions could implement wind generation in favor of oil by 1980 and in favor of coal by 1990.

The nonstorage application is expected to precede operation with short-term battery storage by several years due to lower break-even costs for the fuel-only case and because storage will not be required until wind power generation approaches the stability limit. To evaluate the total potential of wind generation in 1995, the contribution from nonstorage applications is incorporated in the storage application by the addition of battery systems that eliminate network stability problems.

Operation by Nongenerating Utilities. Most of the small municipal, cooperative, and private utility companies in the U.S. do not own their own generating units. They purchase energy from large producers for resale. Large wind turbine generators produce energy on an appropriate scale for such an application. Small companies may be prompted to buy wind turbines due to the high cost of wholesale energy, quotas imposed by the wholesaler because of limited resources which may make it impossible for the small company to satisfy growth in demand, and wholesale rates which are based on peak demand periods in regions where seasonal high winds coincide with the peak demand, i.e., wind generation can "shave the peak" and enable a more favorable wholesale rate.

In Table 1-4, no contribution is indicated for this application since wind generation from nongenerating utilities would reduce the amount produced by generating utilities. It is assumed that the potential contribution from nongenerating utilities has been accounted for in the regional estimates.

Overall Potential for U.S. Utilities in 1995. The overall potential is summarized in Table 1-4. The analysis indicated that almost 19 percent of the high demand estimate (6.4 trillion kWh) of the U.S. demand for electrical energy could be supplied by wind generation if fuel prices rise by an annual rate of 10 percent. A high demand-no fuel cost escalation above inflation scenario was also analyzed. For this case, 4.8 percent of the 1995 demand can be supplied by wind generation. The required investment for the former case is $158 billion (1975 dollars); however, the savings in fossil fuel would offset the investment in wind turbines at an annual rate of 16 percent. The projected utility investments for the new facilities required to meet the demand through 1995 are estimated to be $400 to $800 billion.

The analysis indicates that fuel would be saved at a rate of 2.1 billion equivalent barrels of oil per year, an amount which is slightly greater than the total oil imports for the U.S. in 1975. Hydroelectric applications will be developed first. The implementation of the short-term storage application will occur later.

The Lockheed-California researchers also conducted an analysis of the wind energy resource in U.S. offshore areas. The wind generation potential in offshore areas within 24 km (15 miles) and less than 100 fathoms deep amounts to approximately 6.7 trillion kWh, an amount which exceeds the high demand estimate for 1995. An economic analysis of the trade-off between wind generation in offshore areas and generation from conventional systems was not conducted.

The mission analysis studies have demonstrated that wind generation represents a viable source of renewable energy for the United States. Widespread application of wind power technology, especially in the utility sector, can make a significant contribution to U.S. energy requirements.

Table 1-4. National impact if utilities use wind turbines for the high energy demand-high fuel cost case.

Application	Annual Energy from Wind Turbines (billions of kWh)	% of National Demand 1995	Number of Wind Turbines	Capital Investment ($ billions)	Equivalent Millions of Barrels Per Year	Oil Saved $ Billion Per Year ($12/barrel)
Without storage	126.0	2.0	8,300	16.5	222.8	2.67
With hydroelectric	292.5	4.5	16,800	23.4	517.4	6.21
Pumped hydro storage	80.9	1.3	5,600	10.9	142.9	1.71
Short-Term storage	835.1	12.9	56,500	124.1	1,477.1	17.73
Total	1,208.5	18.7	78,900	158.4	2,137.4	25.65

From Coty (1976).

2
Wind Characteristics and Wind Turbine Siting

Certain wind characteristics are important for the evaluation of the wind resource and the design and performance of wind turbine generators. These include (1) wind speed and direction probability distribution functions, (2) the vertical variation of short- and long-term values of the horizontal wind speed, (3) the mean wind speed and its diurnal, seasonal, and annual variations, (4) wind speed duration (persistence), (5) the probability of extreme winds, and (6) the gustiness of the wind field in both speed and direction. A useful description of wind characteristics in relation to wind energy applications is given in *Initial Wind Energy Data Assessment Study* (Changery, 1975). Justus (1978a) has given a more detailed review of wind characteristics. An understanding of the effect of terrain features and atmospheric conditions on wind characteristics is required for the tentative identification of wind turbine sites. Putnam (1948) was the first to prospect for sites for large wind turbine generators. Golding (1955) has given a useful description of wind characteristics, wind surveys, and wind turbine siting. In this chapter, some of the recent studies of wind characteristics and wind turbine siting are reviewed. Emphasis is placed here on the mathematical description of wind characteristics. In Chapter 3, the relationship between wind characteristics and wind system performance will be examined.

WIND SPEED PROBABILITY DISTRIBUTIONS

The wind speed probability distribution is important in wind energy studies. The speed distribution may be used to evaluate (1) the mean wind power density, (2) the probability that the wind speed lies in certain intervals (e.g., between the cut-in and cut-out values for a particular wind turbine, (3) the energy pattern factor, and (4) the capacity factor for a particular wind turbine. Of course, if time series wind data or data summarized by wind speed class are available, the data may be used directly to calculate these quantities. However, if these quantities are desired for another height or location, an analytical representation of the distribution may provide a useful basis for an extrapolation. One type of distribution function provides the basis for a model which may be used to describe the performance of wind turbine arrays.

Bivariate Distribution

From the viewpoint of statistical theory, it is advantageous to describe the horizontal wind in terms of a bivariate distribution system. One can employ rectangular coordinates such as zonal and meridonal, or one can use the wind speed and direction as variables. Essenwanger (1959, 1976) has studied the former approach. He showed that the elliptical bivariate distribution of two vector components was useful for the calculation of upper-air wind speed probabilities. Brooks and Carruthers (1953) and Crutcher and Baer (1962) have examined the special case of the bivariate Gaussian (normal) distribution.

The bivariate Gaussian distribution $p(u,v)$ of the horizontal wind velocity components u and v is

$$p(u,v) = (2\pi\sigma_u\sigma_v)^{-1} (1-r_{uv}^2)^{-1/2} \exp\left[-\lambda^2/2(1-r_{uv}^2)\right] \tag{2-1}$$

where

$$\lambda^2 = [(u-\langle u\rangle)/\sigma_u]^2 - 2r_{uv}(u-\langle u\rangle)(v-\langle v\rangle)/(\sigma_u\sigma_v)$$
$$+ [(v-\langle v\rangle)/\sigma_v]^2 \tag{2-2}$$

$\langle u\rangle$ and $\langle v\rangle$ are the mean components, σ_u and σ_v are the standard deviations of u and v, respectively, and r_{uv} is the normalized correlation coefficient, i.e., $r_{uv} = \langle uv\rangle/(\sigma_u\sigma_v)$. The five statistics, u, v, σ_u, σ_v, and r_{uv}, are furnished with wind climatology summaries prepared by the National Climatic Center. The univariate speed distribution, which may be derived from the bivariate Gaussian distribution, consists of a summation of Bessel functions (Smith, 1971, 1976). The univariate speed distribution is difficult to use in wind power applications due to its complicated form and parametric requirements.

Rayleigh Distribution

For the special case $\langle u\rangle = \langle v\rangle = r_{uv} = 0$, $\sigma_u = \sigma_v$, the bivariate Gaussian distribution reduces to the one parameter Rayleigh distribution.

$$p(V) = (V/a)^2 \exp[-(V/a)^2/2] \tag{2-3}$$

where V is the wind speed and $a=\sigma_u=\sigma_v$. The Rayleigh distribution function corresponds to the chi distribution for two degrees of freedom. This function has been used extensively in wind power studies (Court, 1974, Baynes, 1974). The first moment of the Rayleigh distribution, the mean wind speed $\langle V\rangle$, may be used to evaluate the parameter a.

$$\langle V\rangle = \int_0^\infty V p(V)\, dV = (\pi/2)^{1/2}\, a. \tag{2-4}$$

The Rayleigh distribution may thus be expressed as

$$p(V) = [\pi V/2\langle V\rangle^2] \exp [-\pi V^2/4\langle V\rangle^2]. \tag{2-5}$$

The measured and Raleigh wind speed probability distribution functions based on five years of data at the Smith-Putnam test site are shown in Figure 2-1. The wind data was collected at the 120-ft level of the Christmas tree on Grandpa's Knob. The distributions correspond to a five-year annual mean wind speed of 7.33 m/s (16.39 mph).

The Rayleigh cumulative probability distribution function is given by the expression

$$p(V \leqslant V_x) = \int_0^{V_x} p(V)dV$$
$$= 1 - \exp[-\pi V_x^2/(4\langle V\rangle^2)]. \tag{2-6}$$

The cumulative probability distribution function gives the probability that $V \leqslant V_x$. The probability $p(V \geqslant V_x)$, i.e., the probability that $V \geqslant V_x$, is $1 - p(V \leqslant V_x)$.

Wind speed distributions are sometimes described in terms of a figure which shows the wind speed versus the number of hours in a year during which the wind speed equals or exceeds the indicated values (Golding, 1955). For the Rayleigh distribution, the number of hours in a year during which $V \geqslant V_x$ ($T(V_x)$) is

$$T(V_x) = 8760 \, p(V \geqslant V_x) = 8760 \exp[-\pi V_x^2/(4\langle V\rangle^2)]. \tag{2-7}$$

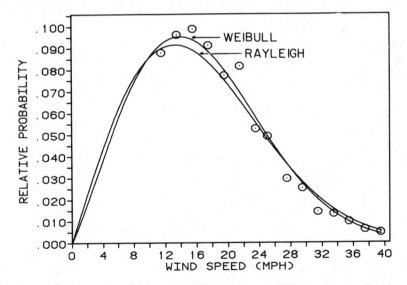

Fig. 2-1. Measured, Rayleigh, and Weibull wind speed probability distributions based on five years of data at the Grandpa's Knob test site. Each distribution corresponds to a mean wind speed of 7.33 m/s (16.39 mph).

Figure 2-2 shows the measured annual wind speed duration curve (wind speed versus $T(V_x)$) for the 120-ft level at Grandpa's Knob.

The second central moment or variance of the Rayleigh distribution is given by the expression

$$\sigma^2 = \langle (V - \langle V \rangle)^2 \rangle = [(4/\pi)-1] \langle V \rangle^2. \tag{2-8}$$

The third moment of the distribution is

$$\langle V^3 \rangle = (6/\pi)\langle V \rangle^3 = 1.9099 \langle V \rangle^3. \tag{2-9}$$

Therefore, the mean wind power density corresponding to the Rayleigh distribution is

$$\langle P_d \rangle = (3/\pi) \rho \langle V \rangle^3. \tag{2-10}$$

If ρ and $\langle V \rangle$ are expressed in kg/m^3 and m/s, respectively, the mean wind power density is obtained in watts. Figure 2-3 shows Rayleigh values of the mean wind power density versus mean wind speed for standard sea level density (1.225 kg/m^3).

Weibull Distribution

The Weibull probability distribution function represents a special case of the generalized gamma distribution. This function has been used for some time in wind load

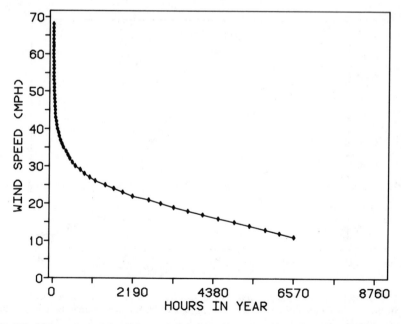

Fig. 2-2. Measured annual wind speed duration based on five years of data at the Grandpa's Knob test site.

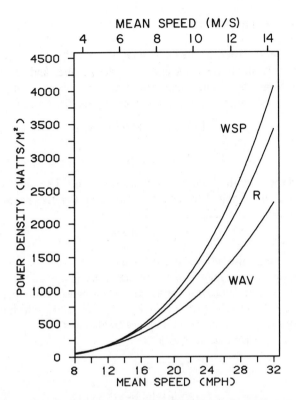

Fig. 2-3. Mean wind power densities for the Rayleigh, Weibull average variance, and Weibull Smith-Putnam reference wind speed probability distribution functions. The values correspond to standard sea-level density (1.225 kg/m^3).

applications. Recently, a number of workers have demonstrated its utility in wind power studies (Justus, 1974, Baynes, 1974, Wentink, 1976, Justus et al., 1976a,b, Hennessey, 1977). The Weibull function is more convenient to use than the uni-variate Gaussian function and more flexible than the Rayleigh function. It is particularly useful for describing the occurrence of high wind speeds (Essenwanger, 1968a, b, 1976).

The Weibull probability distribution function for wind speed may be expressed as

$$p(V) = (k/c) \, (V/c)^{k\text{-}1} \, \exp[-(V/c)^k] \qquad (2\text{-}11)$$

where k is a dimensionless "shape factor" and c is a "scale factor" which has the units of speed. The corresponding cumulative probability distribution function may be expressed as

$$p(V \leqslant V_x) = 1 - \exp[-(V/c)^k] . \qquad (2\text{-}12)$$

The Weibull distribution reduces to the Rayleigh distribution for the special case $k = 2, c = 2 \langle V \rangle / \sqrt{\pi}$.

Justus et al. (1978) have described a number of methods which may be employed to evaluate the Weibull parameters k and c. If the variance and mean wind speed are known, the parameters k and c may be obtained by solving the equations which give the first moment and second central moment of the distribution

$$\langle V \rangle = c \, \Gamma \, (1+1/k) \tag{2-13}$$

and

$$\sigma^2 = \langle V \rangle^2 \left\{ [\Gamma \, (1+2/k)/\Gamma^2 \, (1+1/k)] - 1 \right\} \tag{2-14}$$

respectively, where Γ is the complete gamma function. The parameters c and k may be obtained by solving these equations using an iterative process. The convenient approximate relationship

$$k = (\sigma/\langle V \rangle)^{-1.086} \tag{2-15}$$

and equation 2-13 may also be used to calculate k and c.

The Weibull parameters may also be evaluated using a wind speed probability distribution function in numerical form. If the observed wind speed classes are divided into n speed intervals $O - V_1, V_1 - V_2, \ldots, V_{n-1} - V_n$ which have frequencies of occurrence $f_1, f_2, \ldots, f_n$ and cumulative frequencies $p_1 = f_1, p_2 = f_1 + f_2, \ldots, p_n = p_{n-1} + f_n$, the transformation equations

$$x_i = \ln V_i \tag{2-16}$$

and

$$y_i = \ln [-\ln(1-p_i)] \tag{2-17}$$

may be used to express equation 2-12 in the linear form $y = mx + b$. The linear coefficients may then be evaluated using a least squares process. The Weibull parameters are related to the linear coefficients by

$$c = \exp[-b/m] \tag{2-18}$$

and

$$k = m. \tag{2-19}$$

This method was employed to obtain the Weibull probability distribution function shown in Figure 2-1.

The Weibull parameters may also be evaluated using published values of the monthly mean wind speed and the monthly wind speed corresponding to the fastest one mile run of wind, V_{max}. The values of these wind characteristics are given in climatological summaries such as Local Climatological Data. The probability that $V \geqslant V_{max}$ may be calculated using

$$p(V \geqslant V_{max}) = t/(24d) = 1/(24 V_{max} d) \tag{2-20}$$

where t is the time corresponding to fastest one mile run ($1 \text{ mile}/V_{max}$) and d is the number of days for a given month. The cumulative probability corresponding to V_{max} is

$$p(V \geqslant V_{max}) = \exp[-V_{max}/c]^k. \qquad (2\text{-}21)$$

Equations 2-13, 2-20, and 2-21 may be used to obtain the equation

$$V_{max}/\langle V \rangle = [\ln(24V_{max}d)]^{1/k}/\Gamma(1+1/k) \qquad (2\text{-}22)$$

which can be solved for k using an iterative process. Equation 2-13 can then be solved for the scale factor c.

Justus et al. (1976b) evaluated Weibull parameters for 140 sites in the continental United States. The sites were comprised of surface weather stations with anemometers near the 10-m level. The results were summarized in terms of the equations

$$k = \begin{matrix} 1.05 \ \langle V \rangle^{1/2} & \text{(low variability)} \\ 0.94 \ \langle V \rangle^{1/2} & \text{(average variability)} \\ 0.83 \ \langle V \rangle^{1/2} & \text{(high variability)} \end{matrix} \qquad (2\text{-}23)$$

where $\langle V \rangle$ is expressed in m/s and variability refers to $\sigma/\langle V \rangle$. Since $\sigma/\langle V \rangle \cong k^{-(1/1.086)}$, small k corresponds high variability. The high and low cases were defined by 90 and 10 percentile limits for k for given $\langle V \rangle$, i.e., for a given $\langle V \rangle$, 90 percent of the k values are above $0.83\langle V \rangle^{1/2}$ and 10 percent of the k values are above $1.05\langle V \rangle^{1/2}$. Given a meaningful qualitative estimate of the variability (or variance, σ^2) and the mean wind speed, equations 2-23 and 2-13 may be used to obtain approximate values for k and c.

Figure 2-4 shows Weibull wind speed probability distributions for a mean wind speed of 8.9 m/s (20 mph) and different values of the shape factor k. The distributions for Justus et al.'s low, average, and high variability cases are shown. The shape of the distribution varies from broad to narrow as k decreases.

Figure 2-5 shows the variation of k with mean wind speed for the low, average, and high variability cases. Values of k based on measured monthly wind speed distributions recorded at the Mount Washington Observatory are also shown (Vachon et al. 1979). Mount Washington has an elevation of 6288 ft. The values of k for this high elevation site appear to follow a trend toward higher values at high mean wind speeds.

Values of the Weibull parameters may be obtained on the basis of results shown in Putnam's Figures 37 and 38. Figure 37 shows the relationship between the mean wind speed and the most probable value of the wind speed based on 16 measured wind speed probability distribution functions. Figure 38 shows the relationship between the number of hours during which the wind blows at the most probable wind speed and the mean wind speed for the same set of measured distributions. The most probable value of the wind speed may be obtained by solving the equation

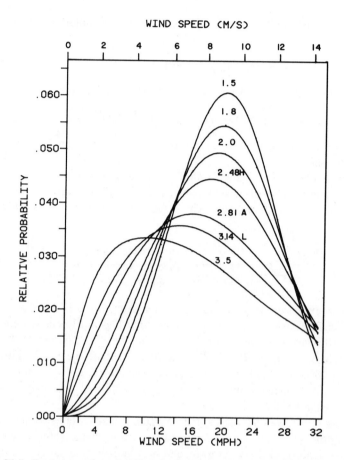

Fig. 2-4. Weibull wind speed probability distributions for a mean wind speed of 8.9 m/s (20 mph) and different values of the shape factor k. The distributions for the Justus et al. low, average, and high variability cases are labeled L, A, and H, respectively.

which results from the condition $dp(V)/dV = 0$ where $p(V)$ is the Weibull probability distribution function. The most probable value of the wind speed (V_{mp}) is

$$V_{mp} = \left\{ [(k-1)/k]^{1-k} / \Gamma(1+1/k) \right\} \langle V \rangle. \qquad (2\text{-}24)$$

Equation 2-24 may be solved by using an iterative process to obtain the value of k for each point (V_{mp}, $\langle V \rangle$) of Figure 37, and equation 2-13 can be used to obtain the corresponding value of c. Weibull parameters may also be evaluated by using point values from both figures. The number of hours during which the wind blows at the most probable wind speed (T_{mp}) is

$$T_{mp} = 8760(k/c)(V_{mp}/c)^{k-1} \exp[-(V_{mp}/c)^k]. \qquad (2\text{-}25)$$

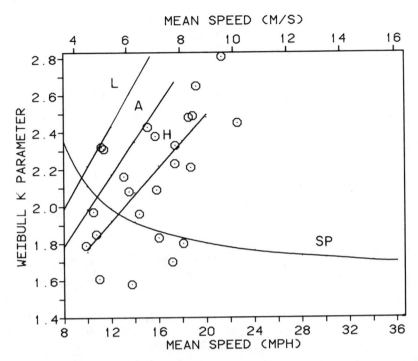

Fig. 2-5. Variation of the Weibull parameter k with mean wind speed based on monthly wind speed probability distributions for Mount Washington. Curves which show the variation of the Weibull parameter k with mean wind speed for the Justus et al. low, average, and high variability cases are labeled L, A, and H, respectively. The curve labeled SP shows the variation based on the Smith-Putnam reference distributions.

Using equations 2-13 and 2-24, the expression for T_{mp} yields

$$T_{mp} = 8760 \left[(k-1)/V_{mp}\right] \exp\left[-(k-1)/k)\right] \qquad (2\text{-}26)$$

Values of T_{mp} and V_{mp} corresponding to the points $(V_{mp}, \langle V \rangle)$ and $(T_{mp}, \langle V \rangle)$ of Putnam's Figures 37 and 38 may be used in an iterative process to determine k, and equation 2-13 can then be used to obtain c. Figure 2-6 shows the values of k obtained by using the former method versus mean wind speed. The curve which gives k versus mean wind speed for the average variability case is shown for comparison.

The Smith-Putnam distributions define a smooth curve of k versus $\langle V \rangle$. The trend is toward small k at high mean wind speed. Figure 2-7 shows V_{mp} versus $\langle V \rangle$ for the 16 Smith-Putnam distributions and the linear regression analysis line. The regression line and equations 2-24 and 2-13 were used to define a reference set of Weibull distribution functions. The curve of k versus $\langle V \rangle$ defined by the reference distributions is shown in Figures 2-5 and 2-6.

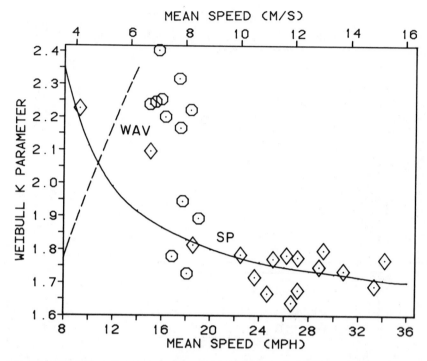

Fig. 2-6. Weibull k values obtained from the analysis of Putnam's Figure 37 versus mean wind speed. Sixteen points are shown. The solid line gives the variation of k based on a linear regression analysis of the points of Figure 37. The dashed line shows the variation of k for the average variability case. The results shown as hexagons are based on the twelve (five-year) monthly distributions for the 120-ft level at the Grandpa's Knob test site.

Kao (1958) and Takle and Brown (1977) have shown how Weibull parameters may be evaluated on the basis of a maximum likelihood solution. Although this method is more cumbersome than the others, it is the method of choice in terms of theoretical statistical considerations.

The third moment of the Weibull distribution is

$$\langle V^3 \rangle = [\Gamma(1+3/k)/\Gamma^3(1+1/k)]\langle V \rangle^3 . \qquad (2\text{-}27)$$

The mean wind power density based on this result is therefore

$$\langle P_d \rangle = (1/2)\, \rho\, [\Gamma(1+3/k)/\Gamma^3(1+1/k)]\, \langle V \rangle^3 . \qquad (2\text{-}28)$$

Figure 2-3 shows the variation of mean wind power density with mean wind speed for the Rayleigh, Weibull average variability, and Weibull Smith-Putnam reference distributions.

Putnam (1948) used the term *cube factor* to describe the ratio of the mean of the wind speed cubed to the cube of the mean wind speed. Golding (1955) called

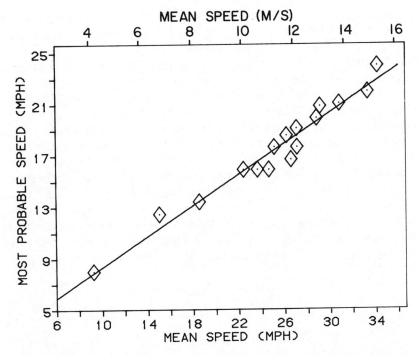

Fig. 2-7. The most probable wind speed versus the mean wind speed for 16 Smith-Putnam distributions. The line is the result of a linear regression analysis.

this quantity the *energy pattern factor*. The Weibull distribution yields the cube factor

$$\langle V^3 \rangle / \langle V \rangle^3 = \Gamma(1+3/k)/\Gamma^3(1+1/k). \tag{2-29}$$

For the Weibull distribution, the ratios $\langle V \rangle / c$, $\sigma / \langle V \rangle$, $\langle V^3 \rangle / \langle V \rangle^3$, and $V_{mp} / \langle V \rangle$ depend upon k alone.

Square Root Normal Distribution

Widger proposed a square root normal model for wind speed probability distribution functions (Widger, 1976, 1977). The model is based on the assumption that the square root transformation of observed wind speeds yields a Gaussian distribution centered on the mean of the square root of observed speeds. Widger performed a linear regression analysis on monthly fastest mile and mean wind speeds for seven New England stations, and found a correlation coefficient of 0.979 for 85 data points. The linear regression equation, the square root normal model, and equation 2-20 were employed to construct a smooth curve of wind power density versus mean wind speed, i.e., of the type shown in Figure 2-3. When this nomogram was employed to estimate mean wind power densities at six of the New England sites, the mean percentage of error was found to be less than 6 percent. The largest error corresponded to an underestimate of 16 percent.

Baker and Hennessey (1977) tested Widger's method via comparisons involving four mountain sites in the Pacific Northwest region. Widger's nomogram underestimated the annual mean power density by as much as 59 percent. At one site the monthly variation of the error ranged from –9 to –60 percent.

Nomograms such as Widger's or those of Figure 2-3 require an energy pattern factor which is either constant or a smooth function of the mean wind speed, e.g., the energy pattern factor for the Rayleigh distribution is 1.91. The monthly variation of error may be understood in terms of the statistical variation of the energy pattern factor. Figure 2-8 shows energy pattern factors based on measured distributions for Mount Washington and Smith-Putnam sites versus mean wind speed. The monthly energy pattern factors for the Mount Washington site show significant variation over small intervals of the mean wind speed over the entire wind speed range. The systematic error associated with the underestimation of wind power

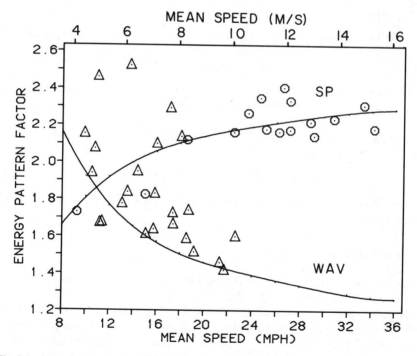

Fig. 2-8. Smith-Putnam and Mount Washington values of the energy pattern factor versus mean wind speed (o, values based on Smith-Putnam distributions and △, values based on monthly wind speed distributions for Mount Washington). The smooth curves show the variation for the Weibull Smith-Putnam (*SP*) and average variability (*A*) reference distributions.

density is caused by the approximate statistical treatment of the fastest mile data and the inaccuracy of the square root normal model for the case of highly skewed distributions.

Widger (1977) calculated square-root-normal distributions using monthly mean wind speed and fastest mile wind data for Concord, New Hampshire. He compared, calculated, and measured distributions for each of 30 months which spanned a five-year period. Justus et al. (1978) used monthly cumulative probabilities and the least-squares-fit procedure to calculate Weibull parameters and distributions for the same 30 months. The differences between the measured and calculated distributions in terms of root-mean-square (rms) errors for cumulative probabilities are given in Table 2-1. This table also gives the errors for the Weibull distributions defined by the fastest mile and mean speed method and the Weibull distributions defined by the linear relationship: $k = 0.87 \langle V \rangle^{1/2}$ where $\langle V \rangle$ is expressed in m/s. The constant of the linear equation is based on the observation that the variability for Concord falls between the average and 90 percentile (high) levels. A comparison of the results in Table 2-1 indicates that the Weibull distributions produce lower rms errors from the observed distributions than the square-root-normal distributions.

Other Distributions

Other functions have been used to provide an accurate description of wind speed probability distributions. Garaty (1977) and Ossenbrugen et al. (1979) have used the gamma distribution. Takle and Brown (1977) and Stewart and Essenwanger (1978) have studied three parameter versions of the Weibull distribution.

Effect of Sampling Rate and Averaging Period

Wind speed probability distributions which are constructed using measured data will depend upon the rate at which wind speed samples are collected (the sampling rate) and the length of the period over which samples are taken. The averaging period for each sample will also affect the statistics since effects associated with

Table 2-1. Root-mean-square errors of distribution fit expressed
as a percentage for Concord, N.H. data

Method description	\multicolumn{8}{c	}{Wind Speed Levels (mph)}
	≤5	≤10	≤15	≤20	≤25	≤30	≤35	≤40
Weilbull least squares	1.40	2.63	1.61	0.47	0.35	0.14	0.09	0.07
Square root normal	11.57	5.20	4.29	1.83	0.54	0.33	0.14	
$\langle V \rangle$ and V_{max} Weibull	5.07	4.46	3.88	1.67	0.55	0.34	0.14	0.07
k versus $\langle V \rangle$ Weibull	3.88	3.44	3.00	1.75	0.47	0.18	0.11	

From Justus et al. (1978).

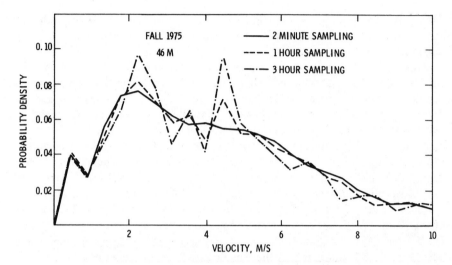

Fig. 2-9. Wind speed probability distributions for two-minute, one-hour and three-hour sampling periods, Fall 1975. Adapted from Figure 1 of Doran et al. (1977).

frequencies greater than the reciprocal of the sampling period will be filtered out by the averaging process.

Doran et al. (1977) have examined the effect of these factors on wind speed probability distributions and output estimates for wind turbine generators. Their analysis was based on data taken from 10- and 46-m levels of a meteorological tower at Hanford, Washington. Wind speeds were averaged for one minute and sampled every two minutes. Data records covering a 54-day period in the fall of 1975 and a 91-day period in the spring of 1976 were analyzed. Wind speed probability distributions were constructed from one-minute average readings taken every two minutes, one hour, and three hours. The distributions for the 46-m level during fall are shown in Figure 2-9.

To simulate the effects of a longer averaging period, e.g., based on wind run data from a strip chart recorder, 30 consecutive readings from each hour of data were averaged and used to construct a wind speed distribution. This distribution is shown with the distribution based on readings taken every two minutes in Figure 2-10. As expected, the smoothness of the distribution is degraded by a decrease in the sampling rate, although the general shape remains the same. A similar loss in smoothness results from increasing the averaging time.

The mean wind speeds and standard deviations for the fall 1975, 46-m data are given in Table 2-2. Despite the irregularities in the curves, the first two moments of the distribution are in good agreement regardless of the sampling rate or averaging period. For both seasons and levels (four cases), the wind speeds in a given season vary by less than 2 percent. The mean speeds based on two-minute samples are al-

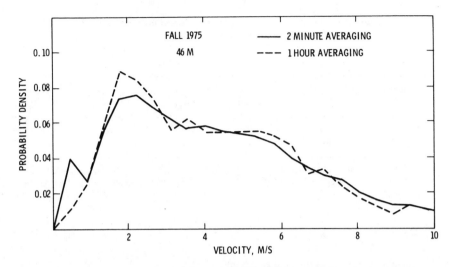

Fig. 2-10. Wind speed probability distributions for two-minute and one-hour averaging periods, Fall 1975. Adapted from Figure 3 of Doran et al. (1977).

ways the highest, while the three-hour sample values are the lowest. For the standard deviations, the two-minute samples yield the highest values in three cases, while the one-hour average values are always the lowest. Differences between standard deviations for two-minute, one-hour, and three-hour samples are generally less than 2 percent (there is one exception). The differences between the two-minute and one-hour average results for all four cases range from 5.7 to 8.2 percent.

Table 2-3 shows the mean wind power density and mean power output for a hypothetical wind turbine generator for distributions at 46-m elevation corresponding to two-minute, one-hour, and three-hour sampling rates, and a one-hour average measurement period. Although the distributions for the lower sampling rates are rough in comparison with the two-minute sample distribution, the sampling rate does not have a significant effect on the power estimates. However, the use of a

Table 2-2. Measured wind characteristics for various sampling and average periods.

Season	Height	Period	$\langle V \rangle$ (m/s)	σ (m/s)
Fall 1975	46 m	2-minute	4.51	3.28
		1-hour	4.50	3.25
		3-hour	4.47	3.25
		1-hour average	4.50	3.08

Adapted from Doran et al. (1977).

Table 2-3. Mean power density and power output estimates from data
for various sampling and averaging periods.[a]

Season	Height (m)	Mean wind power density (W/m^2)			
		2-Minute Samples	1-Hour Samples	3-Hour Samples	1-Hour Average
Fall 1975	46	180	177	178	163
Spring 1976	46	221	227	218	203

Mean power output (kW) estimates for a hypothetical wind turbine generator with cut-in, rated, and cut-out wind speeds of 4.2, 10.4, and 22.4 m/s, respectively, and a rated power of 1.1 MW.[b]

Fall 1975	46	173	172	168	165
Spring 1976	46	261	262	263	252

[a]Adapted from Table 2 of Doran et al. (1977).
[b]The Quadratic power output function and the standard performance model were employed to calculate these results. See Chapter 3.

one-hour averaging period results in consistent underestimates of the mean wind power density and power output. The differences amount to less than 10 percent for the mean wind power density and less than 5 percent for the mean power output. Table 2-4 gives mean power output estimates for a hypothetical wind turbine generator based on wind speed data for two-minute samples and one-hour averaging period and Weibull wind speed probability distributions. The Weibull distribution parameters were calculated by using the mean wind speed and standard deviation for the two-minute and one-hour average wind data. The agreement between the measured distribution and Weibull distribution results is good. The use of one-hour average rather than two-minute sample data for evaluation of the Weibull parameters does not have a significant effect on the power estimates.

Table 2-4. Mean power output estimates based on actual
and Weibull wind speed distributions for 46 m.

	Mean Power (kW)			
	2-Minute Samples	1-Hour Average	Weibull 2-Minute Samples	Weibull 1-Hour Average
Fall 1975	173	165	185	177
Spring 1976	261	252	261	251

From Doran et al. (1977).

VERTICAL WIND SHEAR

General Description

Quantitative methods which may be used to describe the vertical variation of the horizontal wind speed (vertical shear) are important in wind power studies. The two blades of the NASA-Boeing Mod-2 wind turbine generator sweep an area with a diameter of 91.4 m (300 ft) and extend to approximately 107 m (350 ft) above ground level. The power in the wind may therefore vary significantly over the vertical distance spanned by the blades. The instantaneous wind shear profile is important in general for structural design and in particular for the evaluation of the effect of cyclical stresses. A statistical description of the vertical wind shear profile is important for the development of wind turbine generator performance estimates. In the most widely used performance model for horizontal-axis machines, the wind speed frequency distribution referenced to the hub height elevation and the machine power output function expressed in terms of the wind speed at hub height are employed to obtain performance estimates. Given measurements of the wind speed frequency distribution at typical surface weather anemometer heights (9-18 m), it is necessary to extrapolate the distribution or parameters which describe the distribution to much higher levels in order to estimate the output for large machines. Output estimates are sensitive to the method employed to make this extrapolation.

Two methods are frequently used to relate the measured instantaneous (one or two minute average) wind speed V_1 at reference height Z_1 to the measured wind speed V_2 at height Z_2. The power law profile is given

$$\frac{V_2}{V_1} = \left[\frac{Z_2}{Z_1}\right]^n \tag{2-30}$$

where n is an exponent whose value may depend on reference height (Reed, 1974), atmospheric stability (Smith, 1968, Heald, 1979, Sisterson and Hicks, 1979), wind speed (Fales, 1967, American Society of Civil Engineers, 1961, Fichtl and Smith, 1977, Justus and Mikhail, 1976), and surface roughness (Davenport, 1963, Frost et al. 1979, Justus, 1978a). The logarithmic profile is given by

$$V_2 = V_1 \ln (Z_2/Z_0)/\ln (Z_1/Z_0) \tag{2-31}$$

where Z_0 is the surface roughness length (Reed, 1976). The surface roughness is not an actual dimension but rather a parameter which is determined on the basis of measured wind profiles

Review of Theory

The power law profile was proposed first by Hellman (1916). Under certain conditions, n equals 1/7 (Von Kármán, 1921). This value is indicative of a correspondence between wind profiles and fluid flow over flat plates (Schlicting, 1968). The power law profile with $n = 1/7$ (" the one-seventh power law") has been used frequently in wind power studies. Golding (1955) presented a useful discussion of vertical wind shear and cited the utility of the simple expression

$$V_h = \alpha h^{0.17} \tag{2-32}$$

where V_h is the wind speed at height h above flat, open ground and α is a proportionality constant.

Monin and Obukhov (1954) developed a description of short period wind profiles using boundary layer similarity theory. They suggested the relationship

$$\frac{dV}{d\ln Z} = \frac{V^*}{\beta} \; \phi(Z/L) \tag{2-33}$$

which yields

$$V(Z) = (V^*/k^*) \; [\ln(Z/Z_0) - \psi \; (Z/L)] \tag{2-34}$$

where ϕ and ψ are universal functions of the nondimensional ratio of the height Z to the similarity scale L, k^* is the von Kármán constant ($\cong 0.4$), Z_0 is the surface roughness length, V^* is the friction velocity, and $V(Z)$ is the wind speed at height Z. The Monin-Obukhov length L depends on the friction velocity, density, heat capacity, temperature, and the upward eddy flux of heat. Under conditions of neutral stability, $\psi(Z/L) = 0$ and equation 2-34 yields the Prandtl profile:

$$V(Z) = (V^*/k^*)\ln(Z/Z_0). \tag{2-35}$$

Mikhail (1977) and Justus (1978a) have shown how L, ψ, and hence $V(Z)$ may be determined using the surface roughness for a site and the wind speed and net radiation at a given time. Similarity theory is not expected to provide an accurate description of the profile over extended periods due to changes in the atmospheric stability and the variation of surface roughness with wind direction.

Petterssen (1964) reviewed the theoretical description of wind profiles. Spera and Richards (1979) summarized the functional relations which have been proposed for the variation of the power law exponent n with wind speed. These relations are shown in Figure 2-11.

Justus and Mikhail's Equations

The logarithmic relationship between the power law exponent and wind speed proposed by Justus and Mikhail (1976) is noteworthy since it is based on an approach which gives the height variation of both wind speed and the parameters which describe wind speed frequency distributions. Reed (1975) showed that the power law

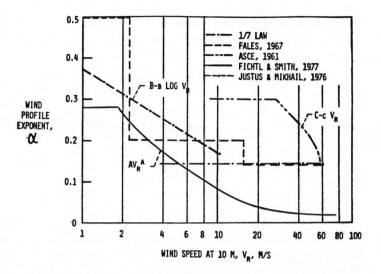

Fig. 2-11. Various recommendations for the relationship between wind profile exponent and wind speed. From Spera and Richards (1979).

profile could be used with an exponent $n(V_1)$ which was a function of the wind speed (V_1) at height Z_1. He proposed an equation of the form

$$V_2 = \alpha V_1^{\beta} \qquad (2\text{-}36)$$

where α and β were functions of Z_1 and Z_2. Justus and Mikhail showed that the functional relations 2-30 and 2-36 are compatible provided that α and β satisfy the equations

$$\alpha = \left(\frac{Z_2}{Z_1}\right)^a \qquad (2\text{-}37)$$

and

$$\beta = 1 + b \ln(Z_2/Z_1) \qquad (2\text{-}38)$$

where the coefficients a and b depend on the reference height Z_1. The exponent n must satisfy the equation

$$n = a + b \ln V_1. \qquad (2\text{-}39)$$

The Weibull wind speed frequency distribution at height Z_1 can be expressed as the cumulative probability function $p(V_1)$ which gives the probability that the wind speed $V \leqslant V_1$:

$$p(V_1) = 1 - \exp\left[-(V_1/c_1)^{k_1}\right] \qquad (2\text{-}40)$$

where c_1 is the scale factor and k_1 is the shape factor of the Weibull distribution for height Z_1. When equation 2-36 is used to eliminate V_1 in equation 2-40, the result will provide $p(V_2)$:

$$p(V_2) = 1 - \exp\left[-(V_2/\alpha)^{k_1/\beta}\, c_1^{-k_1}\right]$$
$$= 1 - \exp\left[-(V_2/c_2)^{k_2}\right] \tag{2-41}$$

provided that the scale factor c_2 and shape factor k_2 for height Z_2 are given by the equations

$$c_2 = \alpha c_1^{\beta} \tag{2-42}$$

and

$$k_2 = k_1/\beta. \tag{2-43}$$

A comparison of equations 2-42 and 2-36 indicates that the functional relationship between the instantaneous wind speed and height and the Weibull scale factor and height is the same, i.e.,

$$\frac{c_2}{c_1} = \left(\frac{Z_2}{Z_1}\right)^n \tag{2-44}$$

where $n = a + b \ln V_1$.

Justus and Mikhail (1976) used data for c and k which was representative of heights up to 100 m to evaluate a, b, α, and β. The results may be expressed as

$$\frac{k_2}{k_1} = \frac{1 - .0881 \ln (Z_1/10)}{1 - .0881 \ln (Z_2/10)} \tag{2-45}$$

$$\frac{c_2}{c_1} = \left(\frac{Z_2}{Z_1}\right)^n \tag{2-46}$$

and

$$n = \frac{0.37 - .0881 \ln c_1}{1 - .0881 \ln (Z_1/10)} \tag{2-47}$$

where the heights Z_1 and Z_2 are expressed in meters. Thus if c_1 and k_1 are known for an arbitrary height Z_1, these equations may be solved for c_2 and k_2 at height Z_2. The constants .37 and $-.0881$ correspond to the values of a and b at 10 m elevation, a_{10} and b_{10} respectively. For the corresponding situation of a known instantaneous wind speed V_1 at an arbitrary anemometer height Z_1, the wind speed V_2 at height Z_2 may be obtained by using the equations

$$\frac{V_2}{V_1} = \left(\frac{Z_2}{Z_1}\right)^n \tag{2-48}$$

and

$$n = \frac{.37 - .0881 \ln V_1}{1 - .0881 \ln (Z_1/10).} \tag{2-49}$$

Equation (2-49 indicates that n will decrease as the wind speed V_1 at arbitrary height Z_1 increases.

The mean wind speed $\langle V \rangle$ and the corresponding Weibull scale and shape factors k and c are related by equation 2-13. For k values in the range $1.4 < k < 3$, the ratio $c/\langle V \rangle$ is in the range $1.1 < c/\langle V \rangle < 1.3$, thus $\langle V \rangle$ is approximately proportional to c with a constant factor. Due to this close correspondence between the mean wind speed and scale factor, equations 2-48 and 2-49 can also be applied to estimate the mean wind speed $\langle V_2 \rangle$ at height Z_2 on the basis of a measured value $\langle V_1 \rangle$ at arbitrary height Z_1, i.e.,

$$\frac{\langle V_2 \rangle}{\langle V_1 \rangle} \cong \left(\frac{Z_2}{Z_1} \right)^n \tag{2-50}$$

$$n = \frac{.37 - .0881 \ln \langle V_1 \rangle}{1 - .0881 \ln (Z_1/10).} \tag{2-51}$$

Spera and Richards' Equations

Spera and Richards (1979) have proposed wind shear profile equations which incorporate both wind speed and surface roughness effects. Their approach is based on a generalization of Justus and Mikhail's equations 2-48 and 2-49. They assumed that n has a normal distribution with standard deviation σ about the mean value $\langle n \rangle$. They also assumed that both $\langle n \rangle$ and σ are functions of the steady, nongusting wind speed and that $\langle n \rangle$ is also dependent on the surface roughness. They proposed the equations

$$V_2 = V_1 (Z_2/Z_1)^{\langle n \rangle},$$

and

$$\langle n \rangle = \frac{n_0 (1 - \log V_1 / \log V_h)}{(1 - n_0 \log (Z_1/Z_r)/\log V_h)} \tag{2-52}$$

where

$$n_0 = (Z_0/Z_r)^{0.2} \tag{2-53}$$

is a surface roughness exponent, V_1 is the steady wind speed at elevation Z_1 in meters per second (m/s), V_h is the homogeneous wind speed in m/s (the value of the wind speed which corresponds to $\langle n \rangle = 0$), Z_r is the reference elevation, 10 m, and Z_0 is the surface roughness length in meters. This equation for the power law exponent is based on the assumption that wind profiles become uniform or homogeneous at high wind speeds for all values of surface roughness. The value of $\langle n \rangle$ converges to zero at V_h for all values of Z_0.

A comparison of equations 2-52 and 2-49 shows that if n is identified with $\langle n \rangle$, the wind data analyzed by Justus and Mikhail yields $n_0 = a_{10} = .37$ and $V_h = \exp[-a_{10}/b_{10}] = 66.7$ m/s. Spera and Richards analyzed wind data from wind turbine sites or potential sites selected by the Department of Energy. Using an assumed value of 67 m/s for V_h at all sites, they found surface roughness exponents n_0 which varied from 0.10 to 0.48 for five sites. The mean exponent for the five sites was 0.32. An analysis of monthly peak wind data from the Clayton, New Mexico site indicated that, at least for this site, the value employed for V_h was reasonable. Equation 2-53 is an empirical relationship between n_0 and Z_0. Literature values of the surface roughness length for five test sites were employed to determine the exponent (0.2) of this equation.

The Weibull parameters c_2 and k_2 at height Z_2 may be obtained from the known values c_1 and k_1 at height Z_1 using the equations

$$c_2 = c_1 (Z_2/Z_1)^{\langle n_{c_1} \rangle} \tag{2-54}$$

$$\langle n_{c_1} \rangle = \frac{n_0(1-\log c_1/\log V_h)}{(1-n_0 \log(Z_1/Z_r)/\log V_h)} \tag{2-55}$$

and

$$\frac{k_2}{k_1} = \frac{(1 - n_0 \log(Z_1/Z_r)/\log V_h)}{(1 - n_0 \log(Z_2/Z_r)/\log V_h)} \tag{2-56}$$

The general utility of Spera and Richards' equations must be established by additional tests.

Illustrative Profiles

Three wind profile equations will be employed in Chapter 3 to illustrate the relationship between vertical wind shear and wind system performance: Justus and Mikhail's equations 2-48 and 2-49, the one-seventh power law, and Justus and Mikhail's equations with modified constants.

These equations will describe cases of high, moderate, and low vertical wind shear. For the high and low cases, the power law exponent decreases as the wind speed increases, although in the latter case n and the range of variation in n are small.

The values of a_{10} and b_{10} for the low wind shear case are based on data for the Grandpa's Knob site. The ratios of the wind speed at the 80- and 185-ft levels of the Christmas Tree to the wind speed at the standard level of 140 ft based on 2000 simultaneous measurements were .954 and 1.030, respectively (Dornbirer and Wilcox, 1945). The wind speed ratio for the 80- and 140-ft levels and the five-year term annual mean wind speed at the standard level were used to obtain estimates for a_{10} and b_{10} of .152 and $-.0362$, respectively.[*]

Figure 2-12 shows height x versus the ratio of the wind speed at height x (V_x) to the wind speed at a reference height of 140 ft (V_{140}) for the illustrative vertical wind shear profiles. The points shown correspond to the measured values at the Grandpa's Knob test site. Figure 2-13 also shows height versus wind speed for the illustrative profiles. The wind speeds are referred to the hub height of the Mod-2 wind turbine (Z_1 = 200 ft). The profiles show the variation of the Mod-2's cut-in, rated, and cut-out wind speeds from their design values at 200 ft to the 30-ft level. Note that the variable exponent of Justus and Mikhail's equations decreases as the wind speed at hub height increases. The exponent value for the unmodified equation is

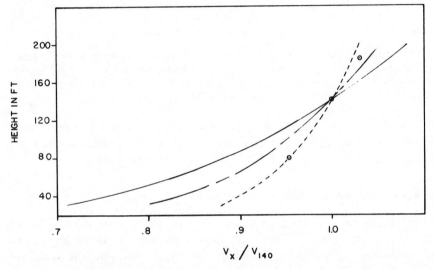

Fig. 2-12. Height versus the ratio of the wind speed at height x (V_x) to the wind speed at the reference height, 140 ft (V_{140}), for three wind shear profiles (_____, Justus-Mikhail equation; _ _ _ _ _, one-seventh power law and _ _ _ _ _ _ _, modified Justus-Mikhail equation). The circled points are the measured values for the Grandpa's Knob test site.

---

[*]The estimate for $a_{10}(n_0)$ was calculated using the expression $\langle n \rangle / [1+(\langle n \rangle \log(Z_1/10) - \log V_1)/\log V_h)]$ where $\langle n \rangle = \ln(.954)/\ln(80/140)$, Z_1 = 44.7 m (140 ft), V_1 = 7.45 m/s (16.66 mph), and V_h = 66.7 m/s. The estimate for b_{10} was calculated using the expression $-a_{10}/(2.303 \log V_h)$. Dornbirer and Wilcox (1945) state that the wind shear factors were determined by using 2000 simultaneous wind speed measurements over a 20-month period. It is likely that wind speeds above a cut-off value, e.g., the 15 mph stated on Putnam's page 63, were employed. The mean wind speed at the 140-ft reference level corresponding to the measured shear factors was not given. It would have been preferable to use this value for the calculation of a_{10} and b_{10} rather than the five-year term annual mean wind speed. However, the constants obtained will serve the purpose of describing a variable exponent power law for the low shear case.

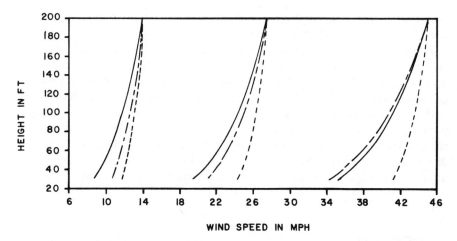

Fig. 2-13. Height versus wind speed for the illustrative vertical wind shear profiles of Figure 2-12. Wind speeds are referred to the hub height of the Mod-2 wind turbine (Z_1 = 200 ft). The profiles show the variation of the Mod-2's cut-in, rated, and cut-out wind speeds from their values at 200 ft (14, 27.5, and 45 mph, respectively) to 30 ft.

0.248 for the cut-in wind speed of 14 mph, and 0.126 for the cut-out wind speed of 45 mph. The one-seventh power law therefore yields a smaller value for the cut-out wind speed at 30 ft than Justus and Mikhail's unmodified equations.

Vertical Wind Speed Profiles at Smith Putnam Sites

The profiles discussed in the previous sections show a monotonic increase of wind speed with height. Similarity theory and simplifications of it pertain to level terrain and for this case site selection is a simple process. In contrast, when the wind flows over a ridge, the "shape" of the profile is more variable and depends upon additional factors, e.g., the angle of incidence of the flow measured from the ridge axis and the geometry of the cross section of the ridge. Putnam (1948) recognized the limitations of the theoretical description of the wind flow over ridges and the need for experimental measurements of profiles at Smith-Putnam sites. Putnam presented an analysis of wind shear data for sites at Grandpa's Knob, three wooded summits, and Mount Washington. The analysis was concerned with the extrapolation of long-term mean wind speeds and the profiles were constructed using hourly mean wind speeds.

Wind speed data was also collected at four levels above the Smith-Putnam site on Scrag Mountain. These data are identified in Putnam's Table IV. The entries for the dates and days of operation are incorrect for the anemometer levels at 53, 66, and 108 ft above ground level. Measurements were actually made during the period

May 31 through August 10, 1940 (62 days). Koeppl (1979a) used the hourly mean wind speeds for coincident records during the 62-day period to construct the vertical wind speed profiles. The results are given in Table 2-5 and Figure 2-14.

The wind speeds for the periods of record in July and August do not show the steady increase with elevation which is shown in Putnam's Figures 39 and 40 for Grandpa's Knob and three wooded summits. A number of interpretations of this finding are possible. The Grandpa's Knob profile was constructed using a different procedure. Hourly mean wind speeds in excess of 6.7 m/s (15 mph) were used to construct the profile, whereas all of the records for the coincident periods in June, July, and August were used to construct the profiles for the Scrag site. The data collection for multiple levels at the Scrag site spanned periods of low winds, particularly during July and August. During the period of relatively high winds in June, the profile for the Scrag site does show a steady increase in wind speed with elevation. Putnam did not describe the procedure for the construction of the profiles of Figure 40 for three wooded summits. Furthermore, the Scrag site is not a mountain summit, but rather a col south of the summit. The col is approximately 91 m (300 ft) lower than the summit. Hence the exposure was markedly different from that at other Smith-Putnam sites where the vertical wind shear profile was measured.

The power law exponent n based on equation 2-30 and the wind speed ratio for the 45- and 100-ft levels in June at the Scragg site is 0.17 (ln(17.88/15.63)/ln (100/45)) . The values of n calculated in this way using the two points of Figure 2-12 for the Grandpa's Knob site are 0.084 and 0.11. A typical value of the exponent for the three wooded summits is 0.13. The vertical wind shear profiles measured during the Smith-Putnam project suggest that vertical wind shear profiles at well-exposed sites in the Green Mountains are characterized by relatively low power law exponents.

Table 2-5. Monthly mean wind speeds for four levels at the Scrag site for periods during which wind speeds were recorded at all levels.

Month	Days of Coincident Data	Level[a] (mph)			
		L (45)	M (58)	S (70)	U (100)
June[b]	15.71	15.63	17.20	17.72	17.88
July	26.91	10.32	11.21	11.01	11.16
August	9.58	10.41	11.55	11.67	11.11

[a]The lower (L), middle (M), standard (S), and upper (U) levels were 45, 58, 70, and 100 ft above the mean height of the trees at the site.

[b]Some of the coincident period records for the month of June were rejected on the basis of statistical criteria.

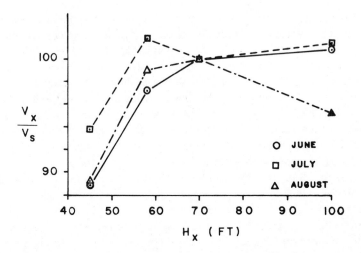

Fig. 2-14. Height (H_x) versus the ratio of the wind speed at height x (V_x) to the wind speed at the standard height, 70 ft (V_s) for coincident periods in June, July, and August at the Scrag site.

Profiles Determined by TALA Measurements

The Smith-Putnam researchers attempted to obtain information about the vertical wind shear and turbulence above mountain summits through pilot balloon measurements. This effort failed to determine the profile because it proved impossible to launch enough balloons over a particular point on the crest with a frequency which produced meaningful results. Balloon launches at a number of sites did show turbulence in the lee; unfortunately, the turbulence frequently caused balloons to drop out of sight. More recently smoke trails from a rocket, the laser scintillation anemometer (Hardy, 1976), and doppler acoustic radar (MacCready, 1977) have been used in such studies. These techniques have proved unsatisfactory due to high cost, low accuracy, or unacceptable reliability. The most promising new technique for the measurement of vertical wind shear profiles involves the use of a Tethered Aerodynamically Lifting Anemometer (TALA).[*] This device is otherwise known as a kite anemometer.

A TALA consists of a sled kite and tail, tethering line, reel, and a tensiometer which consists of either a spring scale or recording strain gauge (Approach Fish Inc., 1979). Figure 2-15 shows a kite anemometer in flight. Lift and drag forces on the kite produce tension in the no-stretch tether line. The tension is calibrated as wind speed. National Bureau of Standards and NASA-Langley wind tunnel calibrations agree within 1 to 2 percent. The calibration has been subjected to field tests by

---

[*]U.S. Patent No. 4,058,010. Manufactured by Approach Fish, Inc., Rt. 1, Box 620B, Ringgold, Virginia, 24586.

Fig. 2-15. A TALA in flight. Courtesy of Approach Fish, Inc., Rt. 1, Box 620B, Ringgold, Virginia, 24586.

flying kites near tower mounted anemometers ranging from 17 to 100 m (Baker et al., 1979). Errors were found which ranged from 1 to 3 percent. The kite will fly in winds as low as 3 m/s and as high as 40 m/s. A peak height of approximately 250 m may be attained in winds as low as 6 m/s. The kite altitude is determined by using the line length and the visual vertical angle to the kite. The vertical angle is measured by using either a clinometer or a tripod-mounted theodolite. A neutral density tail, a narrow balloon filled with helium and air which is attached to the tether line, may be observed to study low-level turbulence, and the shear and flow over surface features.

Wind data collected by using a kite anemometer provides a short-term description of the vertical profile and turbulence of the wind at a site. Wind data of this type has the most significance when it is collected during a period of wind flow from the prevailing direction. The wind speed and atmospheric parameters should represent typical operating conditions for a wind turbine. The quantitative description of the wind flow at a site which is provided by kite measurements may be used to identify problems involving turbulence at a site. For sites with a promising flow pattern, kite data may be used to determine advantageous positions for long-term measurement stations.

Baker et al. (1979) have used kite anemometers to determine short-term vertical wind speed profiles and turbulence intensities ($\sigma/\langle\overline{V}\rangle$) at sites in the West and Northwest. In a typical experiment using a kite furnished with a spring scale, readings are taken at 10-second intervals during a five-minute period at each level. Profiles determined in this way for some sites resemble the long-term profiles shown in Figure 2-14 for the Scrag site.

Vachon et al. (1979) have conducted similar studies in the White Mountains area of New Hampshire. Kite measurements were made at elevations of 15, 30, and 61 m above mountain sites. The wind speed was measured continuously using a portable station of 3 m elevation to obtain temporal corrections for the wind speeds recorded by the kite sensor at different levels. The results were used to calculate power law exponents. The exponents for seven sites ranged from 0.08 to 0.35; the mean value was 0.17. Unfortunately, the measurements were performed during the passage of a weather front and the wind did not flow from the prevailing direction.

STATISTICAL STUDIES

Introduction

Hourly or three-hourly values of the one-minute average wind speed are recorded at National Weather Service sites. These data may usually be obtained from the National Climatic Center (NCC) in digitized form on magnetic tape (TDF data). Computer algorithms may be used to derive statistical data such as the mean, variance, probability distributions of hourly and monthly wind speed and wind power, auto correlation (temporal correlation), persistence, and spectral power density. A number of workers have analyzed data of this type in order to determine the statistical confidence that can be assigned to certain observed quantities as a function of the record length. Attention has also been given to the development of statistical models which may be used to derive required quantities in terms of other more readily available ones.

Treatment of Wind Speed and Wind Power

Corotis (1979) has reviewed the statistical reliability of wind power assessments. The most basic statistics for a site are the mean wind speed and mean wind power. The significance of the observed means for different sites must be known in order to make siting decisions. If the observed means are based on known numbers of independent observations, simple methods may be used to calculate confidence intervals (Benjamin and Cornell, 1970). For example, one is 90 percent certain that the true mean (i.e., the mean for an infinite sampling period) lies within the 90 percent confidence interval. Corotis (1976, 1977) has shown that hourly or three-hourly values of the wind speed have appreciable auto correlation, and thus the assumption of independent sampling is not justified. He used an analysis of variance approach to relate a given number of observed data to a reduced number of equivalent independent observations. The equivalent number was then used in conjunction with standard methods for the determination of confidence intervals.

The analysis of equivalent independent data was performed for both wind speed and wind power on a seasonal basis. The number of independent hours in a day and independent days in a month were first computed for six northern Illinois sites and one Montana site (Corotis et al., 1977). Later, data from 21 sites across the U.S. were analyzed. Although some variation among sites and regions was noted, generally there were 2 to 3 independent hours per day and 10 to 25 independent days per month for the case of the mean wind speed. For all 28 sites, the bounds on the number of equivalent independent hours per month, when averaged over the four calendar seasons, varied between 30 and 75. The results for the mean wind power were similar. Using these statistics it is possible to determine the duration of a survey which gives a desired confidence level.

Table 2-6 gives minimum survey durations in terms of independent months which correspond to stated confidence levels and accuracies of the mean wind speed. To obtain these results, it was assumed that the hourly wind speed obeys the Rayleigh frequency distribution. The hourly coefficient of variation for this distribution is 0.52. In order to convert the results of Table 2-6 to calendar months, the values must be increased slightly. The increase depends upon the month-to-month correlation at the site. In no case was the increase more than 50 percent. Thus, the values of Table 2-6 should be multiplied by a factor between 1.0 and 1.5 to obtain calendar months.

Although this procedure was derived for seasonal mean wind speeds at a site, it can be applied directly for the case of annual mean wind speeds as long as the data collection period is a multiple of whole years. Otherwise the effect of the annual cycle would distort the results. An inspection of Table 2-6 shows that one can be about 90 percent confident that the mean wind speed determined by a season of data is within about 10 percent of the true long-term wind speed for that season.

Table 2-6. Minimum survey duration in terms of independent months for specified confidence and accuracy levels of the mean wind speed.

Independent Hours Per Month	90% Confidence			95% Confidence		
	Accuracy			Accuracy		
	± 5%	± 10%	± 15%	± 5%	± 10%	± 15%
30	9.8	2.4	1.1	13.9	3.5	1.5
75	3.9	1.0	0.4	5.5	1.4	0.6

From Corotis (1979).

The same approach was applied to the cube of the wind speed to obtain results for available wind power. The hourly correlation for wind power was observed to be about the same as for wind speed. Thus, the number of equivalent independent hours in a month is bounded approximately by 30 and 75. However the minimum survey duration is much longer for wind power because the hourly coefficient of variation is larger, i.e., 1.55 for the Rayleigh cubed wind speed distribution. Table 2-7 gives typical results for the minimum survey duration in independent months. An inspection of this table shows that it takes almost nine times as long to obtain the wind power at a given confidence and accuracy as it does to obtain the mean wind speed. From a practical viewpoint, this difference may be unimportant. In Chapter 3, model wind system performance calculations show that the output from wind turbine generators obeys an approximate linear relationship with mean wind speed over an intermediate speed range that includes most viable wind power sites. Nonlinear variation is found for low and high values of the mean wind speed. Therefore Table 2-6, and not 2-7, has greater relevance in terms of the evaluation of wind power sites.

An inspection of Tables 2-6 and 2-7 indicates that a four-fold increase in survey duration yields a doubling of accuracy (e.g., from ±10 percent to ±5 percent). The reliability can be doubled (e.g., from 90 percent confidence to 95 percent) with an increase in survey duration of 40 percent.

Corotis (1979) has discussed the statistical treatment of daily measurements of the wind speed. Wind speeds at National Fire Weather stations are recorded once per day, typically at 2 P.M. local time. Since consecutive hourly wind speed data from National Weather Service stations have been shown to be equivalent to only two to three independent readings per day, a single daily reading produces a confidence interval which is no more than twice as large as that for 24 consecutive hourly readings. Two to three years of once-daily readings can therefore provide the same confidence and accuracy as one year of hourly readings. For the case of once-daily readings, it is necessary to either make an adjustment for the diurnal cycle or record data at different hours on different days so that the effect of the diurnal cycle cancels in the computation of the mean wind speed.

Table 2-7. Minimum survey duration in terms of independent months for specified confidence and accuracy levels for site mean wind power.

Independent Hours Per Month	90% Confidence			95% Confidence		
	Accuracy			Accuracy		
	± 5%	± 10%	± 15%	± 5%	± 10%	± 15%
30	86.7	21.7	9.6	123.1	30.8	13.7
75	34.7	8.7	3.9	49.2	12.3	5.5

From Corotis (1979).

Annual and Seasonal Variation of the Mean Wind Speed at Grandpa's Knob

Putnam's Figures 51 and 52 show the variation of calculated annual and seasonal outputs for the Smith-Putnam turbine at three sites. He did not show the variation of annual and seasonal mean wind speeds. Figures 2-16 and 2-17 show the variation of the annual and seasonal mean wind speeds at Grandpa's Knob (Koeppl, 1980). The values correspond to the 120-ft level of the Christmas tree during the five-year test period. The maximum positive and negative deviations of the mean annual wind speed from the five-year mean (16.39 mph) are +13.4 and −7.4 percent, respectively.[*] The maximum positive and negative deviations of the mean monthly wind speed from the corresponding five-year mean are +28.6 percent (January, 1942) and −28.6 percent (January, 1941), respectively.

Climatological Adjustment Procedures

Putnam (1948) used a climatological adjustment procedure to convert short-term measured mean wind speeds for survey sites to estimates of the long-term means. The method relies on the availability of long-term wind data from nearby "clima-

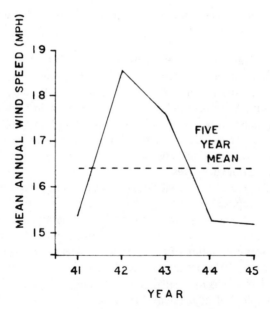

Fig. 2-16. Variation of the annual mean wind speed at the 120-ft level of the Christmas tree at Grandpa's Knob.

[*]The variation of the mean annual wind speed does not follow the output variation of Figure 51. The differences may be due to the dispersion of the calculated output at a given mean wind speed which is illustrated in Putnam's Figure 33.

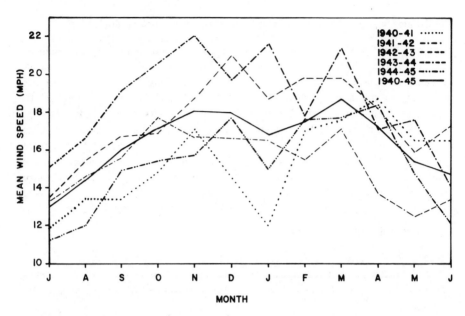

Fig. 2-17. Variation of the monthly mean wind speed at the 120-ft level of the Christmas tree at Grandpa's Knob.

tological" sites. Grandpa's Knob served as the climatological site for the Smith-Putnam measurement program. Putnam's method has become known as climatological reduction by the method of ratios. Conrad and Pollak (1962) have described the prerequisites for the application of this method in terms of deviations from the climatic mean which are both "quasi-constant" and "relatively homogeneous." The former prerequisite is mathematically equivalent to the requirement of a high spatial cross-correlation between the deviations from the climatic means at the survey and climatological sites.

Justus et al. (1979) have examined Putnam's adjustment procedure and another method which was proposed recently by Corotis. Corotis's method is based on the work of Feller (1966) and statistical principles which he discussed in connection with earlier work (Corotis, 1974). Corotis's adjustment procedure makes explicit use of the spatial cross-correlation of the mean wind speeds at the survey and climatological sites. Justus (1976a) and Corotis (1977) have shown that correlations based on hourly wind speeds for nearby stations are high and therefore support the concept of a climatological adjustment. However, the cross-correlation which is required in the adjustment procedure should correspond to the average over the full short-term collection period at the survey site (Corotis, 1979). Because this collection period will usually be about one year, or at least several months, the monthly and annual average spatial cross-correlation provide a better assessment of the relevance of the adjustment.

Justus et al. (1979) studied the behavior of spatial cross-correlation by analyzing monthly mean wind speeds obtained from Local Climatological Data publications for 40 sites which span regions that include 21 states. Ten or more years of data for a fixed location and a constant anemometer level were available for each site. In order to examine regional effects and facilitate the study of the relationship between the magnitude of the spatial cross-correlation and site separation, the sites were divided into six subgroups of regional sites. Data for concurrent periods were analyzed for each of these groups. The spatial cross-correlations were evaluated for each regional group and the resulting correlations were averaged according to inter-site separation distance intervals of 0–200 km, 200–400 km, and so forth. Figure 2-18 shows average spatial cross-correlation for monthly and annual mean wind speeds versus site separation. The correlations correspond to averages over the six regional groups. The error bars correspond to the standard deviation of the mean spatial correlation. Due to the large scatter of individual values of the spatial cross-correlation, no significant differences were detected among the six regional groups. The 10 and 90 percentile values of the monthly correlation in the separation range 0–400 km were found to be 0.14 and 0.62, respectively; the corresponding values for the annual correlation are –0.30 and 0.66, respectively. Therefore, although the average monthly and annual spatial cross-correlations for 0–200 km separation are 0.5 and 0.3, respectively, these values are not defined with the precision which could justify their routine use for a climatological adjustment.

The results of Justus et al.'s investigation of the accuracy of the Putnam and Corotis methods for making a climatological adjustment are given in Tables 2-8 and 2-9. The test of accuracy was based on the analyses of data for the site pair in each regional group which had the smallest separation; the climatological site was chosen on an alphabetical basis. Table 2-8 gives the average monthly and annual spatial cross-correlations for each site pair, the years for the comparison, the mean annual wind speed ($\langle \overline{V} \rangle$), and the ratio of the standard deviation to the mean wind speed for the comparison period ($\sigma / \langle \overline{V} \rangle$). Table 2-9 gives the annual climatic mean wind speed (for the cross-correlation period), the mean wind speed for the last year of the study

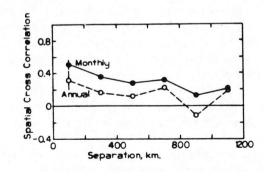

Fig. 2-18. Monthly and annual spatial correlations of wind speed versus site separation (averaged over all sites). From Justus et al. (1979).

Table 2-8. Site pairs used in the climatological site to candidate site conversion test.

Sites	Separation (km)	Average Cross-Correlation Monthly	Average Cross-Correlation Annual	Years for Cross-Comparison	Candidate Site Winds $\langle V \rangle$ (mph)	Candidate Site Winds $\sigma/\langle V \rangle$ (%)
JFK–LGA	16	0.465	0.091	1963–1976	11.6	4.3
ABR–HON	120	0.591	0.572	1965–1976	11.4	5.3
CNK–GRI	168	0.562	0.367	1964–1976	11.8	4.2
CPR–CYS	237	0.144	−0.195	1965–1976	12.9	3.9
AKN–BET	391	0.349	0.470	1963–1974	12.4	4.8
HNL–HOG	162	0.584	0.524	1965–1976	12.2	7.4

period, and the value for the last year minus the climatic mean, divided by the climatic mean (designated $\triangle$%) for the climatological and candidate site of each pair.

The climatological estimate for Putnam's method of ratios ($\langle \overline{V}_c \rangle$) was calculated using

$$\langle \overline{V}_c \rangle = \langle \overline{V}_1 \rangle (\langle \overline{V} \rangle / \langle \overline{V}_o \rangle) \tag{2-57}$$

where $\langle V_1 \rangle$ is the observed one-year mean at the survey site, $\langle V \rangle$ is the long-term mean at the climatological site for the study period, and $\langle V_o \rangle$ is the one-year mean at the climatological site (for the period concurrent with $\langle V_1 \rangle$). The climatological estimate for Corotis's method was calculated using

$$\langle \overline{V}_c \rangle = \langle \overline{V}_1 \rangle + \rho (\langle \overline{V} \rangle - \langle \overline{V}_o \rangle) \, \sigma_c / \sigma \tag{2-58}$$

where σ_c and σ are the standard deviations for the climatological and survey sites and ρ is the spatial correlation between these sites.

The results in Table 2-9 show that neither of the methods yields a reduction in the root-mean-square (rms) deviations between estimated and observed values of the climatic mean wind speed. The results also show that neither of the methods yields estimates which represent an improvement upon the values which correspond to one year of on-site data for more than two of the six cases examined. The results in Table 2-9 indicate that some improvement is obtained when estimates are calculated for the monthly climatic means. This finding suggests that an improved adjustment for the annual mean may be obtained through the separate adjustment of the monthly means. Table 2-9 shows that no improvement is gained by using the average of 12 adjusted monthly means to estimate the long-term annual mean.

In the application of the method of ratios by Justus et al. (1979), the four mean wind speeds of equation 2-57 correspond to the same (annual) period. Putnam applied a different version of the method of ratios. The mean wind speeds recorded during a concurrent period of from 22 to 301 days at the survey and climatological (Grandpa's Knob) sites were employed for $\langle V_1 \rangle$ and $\langle V_o \rangle$, respectively, and a long-term annual mean for the wind speed at the climatological site was used to obtain an estimate for the long-term annual mean at the survey site. The use of different

Table 2-9. Climatological site to candidate site conversion test data.

Climatological Site	Long-Term Mean ⟨V⟩ (mph)	Last Year Average ⟨V₀⟩ (mph).	(Δ %)	Candidate Site	Long-Term Mean (mph)	Last Year Average ⟨V₁⟩ (mph)	(Δ %)	Eq. (1) Long-Term Mean Calculated ⟨Vc⟩ (mph)	(Δ %)	Eq. (2) Long-Term Mean Calculated ⟨Vc⟩ (mph)	(Δ %)	Last Year Monthly rms Δ %	Estimated Long-Term Mean Monthly rms Δ %	Annual Average from Monthly Long-Term Estimates (mph)	(Δ %)
JFK	12.5	12.7	1.6	LGA	11.6	12.4	6.9	12.2	5.2	12.4	6.9*	8.2	5.7	12.2	5.0
ABR	11.2	11.8	5.4	HON	11.4	11.7	2.6	11.1	-2.6*	11.4	0.0	8.3	5.4	11.1	-2.6*
CNK	12.2	12.2	0.0	GRI	11.8	12.4	5.1	12.4	5.1*	12.4	5.1*	9.0	7.1	12.5	5.7*
CPR	12.8	11.9	-7.0	CYS	12.9	12.6	-2.3	13.6	5.4*	12.5	-3.1*	8.5	8.7*	13.6	5.4*
AKN	10.9	10.3	-5.5	BET	12.4	12.0	-3.2	12.7	2.4	12.2	-1.6	10.2	15.7*	12.9	4.1*
HNL	11.8	11.5	-2.5	HOG	12.2	12.9	5.7	13.2	8.1*	13.1	7.4*	10.6	8.6	13.2	8.4*
							4.6 rms		5.2 rms		4.8 rms				

*These data not improved by the adjustment process using the climatological site data.

275

periods in this way introduces an uncertainty which was not examined in the study by Justus et al. (1979). Bortz et al. (1979) used a similar version of Corotis's method to convert short-term mean wind speeds for seven mountain sites in New Hampshire to estimates of the long-term annual means. Measurements of the wind speed were made three times a day at each site during a period of 14 days in August. Mount Washington was employed as the climatological site. The spatial cross-correlations during the brief period ranged from 0.34 to 0.82; the average for the seven sites was 0.57.

Putnam (1948) did not justify the application of the method of ratios on the basis of calculated values of the spatial cross-correlation for Smith-Putnam sites. Koeppl (1979a) evaluated the cross-correlation for the Scrag and Grandpa's Knob sites. The cross-correlation (r) based on six values of the monthly mean wind speed for these sites was high (0.946). The Grandpa's Knob site is located approximately 60 km to the southwest of the Scrag site. The unobstructed prevailing winds at Grandpa's Knob are from the southwest. The difference in elevation between the two sites is approximately 183 m and the Scrag site lies on a ridge which is downwind from the higher Lincoln Ridge. The regular relationship between the direction of the prevailing winds and the topographical features in the Green Mountain region may yield the conditions which make climatological adjustments useful.

Putnam did not analyze the six months of wind speed data collected at the Scrag site. Koeppl (1979a) used the method of ratios to estimate the long-term (five year) annual mean for the Scrag site. An approach suggested by Corotis (1979) and five-year monthly mean wind speeds for the Grandpa's Knob site were used to adjust the six-month mean for the annual cycle. The estimate obtained for the five-year annual mean wind speed at the Scrag site referenced to 42.7 m (140 ft) above ground level was 7.15 m/s (16.0 mph). The Green Mountain Power Corporation proposed the Scrag site in the DOE's second wind turbine candidate site competition.

Run Duration Models

Corotis and coworkers have analyzed the durations of runs both above and below various wind speed levels (Corotis et al., 1978, Sigl et al., 1979, Corotis, 1979). A simple model was developed which gives the probability distribution of wind speed persistence above and below fixed reference speeds. The statistical description of the persistence at a site is important in connection with studies of power reliability and energy storage. A composite distribution has proven useful for the description of runs. The probability that a particular run has a duration of time t or less ($F(t)$) is expressed as

$$F(t) = \begin{cases} 1 - (t/t_0)^{1-b} & \text{for } t_0 \leqslant t \leqslant t_1 \\ 1 - Ae^{-\lambda t} & \text{for } t_1 < t < \infty \end{cases} \qquad (2\text{-}59)$$

where the time t_0 is the duration of the shortest observed run, i.e., 30 minutes for hourly data since, on the average, a run shorter than this is most likely not observed (Bendat and Piersol, 1971), time t_1 is the partition parameter between the power and exponential functions, and the parameters A and λ are expressed in terms of the other parameters to effect a smooth transition from the power function (which describes runs of short duration) to the exponential function (which describes runs of long duration). The analysis of observed data showed that while t_1 varies over a relatively large range, the complementary cumulative $G(t_1) = 1 - F(t_1)$ is relatively constant at 0.25. For the case $t_0 = 0.5$ and $G = 0.25$ the continuity equations for A and λ yield:

$$t_1 = 0.5 \, (0.25)^{1/(1-b)}, \tag{2-60}$$
$$A = 0.25 \, e^{b-1} \tag{2-61}$$

and

$$\lambda = 2 \, (b-1) \, (0.25)^{1/(1-b)} \tag{2-62}$$

The mean run duration m_t may be expressed as

$$m_t = 0.5 \left\{ \alpha^{-1} \, [1 - (0.25)^{\alpha}] + [b/(b-1)] \, (0.25)^{\alpha} \right\} \tag{2-63}$$

where $\lambda = (2-b)/(1-b)$ and the values given by equations 2-60 through 2-62 assume that the run duration is measured in hours. The values of the parameter b can be computed from hourly data by using the method of maximum likelihood.

SITING

Introduction

The importance of wind turbine siting was emphasized by the experience of the Smith-Putnam researchers. The project's outstanding meteorologists and aerodynamicists predicted that the annual mean wind speed at the Grandpa's Knob site was 10.7 m/s (24 mph). The five-year measured mean was instead 7.45 m/s (16.66 mph) and the difference meant that the turbine would produce 30 percent of the anticipated output at the test site. Putnam summarized the findings of the Vermont wind survey program with the remark, "After five years of increasing familiarity with the problem of site selection, we can point to no analogy between the profiles of mountains and the profiles of airfoils by which one can predict mean wind velocities at hub height within limits which will be useful." Golding (1955) performed an extensive British wind survey. He concluded that, "The theory of wind flow over hills is complex and evidence is still lacking on the precise shape of hill for maximum accelerating effect."

Although a great deal has been learned since these pioneering efforts in specific studies of the atmospheric flow, work performed to this date has not produced a useful description of the combined influence of terrain aspect ratio, roughness, insolation, and stratification.

This section is concerned with the description of general guidelines for siting based on the Smith-Putnam and other field measurement programs, a brief review of the literature on siting studies and results, the use of ecological indicators of wind energy for wind prospecting, and the use of numerical and physical modeling for site selection. We are concerned here with the identification of energetic wind turbine sites and not the evaluation of sites on the basis of other important practical criteria (e.g., the number of machines and their spacing at the site, proximity to power lines, site accessibility, environmental impact, and so forth).

Smith-Putnam Studies

The results from Putnam's measurement program show that elevation provides the best indication of the annual mean wind speed at the survey sites. Figure 2-19 shows the variation of the estimated five-year mean annual wind speeds at nine Smith-Putnam survey stations with elevations in the range from 457 to 1252 m (elevation above mean sea level).[*] The wind speeds correspond to "trees removed"

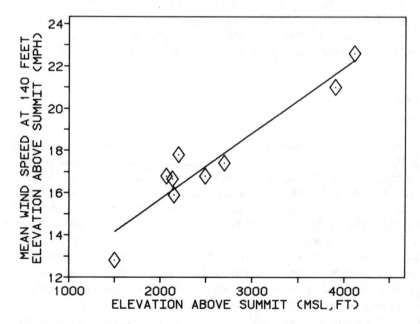

Fig. 2-19. Variation of mean wind speed with elevation for nine Smith-Putnam survey stations.

---

[*]See Putnam's Table VII. The points for Mount Washington (elevation 6288 ft) and Crown Point (elevation 100 ft) are not shown. The five-year term wind speeds at the survey sites were calculated by making a climatological adjustment using the method of ratios.

values which have been scaled to 140 ft elevation above ground level using the wind shear profile for Grandpa's Knob. The coefficient of determination (r^2) for these data is 0.91. The exposure to the prevailing winds was good at all of the Smith-Putnam sites. However, there was a great deal of variation in the geometry of the mountain sites in the direction of the prevailing winds. Consequently, the high correlation may surprise those who emphasize the significance of profiles in conjunction with speedup factors.

As indicated earlier, Koeppl (1979a) estimated the mean annual wind speed for the site near Scrag mountain. The value obtained for 140 ft elevation above ground level was 7.15 m/s (16.0 mph). The value predicted using the linear correlation based on the results of Figure 2-19 and the elevation for the Scrag site (2750 ft at 140 ft above ground level) is 8.0 m/s (18 mph). The "measured" result is thus 2 mph lower than the value predicted by the elevation correlation. However, the exposure at the Scrag site was not as good as that at most Smith-Putnam sites. Due to a property line restriction, the Scrag site was not located at the mountain summit. The tower was installed on a col 300 ft below the mountain summit. Observations of the trees at the summit of Scrag Mountain are consistent with the higher mean annual wind speed predicted by the elevation correlation at the 140-ft level above the summit (8.5 m/s–19 mph–at 3050 ft).

The regularity of the relationship between the topographic features of the Green Mountain ridges and the prevailing winds yields a high correlation between elevation and mean wind speed which is not found in regions of more complex terrain (Wahl, 1966, Pristov, 1959).

Other Studies and Results

Meroney et al. (1976a,b) have presented a review of wind power site selection studies. They describe a consensus of experience for wind site evaluation as follows:

1. Ridges should be athwart the principal wind direction, but high speeds are not likely on upwind foothills.
2. Hilltops should not be too flat and slopes should extend all the way to the summit.
3. A hill on the coast as opposed to an inland hill surrounded by other hills is more likely to provide high winds.
4. Speedup is greater over a ridge of given slope than over a conical hill of the same slope.
5. Speedup over a steep hill decreases rapidly with height.
6. The optimum hill slope is probably between 1:4 and 1:3 with 1:3.5 best.
7. Topographical features in the vicinity of the hill produce the structure of the flow over it.
8. Hills with slopes greater than 1:3 should probably be avoided.

Table 2-10. Frenkiel's classification of wind turbine sites on hills.[a]

Quality	$R = V_{40m}/V_{10m}$	n[b]	Slope
Optimum	$R < 1.05$	0.0	1:3.5
Very Good	$1.05 < R < 1.1$	0.07	1:6
Good	$1.1 < R < 1.15$	0.1	1:10
Fair	$1.15 < R < 1.21$	0.14	1:20
			1:6
Avoid	$1.21 < R$	> 0.14	$> 1:20$
			$< 1:2$

[a]Adapted from Frenkiel (1962).
[b]Simple power law exponent.

9. Vertical wind speed above a summit does not increase as much with height above ground as over level terrain.

Frenkiel (1964) carried out a comprehensive study of the wind flow over hills in Israel. He described measurements of the wind speed, direction, and temperature at different heights in order to characterize the effect of topography on the vertical wind shear profile. He showed that for any particular wind direction, the profile was independent of the speed within the significant speed range for wind turbines. Frenkiel (1962) ranked sites on the basis of the uniformity of the wind profile at the summit. His ranking criteria are given in Table 2-10. He suggests that flat terrain (slope 1:20) and very steep terrain (slope 1:2) should be avoided and that a hill of slope 1:3.5 is best.

A number of siting studies and techniques are described in volume seven of the *Proceedings of the United Nations Conference on New Sources of Energy* (1964). Wegley et al. (1978) have prepared a siting handbook for small wind turbines which gives a useful review of siting principles. Justus (1978a) has given a brief review of siting principles. Frost and Nowak (1979) have developed guidelines for siting wind turbine generators relative to small-scale, two-dimensional terrain features based on experimental data and analytical flow field models.

The diurnal variation of wind power is particularly important in connection with the economic evaluation of wind generation. Figure 2-20 shows diurnal variation models for low, moderate, and high wind regimes which the General Electric researchers used in their mission analysis study (Garate, 1977). The models are consistent with observations which show that over level terrain, near-surface winds usually show a maximum during the early afternoon whereas upper winds show a minimum during this period. This behavior is understandable in terms of downward momentum transfer under unstable conditions during the day and low momentum transfer downward under stable conditions during the evening. The high wind model is based on wind data from Whiteface Mountain in New York's Adirondack

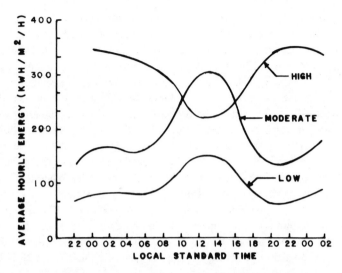

Fig. 2-20. Diurnal variation model of wind energy in favorable wind regimes. From Garate (1977).

Mountain region. If the electrical demand peaks during the afternoon (e.g., due to air conditioning loads), a site in the moderate wind regime may be preferable to a high wind site. Putnam's Figure 53 shows the diurnal variation of the output for the Smith-Putnam turbine based on wind data for Grandpa's Knob, Mount Washington, and Blue Hill. The outputs for the high wind sites at Grandpa's Knob and Mount Washington are in qualitative agreement with the General Electric model. However, Coty (1976) showed that the diurnal variation of winds at mountain sites can parallel either the surface or upper winds due to separation from the surface winds which may depend upon atmospheric stability.

Prospecting with Ecological Indicators of Wind Energy

Trees. Putnam recognized that trees might provide a useful description of the local wind climatology. He invited ecologist Robert F. Griggs to participate in the Smith-Putnam project's siting studies. He and Griggs developed a method which permitted estimates of mean wind speeds to be derived from observations of wind-deformed trees. Putnam called this method quantitative ecology. The method was developed on the basis of correlations involving the characteristics of wind-deformed coniferous trees and measured mean wind speeds at Smith-Putnam sites.

Important applications of quantitative ecology were made in 1945 and 1979. In Chapter XI, Putnam describes Jackson and Moreland's determination of the worth to the Central Vermont Public Service Company of a farm of six 1500 kW units on the Lincoln Ridge. The Lincoln Ridge sites are shown as anemometer sites in Put-nam's Figure 25; however, no measurements were actually made at these locations. Jackson and Moreland relied on estimates for the annual mean wind speeds based on quantitative ecology. The average value for the six sites was 12 m/s (27 mph) (Dornbirer and Wilcox, 1945). This value corresponds to an estimate for an eleva-tion of 140 ft above the ridge which is based on the vertical wind shear at the Grandpa's Knob site. This result was used to obtain a conservative estimate of the specific output for use in the economic analysis. In 1979, the Vermont Public Ser-vice Board measured the wind speed at the Lincoln Peak ridge site. The measured mean wind speed for a three-month period was corrected for the effect of the annual cycle using measured data for the Grandpa's Knob site. When this result was scaled to 140 ft using the wind shear for Grandpa's Knob, the estimate obtained for the annual mean wind speed was 11.7 m/s (26 mph) (Koeppl, 1979b). The agree-ment between the predicted and measured wind speed is good; however, additional measured wind data at the Lincoln Ridge and other sites will be required to define the accuracy of quantitative ecology for applications in the interior New England region.

Additional wind data will be collected at Lincoln Peak and another site on the Stratton Mountain Ridge. In 1979, the Green Mountain Power Corporation devel-oped a successful proposal for the Lincoln Peak site in the DOE's second wind tur-bine candidate site competition. Quantitative ecology was used to estimate the annual mean wind speed at sites on the Stratton Mountain ridge. The Stratton Mountain Ridge is a monadnock of approximately 1220m(4000-ft) elevation which is located approximately 120 km south of the Lincoln Ridge. The indicated values of the annual mean wind speeds referenced to the 140-ft level above the Stratton Mountain Ridge range from 8.8 to 11.2 m/s (Koeppl, 1979c). The Vermont Elec-tric Cooperative developed a successful candidate site proposal for this site. Wind velocity (speed and direction) measurements will be made at two or three levels of a 50-m meteorological tower to determine the viability of these locations for future demonstration projects.

Putnam claimed that quantitative ecology could be used to rank sites according to their order of merit for a measurement program, indicate sites of submarginal economic interest, and produce estimates of the long-term mean wind speed with an uncertainty which may be on the order of ±20 percent. My own observations of the vegetation at Smith-Putnam sites support this claim.

Hewson and coworkers (Hewson et al. 1977, 1979a, Wade and Hewson, 1979b) have reviewed studies of the effects of wind climatology on trees and extended Putnam and Griggs's pioneering work. In these studies, three indices of wind defor-mation were defined and "calibrated" against wind characteristics such as the mean annual wind speed, mean growing season wind speed, mean nongrowing season wind speed, and percentage of winds from the prevailing direction. They employed two species, the Douglas fir and Ponderosa pine, which are common in their Pacific Northwest study area.

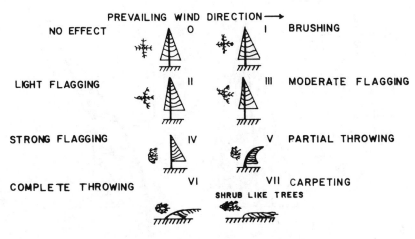

Fig. 2-21. The Griggs-Putnam index of wind deformation. From Hewson et al. (1979a).

The Griggs-Putnam index G is a rating on a subjective scale. The value of this index is assigned on the basis of the permanent bending of needles, twigs, branches, and trunks of coniferous trees as shown in Figure 2-21. The deformation ratio D measures the effect of wind on crown asymmetry and trunk deflection. The index D is defined by

$$D = \alpha/\beta + \gamma/45 \qquad \left\{ 1 \leqslant \alpha/\beta \leqslant 5 \right\} \qquad (2\text{-}64)$$

where α is the angle between the crown and the stem on the leeward side of the tree, β is the angle between the crown and stem on the windward side of the tree, and γ is the angle of inclination of the trunk from the vertical. The characteristics α, β, and γ are shown in Figure 2-22. The compression ratio C measures the effect of the wind on reaction wood. Reaction wood forms on a particular side of the tree in response to a force which displaces the tree from the vertical. For trees exposed to the prevailing winds, reaction wood forms on the leeward side of conifers and the windward side of deciduous trees. Figure 2-23 illustrates the eccentric growth pattern of a wind-deformed coniferous tree. The compression ratio measures the ratio of leeward growth to windward growth; its value may be determined by measurements on tree core samples taken from the leeward and windward sides of a tree.

The indices G, D, and C were calibrated against the wind characteristics cited earlier by using data from 24 locations which had either or both of the tree species and a year or more of wind data. The wind characteristic which showed the highest correlation with the indices was the mean wind speed. Figure 2-24 shows the relationship between the Smith-Putnam index and the mean annual wind speed. The indices were validated by using them to predict mean animal wind speeds for locations which were not employed in the calibration process. Prediction errors of 15,

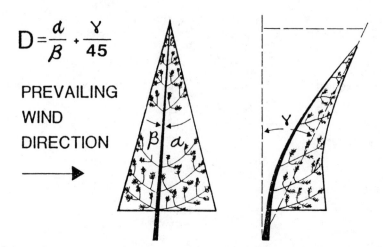

Fig. 2-22. Illustration of the deformation ratio. A photograph is taken from the direction perpendicular to the direction of maximum asymmetry. The angles are measured on the photographs. From Hewson et al. (1979a).

18, and 20 percent were cited for the indices G, D, and C, respectively (Wade and Hewson, 1979b).

There are a number of practical limitations on the use of trees as an indicator of the local wind climate. The indices studied by Hewson and coworkers must reflect both wind speed and direction. The directional variation of the strong winds during the significant period will influence the characteristics of the deformed trees. The

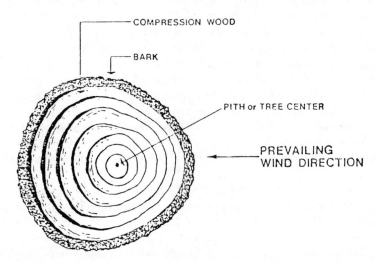

Fig. 2-23. A typical cross-section of a conifer tree growing in a windy location. From Hewson et al. (1979a).

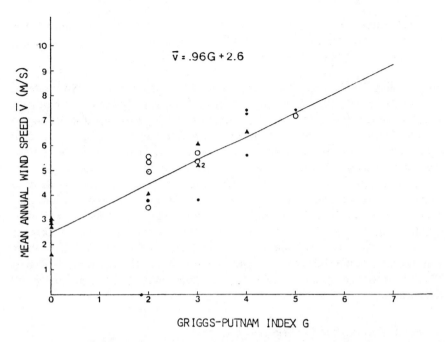

Fig. 2-24. The relationship between the Griggs-Putnam index and mean annual wind speed. Open circles represent locations where contact anemometer data were used. Closed circles represent locations with less than four years of data, and triangles denote more than four years of wind data. From Hewson et al. (1979a).

scatter in Figure 2-24 may be understood in terms of this effect alone. The response to the wind force is species-dependent. For example, Putnam and Griggs found that the white pine showed minimal flagging at a specimen height mean annual wind speed of 4.7 m/s (10.6 mph), whereas the Balsam shows comparable flagging at 7.7 m/s (17.3 mph).* A tree must be exposed to the prevailing winds in order to show the effect of the wind force. The response of trees in a continuous forest canopy will depend on the packing density. Factors other than the wind which may deform trees in a similar fashion include phototropism, ice and snow damage, and ground orientation (asymmetric rings may be found in trees growing on slopes).

Hennessey (1980) has presented a critique of the "jackknife technique" used by Wade and Hewson (1979b) to estimate prediction errors for mean annual wind speeds. He reanalyzed their data and concluded that "wind deformed trees will prove to be unreliable as well as imprecise estimators of either mean annual wind speeds or more complicated quantities such as wind power potential."

Noguchi (1979) has studied the reliability of the use of wind-deformed trees as indicators of wind direction and magnitude at 590 sites in Hawaii where the north-

---

*See Putnam's Table IX.

east tradewinds are persistent throughout the year. He found that wind patterns based on the observation of wind-deformed trees were in reasonable agreement with those based on instrumental observations.

Eolian Land Forms. Marrs and Marwitz (1978) have used LANDSAT satellite imagery and aerial photography to study the relationship between wind characteristics and eolian land forms. They and other workers have identified several important relationships between dunes and wind flow. The alignment of dunes and dune fields allows the interpreter to draw the streamlines of the horizontal flow pattern; the regular spacing of dunes indicates gravity waves in the near-surface airflow; the rate of dune movement appears to be related to the average speed of winds via an empirical equation; an analysis of sediment size yields information on the local variation of wind energy and, although dune form is not a good indicator of wind speed, certain forms are characteristic of unidirectional wind flow. Playa lakes and their associated sediment plumes are formed by wind deflation. These land forms also provide a useful indication of wind patterns. Marrs and Kopriva (1978) have mapped the arid and semi-arid areas of the conterminous U.S. which are susceptible to eolian action.

Numerical Modeling for Site Selection

Introduction. Putnam (1948) described von Kármán's numerical studies of the speedup of air over ridges. Results from this simple uniform potential flow model, together with an analysis of meteorological data, led Putnam to expect a speedup factor of 1.2 at the Grandpa's Knob test site. The measured factor based on five years of data was 0.88. The accuracy of the measured speedup factors for Smith-Putnam sites* depends upon the accuracy of the values used for the free-air wind speed at each site. The values adopted for the mean free-air wind speed at all of the Smith-Putnam survey stations were based on the height profile of the free-air wind speed at Burlington. The Burlington profile was based upon Lange's analysis of pilot-baloon data. Petterssen (1964) expressed a lack of confidence in the speedup factors determined using this approach:

In retrospect it appears that the observed speed-up factors . . . . . are much in doubt, the uncertainties being due to difficulties in determining the undisturbed wind [speed] at the level concerned . . . . . In the opinion of the writer there is really no firm observation to show that speed-up factors in excess of unity are obtainable over large mountain ridges.

The development of accurate numerical models and the acquisition of relevant measured data will help resolve this uncertainty.

---

*These factors are given in column 14 of Putnam's Table VII.

Freeman et al. (1976) have presented a useful classification of wind field models—
primitive equation models, simplified physics models, and *objective analysis models*—
which may be applied in site selection programs.

Primitive Equation Models. In the primitive equation models, the Navier-Stokes
equations of motion for fluid flow are solved. All physical processes which effect
the near surface winds are included. The SIGMET computer codes developed by
Freeman et al. (1976) and Traci et al. (1977, 1978) solve the conservation equations
for the wind components, temperature, and moisture. The equations account for
topography, advection, turbulent heat, radiation, momentum and moisture transport, and Coriolis effects. Primitive equation models are applied by performing a
three-dimensional time-dependent calculation starting from initial conditions which
are determined by the synoptic or larger scale weather regime. Many sets of initial
conditions must be examined in order to provide an adequate description of the
climatology of a mesoscale (100 km) region. Anthes and Warner (1974), Orville
(1968), and Fosberg et al. (1976) have prepared models of this type. Primitive
equation analysis may prove useful for the study of certain conditions which are
important in wind system siting. However, due to the complicated and expensive
nature of the calculations required by this class of models, it is unlikely that the
approach will find general use in site evaluation work.

Simplified Physics Models. The simplified physics models are based on the simplification of the Navier-Stokes equations. Freeman et al. (1976), Meroney et al.
(1976a,b), and Eagan (1975) have reviewed the development of models of this
type. O'brien and Hurlburt (1972) has described a general *n*-layer model based on
the Lagrangian approach to the primitive equations. The two-layer approximation
has been applied by Kasahara et al. (1965) and the one-layer approximation (known
as the shallow fluid or shallow water approximation) has been applied by Lavoie
(1972) and others. Hino (1968) has developed a model for incompressible flow
over complex terrain which assumes that the geostropic flow is uniform and that a
perturbation can be calculated in a two-layer approximation. The flow in the free
layer is incompressible and inviscid. The normal component of the flow in the
boundary layer near the surface is neglected. Hino claims that the results are as
reliable as those from conventional experiments in wind tunnels. Frost (1974) has
solved the turbulent boundary layer equations and an additional transport equation
for eddy viscosity. Jackson and Hunt (1975) and Jackson (1975) have developed
analytical solutions for the flow in an adiabatic turbulent boundary layer over a
two-dimensional hill of uniform roughness and small curvature. Jackson (1975)
suggested that the flow perturbation caused by the change in surface topography has
the same distribution as the perturbation to a uniform inviscid flow produced by
the same surface shape. This suggestion led Meroney et al. (1976a,b) and Derickson
and Meroney (1977) to develop a model based on a coupled, nonlinear system of

steady state momentum and energy equations. They employed the simplification provided by neglecting viscosity and turbulence effects. Traci et al. (1979) have verified a simplified physics, terrain conformal wind field model through detailed comparisons of model results and field data for sites on Oahu, Hawaii.

Very few comparisons have been made between results based on simplified physics models and field measurements. Frost and Nowak (1979) have compared Golding's (1955) measured speedup factors for two hills in Great Britain with results from a simple, analytical model for the flow over a low hill. The results are shown in Table 2-11. The agreement between the results amounts to approximately 10 and 4 percent for Costa Hill and Vestra Fiold, respectively. Given the model's assumptions and the additional assumptions required for its application to these sites, the agreement is good. If simplified physics models are to play an important role in the evaluation of wind turbine sites, the results from promising models must be compared with field measurements in order to identify useful approaches and approximations.

Agopian and Crow's Simplified Physics Model. Putnam recognized that effects such as atmospheric density stratification and temperature inversions which could not be modeled in wind tunnels could play an important role in wind turbine siting. Lange (1964) discussed the difficult aspects of the theoretical description of wind flow in complex terrain. He showed that the energy required for the vertical motion of air under stable atmospheric conditions was large and that this effect acts to retard the flow of air over summits. By equating the energy required to overcome the stability of the standard atmosphere with the kinetic energy of the wind, he showed that an 11-m/s wind flow would be reduced to zero during the ascension of a 1000-m obstacle.

Agopian and Crow (1978) used a simplified physics model to study the effect of atmospheric density stratification on the flowfield over isolated terrain features. The flowfield is predicted using the linearized Boussinesq wave equation for the vertical component of the wind velocity. The effect of friction, and hence the earth's rotation, is neglected. A two-layer model is employed for the atmosphere. Each layer is characterized by a constant lapse rate, i.e., temperature decrease with altitude, and a temperature jump across the interface is introduced in order to simu-

Table 2-11. Comparison of model results with Golding's measured values.[a]

Site	V_{11}/V_3[b]		V_{20}/V_3[b]	
	Measured	Calculated	Measured	Calculated
Costa Hill	1.06	1.15	1.11	1.22
Vestra Fiold	1.18	1.14	1.25	1.23

[a]From Frost and Nowak (1979).
[b]Speedup factor. The subscripts refer to the height in m.

late inversions. The freestream wind is assumed to be uniform in height far upstream of the mountain. The linearization of the equations of motion permits a solution by the Fast Fourier Transformation technique.

Calculations were performed for two-dimensional (bell-shaped) and three-dimensional (Gaussian) mountains. The two-dimensional mountain is described by

$$\eta(x) = \frac{h}{1 + (x/a)^2} \qquad (2\text{-}65)$$

where h is the height of the mountain, a is the half-width of the mountain, and x is the horizontal stream coordinate. Figures 2-25a, 2-25b, and 2-25c show the shape of the mountain η, the surface wind speed U, and the surface power flux $W((1/2)\rho U^3))$ for three different stratifications in the troposphere (the lower layer). Figure 2-25a shows results for a neutrally stable (adiabatic) troposphere where the lapse rate is

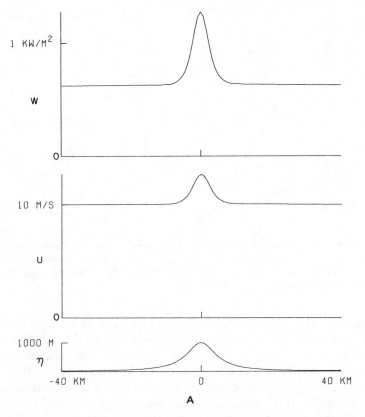

Fig. 2-25. Surface wind speed and power flux: (a) neutrally stable (adiabatic) troposphere, U_∞ = 10 m/s (b) standard troposphere, U_∞ = 10 m/s (c) isothermal troposphere, U_∞ = 10 m/s and (d) isothermal troposphere, U_∞ = 5 m/s. The mountain height h is 1 km. From Agopian and Crow (1978).

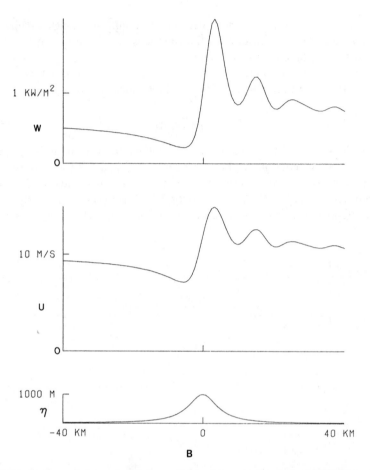

Fig. 2-25. Surface wind speed and power flux: (a) neutrally stable (adiabatic) troposphere, U_∞ = 10 m/s (b) standard troposphere, U_∞ = 10 m/s (c) isothermal troposphere, U_∞ = 10 m/s and (d) isothermal troposphere, U_∞ = 5 m/s. The mountain height h is 1 km. From Agopian and Crow (1978). (Continued)

9.8° K/km. Figures 2-25b and 2-25c show results for stable stratifications, the former for a lapse rate of 6.7° K/km (standard troposphere) and the latter for a zero lapse rate (isothermal troposphere). In these calculations, the lapse rate for the stratosphere (the upper layer) is taken as zero (isothermal stratosphere). The height of the tropopause in the freestream is 10 km and there is no temperature jump across it. The height of the mountain is 1 km and the freestream wind speed (U_∞) is 10 m/s. The direction of the airflow is from left to right and the ground temperature is 288°K. The value of the half-width of the mountain is $5h$.

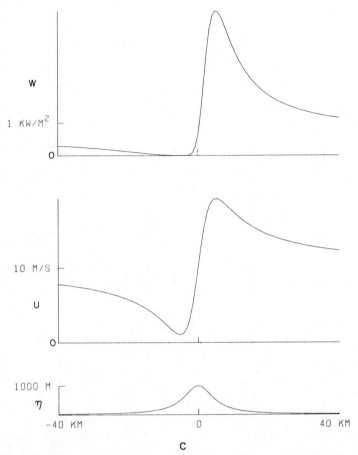

Fig. 2-25. Surface wind speed and power flux: (a) neutrally stable (adiabatic) troposphere, U_∞ = 10 m/s (b) standard troposphere, U_∞ = 10 m/s (c) isothermal troposphere, U_∞ = 10 m/s and (d) isothermal troposphere, U_∞ = 5 m/s. The mountain height h is 1 km. From Agopian and Crow (1978). (Continued)

The maximum wind speed occurs at the peak of the mountain for the case of a neutrally stable (adiabatic) troposphere. For the two stable cases, the location of the maximum wind speed moves to the lee side of the mountain. For the case of an isothermal troposphere, the wind speed falls sharply on the lee side of the mountain due to the energy required for the climb. For a freestream wind speed of 5 m/s, the wind speed falls to a negative value on the lee side for both cases of stable flow, i.e., the flow is reversed because the air near the ground does not have enough energy to surmount the obstacle. Figure 2-25d shows the results for the case of an isothermal troposphere and a freestream wind speed of 5 m/s.

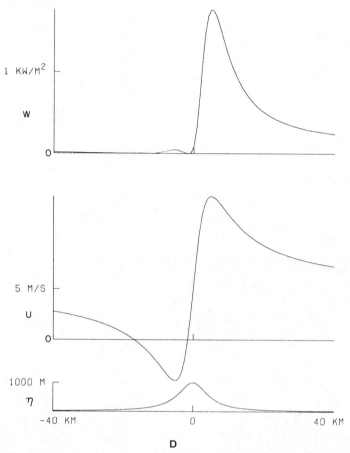

Fig. 2-25. Surface wind speed and power flux: (a) neutrally stable (adiabatic) troposphere, U_∞ = 10 m/s (b) standard troposphere, U_∞ = 10 m/s (c) isothermal troposphere, U_∞ = 10 m/s and (d) isothermal troposphere, U_∞ = 5 m/s. The mountain height h is 1 km. From Agopian and Crow (1978). (Continued)

Figure 2-26 shows an example of a resonant wave in a neutrally stratified atmosphere with an inversion layer. The height of the lower layer in the freestream is 0.7 km, the temperature jump across the layers is 10° K, and both layers are neutrally stable.

Figure 2-27 shows the maximum speedup factor S_{max} (U_{max}/U_∞) as a function of the ratio of the mountain half-width to mountain height (a/h). Figure 2-28a shows the maximum speedup factor as a function of the freestream wind speed for a mountain that is 1 km high. Results are shown for the three cases of atmospheric

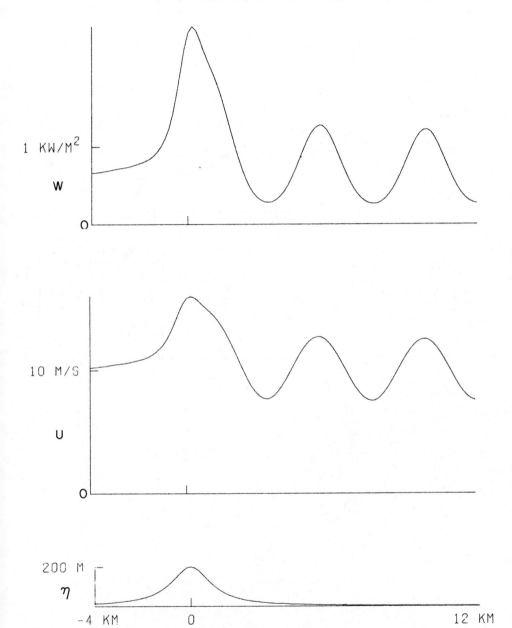

Fig. 2-26. Resonance waves in a neutrally stratified atmosphere with an inversion layer. $H = 700$ m and $\triangle T = 10°$ K. From Agopian and Crow (1978).

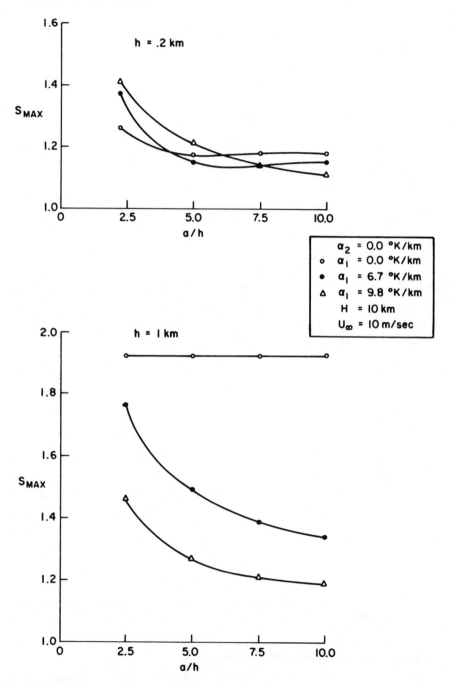

Fig. 2-27. Maximum speedup factor as a function of the ratio of mountain half-width to mountain height. From Agopian and Crow (1978).

stratification described earlier. The parameters α_1 and α_2 of the figure legend give the lapse rates for the troposphere and stratosphere, respectively, and H is the height of the tropopause. Note that S_{max} may not be relevant in terms of siting considerations based on stable stratifications since the location of the maximum is on the lee side of the mountain (see Figure 2-25). Figure 2-28a shows a decreasing relationship between the speedup factor and freestream wind speed for the isothermal troposphere, resonance behavior for the standard troposphere, and constant behavior for the adiabatic troposphere (neutral stability). Figure 2-28b shows the speedup factor as a function of freestream wind speed and location on the mountain for the standard troposphere.

Calculations were performed for the three-dimensional motion of air over a Gaussian mountain of the form

$$\eta(x,y) = h \ \exp\left\{-(x'/a)^2 - (y'/b)^2\right\} \qquad (2\text{-}66)$$

where x' and y' are the principal coordinates, h is the height of the mountain, and a and b are the lengths of the semiminor and major axes. Figure 2-29a shows the flow geometry; the ellipse indicates the contour of elevation $h \ \exp(-1)$. The freestream wind speed is uniform in the vertical direction Z. Figure 2-29b shows the contours of constant elevation for a 1-km mountain. Figures 2-29c and 2-29d show the results of flow calculations performed for the standard troposphere ($\alpha_1 = 6.7°\text{K/km}$, $\alpha_2 = 0.0°\text{K/km}$), a tropopause height of 10 km, a freestream wind speed of 8 m/s, a ground temperature of 288°K, and no temperature jump across the layers. Contours of constant surface wind speed are shown for flow inclinations of 0 and 60°, respectively. For both inclinations, the location of the wind speed maximum is approximately one mountain half-width downstream from the mountain peak.

Figure 2-30 shows the location of the maximum wind speed as a function of the aspect ratio (b/a) for the same conditions and mountain height. For flow inclinations of 60 and 90°, the maximum wind speed occurs on the side of the mountain. The occurrence of the maximum on the side of the mountain indicates considerable flow around rather than over the mountain. Additional calculations showed that trapped resonance waves occur in neutral conditions with an inversion. The maximum surface wind speed attains a value which is considerably higher than it does in the absence of an inversion. Although the maximum occurs on the lee side of the mountain, its location is closer to the peak than for the case of stable flow with no inversion.

A separation criterion was applied to the results for the flow over a three-dimensional mountain in the standard atmosphere. According to the criterion used, regions of separation may occur on both the windward and leeward sides of the mountain. On the leeward side, separation occurs just downstream of the location of the maximum wind speed.

Objective Analysis Models. These models interpolate observed wind velocity data on the basis of constraints imposed by topography and the conservation of mass.

Three-dimensional mass consistent wind field models have been developed by Seltari and Lantz (1974), Dickerson and Orphon (1975), Sherman (1975), Freeman (1976), and Knox et al. (1976). Hardy and Walton (1977) have developed a technique known as principal component analysis. This method may be used to iden-

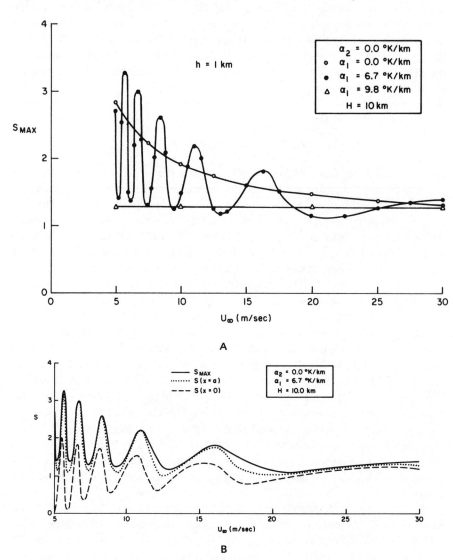

Fig. 2-28. (a) Maximum speedup factor as a function of freestream wind speed, $h = 1$ km, and (b) speedup factor as a function of freestream wind speed and location on the mountain, $h = 1$ km. From Agopian and Crow (1978).

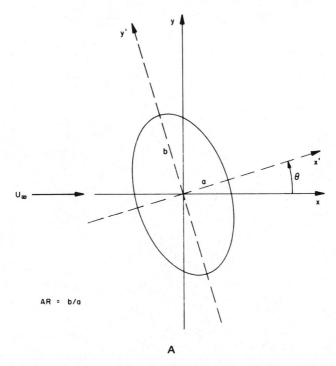

A

Fig. 2-29. (a) Three-dimensional flow geometry, (b) contours of constant elevation for a 1-km high Gaussian mountain, (c) contours of constant surface wind speed for U_∞ = 8 m/s and θ = 0°, and (d) contours of constant surface wind speed for U_∞ = 8 m/s and θ = 60°. From Agopian and Crow (1978).

tify significant wind patterns in a region and perhaps reduce the number of model calculations required to provide a useful description of regional climatology. Bhumralkar et al. (1980) have developed a similar model which allows input data sets from three to five stations to be resolved into orthogonal components along a set of eigenvectors. The solution for each eigenvector is obtained, and hourly interpolated winds are then formed from the linear combinations of these solutions.

Vukovich and Clayton (1977) have developed a statistical model which may be used to predict wind statistics at remote locations. Synoptic data and a primitive equation model are employed to obtain the parameters of the statistical model.

Physical Modeling for Site Selection

Putnam described von Kármán's studies of the airflow over models of Smith-Putnam sites in wind tunnels. The results from these studies did not convince the project's researchers that they could use these data as the basis for extrapolations to full-scale

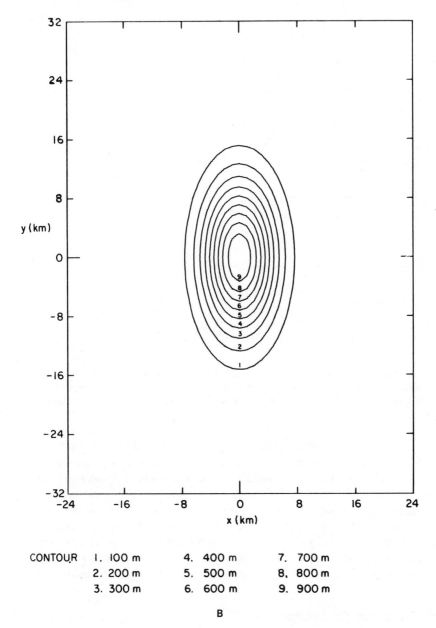

CONTOUR   1. 100 m   4. 400 m   7. 700 m
          2. 200 m   5. 500 m   8. 800 m
          3. 300 m   6. 600 m   9. 900 m

B

Fig. 2-29. (a) Three-dimensional flow geometry, (b) contours of constant elevation for a 1-km high Gaussian mountain, (c) contours of constant surface wind speed for U_∞ = 8 m/s and θ = 0°, and (d) contours of constant surface wind speed for U_∞ = 8 m/s and θ = 60°. From Agopian and Crow (1978). (Continued)

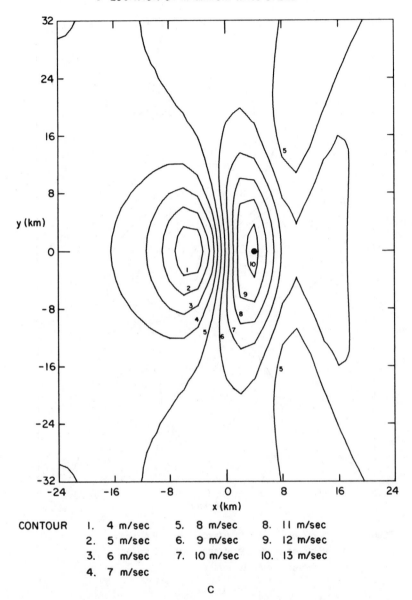

● LOCATION OF MAXIMUM WIND SPEED

CONTOUR	1.	4 m/sec	5.	8 m/sec	8.	11 m/sec
	2.	5 m/sec	6.	9 m/sec	9.	12 m/sec
	3.	6 m/sec	7.	10 m/sec	10.	13 m/sec
	4.	7 m/sec				

c

Fig. 2-29. (a) Three-dimensional flow geometry, (b) contours of constant elevation for a 1-km high Gaussian mountain, (c) contours of constant surface wind speed for U_∞ = 8 m/s and θ = 0°, and (d) contours of constant surface wind speed for U_∞ = 8 m/s and θ = 60°. From Agopian and Crow (1978). (Continued)

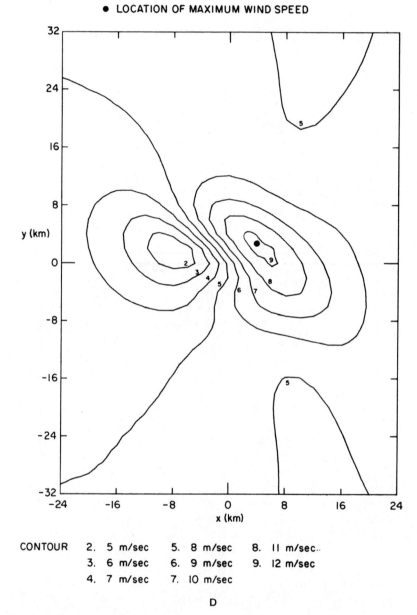

● LOCATION OF MAXIMUM WIND SPEED

CONTOUR	2. 5 m/sec	5. 8 m/sec	8. 11 m/sec
	3. 6 m/sec	6. 9 m/sec	9. 12 m/sec
	4. 7 m/sec	7. 10 m/sec	

D

Fig. 2-29. (a) Three-dimensional flow geometry, (b) contours of constant elevation for a 1-km high Gaussian mountain, (c) contours of constant surface wind speed for U_∞ = 8 m/s and θ = 0°, and (d) contours of constant surface wind speed for U_∞ = 8 m/s and θ = 60°. From Agopian and Crow (1978). (Continued)

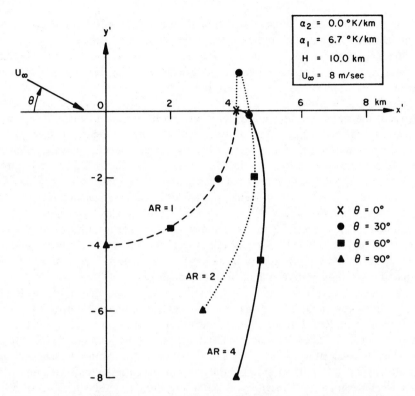

Fig. 2-30. Location of maximum surface wind speed as a function of mountain aspect ratio and direction of the freestream. From Agopian and Crow (1978).

in nature, e.g., the wind tunnel result for the speedup factor for Mount Washington was 1.3 whereas the actual value was thought to be 1.5, and the effect of reverse flagging of Balsams on the lee side of Pond Mountain could not be reproduced in the wind tunnel. The relevance of the wind tunnel data was diminished because the approach profiles were not simulated, i.e., the approach vertical wind speed profiles may have had power law coefficients near zero.

Meroney et al. (1976a,b) have reviewed the status of laboratory stimulation experiments. Simularity refers to a relation between model and full-scale flow systems which ensures that proportional alterations of the units of length, mass, and time will scale measured quantities from one system to the other. For airflow over mountains, geometrical, dynamical, and thermal simularity must be achieved. The full-scale region must be small enough (~ 150 km) to ensure that the effect of Coriolis acceleration is negligible. Other effects which are neglected in the simulation of atmospheric flow are the variation of hydrostatic pressure with elevation, condensation and evaporation processes, compressibility, and the unsteady behavior of the full-scale winds. Despite these limitations, useful results have been obtained from laboratory simulation studies of the airflow in the troposphere.

Meroney et al. (1976a,b), Bouwmeester et al. (1978), and Meroney (1979) have performed wind tunnel measurements on numerous triangular and sinusoidal hill models. Static pressure holes, surface hot wires, and preston tubes were used to monitor the flow under neutral conditions. Figure 2-31 shows typical measurements of the static pressure, speed, speedup, and turbulence intensity distribution for a 1:4 triangular "hill." The pressure distributions for such models show that the speed is reduced near the foot of the hill; strong adverse pressure gradients can produce local separation near the foot of steep hills. Figure 2-32 shows contour plots of the speed for flow over 1:4 triangular and sinusoidal hills. The speedup is somewhat greater for the triangular-shaped hill. The sine-shaped hill has a wider range near the crest where the speed distribution remains independent of the horizontal (x) position. Figure 2-33 shows wind tunnel results for the speedup ratio obtained using a 1:4 triangular hill model with and without upstream roughness. The results are preliminary in nature due to problems concerning the technique of matching the start of a hill with the roughness.

Meroney et al. (1978) have simulated the wind characteristics of the Rakaia River Gorge region of New Zealand. The validity of the laboratory simulation methods was studied through a limited field measurement program. On two spring days that were selected for strong neutral airflow through the valley, three teams surveyed up to 27 sites on either side and within the gorge. Some of the conclusions reached by the researchers were: (1) the atmospheric shear layer flowing over complex terrain can be reproduced by physical modeling to within the limit in which the atmosphere produces stationary results, (2) the individual day-to-day wind speeds found over complex terrain were reproduced by physical modeling to sample correlation coefficient levels of from 0.70 to 0.76, and (3) the individual day-to-day site wind directions over complex terrain were reproduced by physical modeling to sample correlation coefficient levels of from 0.65 to 0.67. More such studies are needed in order to establish the range of conditions over which physical modeling can reproduce an accurate description of full-scale flow.

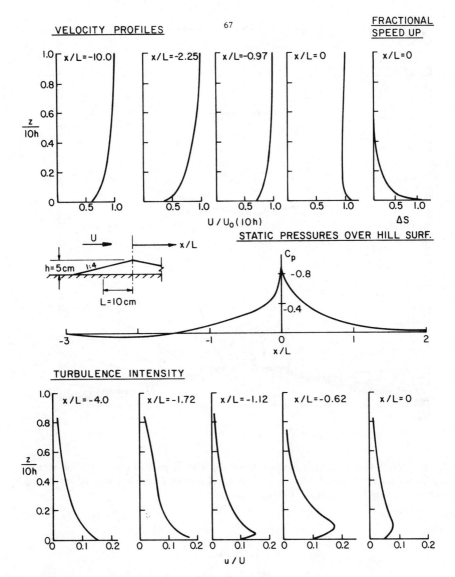

Fig. 2-31. Speed. static pressure, and turbulence measurements over a 1:4 triangular hill. From Meroney et al. (1976a).

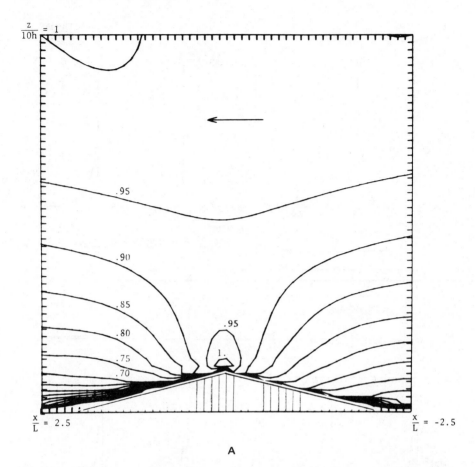

A

Fig. 2-32. (a) Speed contours (nondimensionalized with respect to U_0 $(10h)$) for a 1:4 triangular hill model, $U_0 = 15.24$ m/s, and (b) speed contours for a 1:4 sine hill model, $U_0 = 15.24$ m/s. From Meroney et al. (1976a).

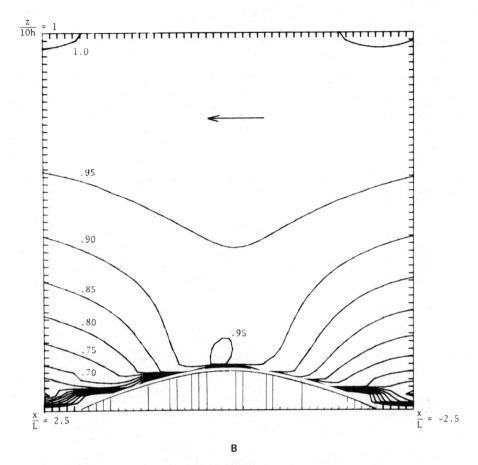

B

Fig. 2-32. (a) Speed contours (nondimensionalized with respect to U_0 $(10h)$) for a 1:4 tri-angular hill model, U_0 = 15.24 m/s, and (b) speed contours for a 1:4 sine hill model, U_0 = 15.24 m/s. From Meroney et al. (1976a). (Continued)

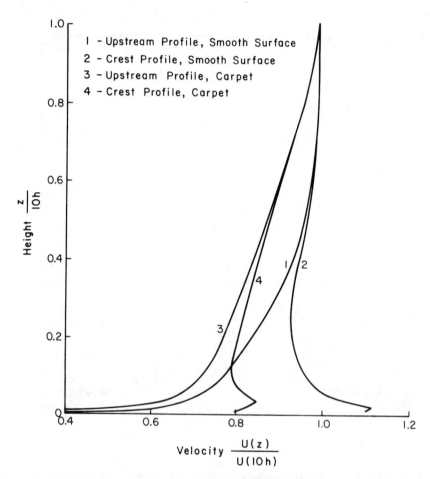

Fig. 2-33. Speed profiles above crest with different roughness features for a 1:4 triangular hill model, U_0 = 9.14 m/s. From Meroney et al. (1976a).

3
Wind System Performance

GENERAL DESCRIPTION

An estimate for the mean power output ($\langle P \rangle$) from a wind turbine generator may be computed using the equation

$$\langle P \rangle = \int_0^\infty P(V) f(V) \, dV \tag{3-1}$$

where $P(V)$ is the power output at the blade center* wind speed V, and $f(V)$ is the wind speed probability distribution function corresponding to blade center elevation. The actual output will be different from the result given by this simple equation since a number of significant factors were not considered in its derivation. These factors include the profile of the horizontal component of wind velocity, the directional variability of wind flow, and turbulence. Nevertheless this equation can provide a useful approximate description of wind system performance. Putnam used this equation to describe the performance of the 1250-kW test unit at Grandpa's Knob.

The wind turbine generator will begin to produce power at the cut-in wind speed V_o. The power increases monotonically from zero at V_o to the maximum or rated power (P_r) at the rated wind speed (V_r). Above the rated wind speed, the power remains constant until the machine's cut-out wind speed (V_c) is reached. The machine's blades are feathered or designed to stall at wind speeds above the cut-out value to prevent damage to the machine. The limits of zero and infinity for the integral of equation 3-1 may thus be replaced by V_o and V_c, respectively.

Approximate power output functions may be constructed on the basis of design values of V_o, V_r, V_c and P_r. Figure 3-1 shows approximate power output versus wind speed for the U.S. Department of Energy's Mod-0A, Mod-1, and Mod-2 wind machines. These machines are of the horizontal-axis type and operate at a fixed number of revolutions per minute (rpm). Above the rated wind speed, the power is maintained at a constant level up to the cut-out wind speed by varying the pitch of

---

*The hub height and blade center line elevations are employed for horizontal- and vertical-axis machines, respectively.

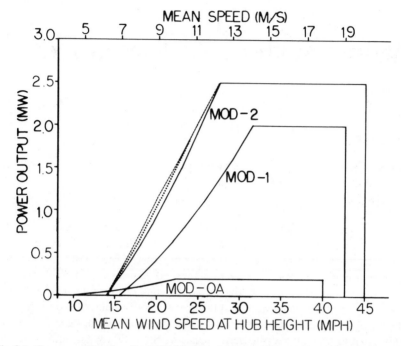

Fig. 3-1. Power outputs for the DOE Mod-0A, Mod-1, and Mod-2 wind turbines. The lines labeled 2 were calculated on the basis of the machine parameters given in Table 3-1 and equations 3-11. Two additional curves are shown for the Mod-2. The line labeled 1 shows linear variation between cut-in and rated wind speeds. The line labeled 3 shows the cubic variation of equations 3-18 based on the parameters in Table 3-1 and C_{pm} = .382, C_{pr} = .334, and V_m = 9.83 m/s (22 mph) at hub height.

the blades, or blade tips, toward the feathered position. The power output curves shown were calculated on the basis of the machine parameters given in Table 3-1 and on models which will be described later in this chapter. Accurate power output functions may be established during operational tests on the basis of field measurements of power and wind speed under various conditions. Figure 3-2 shows measured and model values of the alternator output versus wind speed for the Mod-0A machine at Clayton, New Mexico.

Wind speed probability distribution functions commonly used in performance studies are measured distributions obtained under various conditions, Rayleigh distributions, and reference sets of Weibull distributions.

The convolution and integration of the power output and wind speed probability functions yields a mean power output which is nonlinear, i.e., the value of $\langle P \rangle \neq P(\langle V \rangle)$ where $\langle V \rangle$ is the mean wind speed for a given averaging period.

Table 3-1. Machine characteristics for the Mod Series.

Type	Hub Height m (ft)	Rotor Diameter m (ft)	Rated Power MW	Wind Speeds at Hub Height (mph)			C_{pr}[d]
				m/s V_o	V_r	V_c	
Mod-0A[a]	30.5(100)	38.1(125)	0.2	4.25(9.5)	10.0(22.4)	17.9(40.0)	0.29
Mod-1[b]	42.7(140)	61.0(200)	2.0	6.97(15.6)	14.1(31.5)	19.0(42.6)	0.40
Mod-2[c]	61.0(200)	91.4(300)	2.5	6.26(14.0)	12.3(27.5)	20.1(45.0)	0.33

[a]Neustadter (1979).
[b]Poor and Hobbs (1979b) provide the cut-in, rated, and cut-out wind speeds referenced to an elevation of 30 ft; the values are 11, 24.6, and 35 mph, respectively. Justus and Mikhail's equations 2-48 and 2-49 were used to obtain the values of the cut-in and cut-out wind speeds at hub height. The value of the rated wind speed was calculated using $C_{pr} = 0.40$, standard sea level density (1.225 kg/m^3), and equation 3-8. This procedure was recommended by R. Donovon of the NASA-Lewis Research Center (1978, private communication).
[c]Douglas (1979).
[d]The values of C_{pr} cited here were calculated using the rotor diameters and rated wind speeds of columns (3) and (6), standard sea level density 1.225 kg/m^3 and equation 3-8.

PERFORMANCE COEFFICIENTS

The efficiency of a wind turbine generator is often described in terms of a performance coefficient which describes the relationship between the power in the wind, the power at the rotor shaft, and the system electrical output. The rotor or aerodynamic power coefficient (C_p^R) is the decimal fraction of the power contained in a stream of wind flowing perpendicular to the rotor disk which is transformed to power at the rotor shaft. Rotor theory and experimental observations have shown that the rotor power coefficient depends upon the tip speed ratio λ and the blade pitch angle β. The tip speed ratio is

$$\lambda = R\omega/V \qquad (3\text{-}2)$$

where R is the rotor radius, ω is rotor angular speed, and V is the wind speed. The blade pitch angle is defined as the angle between the chord of the blade (at the $0.75R$ station if the blade has pitch) and the plane of the cone of rotation of the rotor (Wilson and Lissaman, 1974). The rotor power coefficient is defined by the equations

$$C_p^R (\beta, \lambda) = P^R(V)/P_w(V) \qquad (3\text{-}3)$$

and

$$P_w(V) = (1/2) \rho A V^3 \qquad (3\text{-}4)$$

where $P^R(V)$ is the power at the rotor shaft for wind speed V, ρ is atmospheric density, A is the area swept by the blades, and $P_w(V)$ is the power of the wind

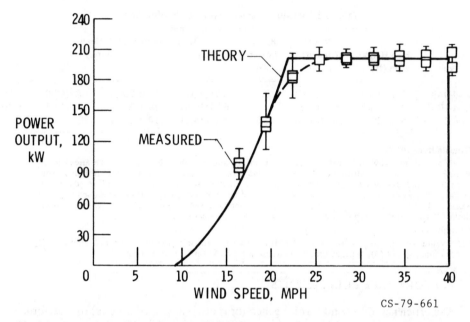

Fig. 3-2. Comparison of measured and predicted power for the Mod-0A 200 kW wind turbine at Clayton, New Mexico. From Glascow and Robbins (1979).

which flows at normal incidence to the swept area. The wind speed V corresponds to that of the undisturbed wind, upwind of the machine, at blade center elevation. Figure 3-3 shows how C_p^R varies with β and λ for a typical high speed rotor. For the case of a variable rpm machine which adjusts β and λ to maintain C_p constant at the maximum value (C_{pm}) the power output may be expressed as

$$P^R(V) = C_{pm} \, P_w(V). \tag{3-5}$$

Betz (1920) used an idealized laminar-flow model to derive an upper limit of $16/27 = 0.593$ for the fraction of the kinetic energy of the wind impinging on the area swept by a wind turbine which the machine can convert to useful power. Inglis (1979) has shown how deviations from the idealized model involving rotational kinetic energy of the downwind stream and turbulent mixing from outside the boundaries of the idealized stream can either increase or decrease the available power. The Betz limit is therefore not a strict upper limit. Wilson and Lissaman (1974) and Shepherd (1978) have reviewed the relationship between the aerodynamic theory of rotors and wind system performance.

The system power coefficient (C_p) is the fraction of the power contained in the wind which is transformed to electrical power $(P(V))$.

$$C_p(V) = P(V)/P_w(V). \tag{3-6}$$

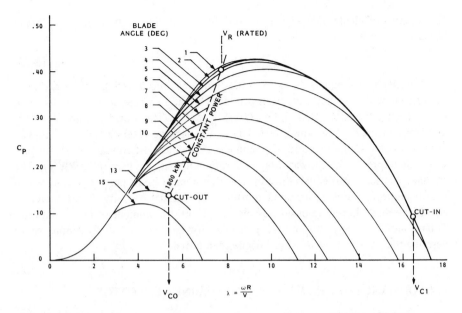

Fig. 3-3. Variation of power coefficient with blade angle and tip speed ratio for the Mod-1 wind turbine generator. From Poor and Hobbs (1979a).

The value of this performance coefficient depends upon β, λ, and other machine parameters γ. If a complete and accurate description of C_p (β, λ, γ) is unavailable, an approximate function may be developed by using a model and certain significant machine parameters.

Equations 3-4 and 3-6 may be used to express equation 3-1 as

$$\langle P \rangle = 1/2 \ \rho A \int_{V_o}^{V_c} C_p(V) \ V^3 \ f(V) \ dV. \tag{3-7}$$

CAPACITY FACTOR

The capacity or plant factor C_f is defined as the ratio of the mean power ($\langle P \rangle$) to the rated power. The rated power may be expressed in terms of the system performance coefficient at rated wind speed (C_{pr}) and the rated wind speed:

$$P_r = C_{pr}(1/2)\rho A \ V_r^3 \tag{3-8}$$

The capacity factor may be obtained by dividing equation 3-7 by equation 3-8.

$$C_f = \frac{\langle P \rangle}{P_r} = \int_{V_0}^{V_c} \frac{C_p(V)}{C_{pr}} \left(\frac{V}{V_r}\right)^3 \ f(V) \ dV. \tag{3-9}$$

The amount of electrical energy produced during a period of duration t_0 (E_{t_0}) may be calculated using the equation:

$$E_{t_0} = t_0 \langle P \rangle = t_0 \, C_f P_r. \tag{3-10}$$

The capacity factor equals the ratio of the energy produced during a given period to the energy which would have been produced if the machine operated at its rated power during the entire period, i.e., $C_f = E_{t_0}/(t_0 P_r)$. The capacity factor therefore provides a measure of the energy efficiency of a machine in terms of its rated power. Putnam used a similar measure of machine performance, the annual specific output, which he expressed as kilowatt hours per rated kilowatt per year (kWh/kW/y). The annual specific output equals the equivalent number of hours per year during which the machine operates at its rated output. The annual capacity factor therefore equals the annual specific output divided by 8760, the number of hours in a year. Today, annual energy outputs (E_A) or capacity factors are usually employed to describe the performance of wind turbine generators.

PERFORMANCE OF THE MOD-2 WIND TURBINE GENERATOR

Boeing (1978) has described the design study of the performance of the Mod-2. Figure 3-4 shows the factors which are responsible for system performance losses at the rated wind speed. The Mod-2 is a horizontal-axis machine with an upwind two-bladed rotor which has a diameter of 91.4 m (300 ft). This machine produces its rated power of 2.5 MW at a hub height wind speed of 12.3 m/s (27.5 mph). An artist's drawing of a group (farm) of Mod-2 wind turbine generators is shown in Figure 3-5. The power in the wind at the rated wind speed for standard sea level density (1.225 kg/m^3) is 7481.7 kW. The power at the rotor shaft is 2715.3 kW and thus the value of the rotor performance coefficient $C_p^R (V_r)$ is 0.363. The value of the system power coefficient $C_p(V_r) = C_{pr}$ is 0.334, i.e., 2500/7481.7.

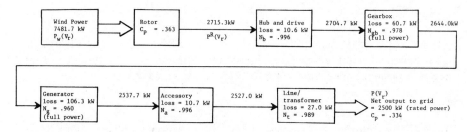

Fig. 3-4. System performance losses for the Mod-2. The results correspond to the rated wind speed of 12.3 m/s (27.5 mph) at hub height elevation and standard sea level air density. The factors N give efficiencies for the system components. Adapted from Boeing Engineering and Construction Company (1978).

Fig. 3-5. An artist's drawing of a farm of Mod-2 wind turbines. Courtesy of NASA Lewis Research Center.

The hub and drive losses are the result of friction in the bearings of the Mod-2's teetered rotor, rotor shaft bearings, and a high speed flexible coupling. For operation at partial power, this loss is approximately constant. The loss of power in the gearbox is due mainly to mesh losses. The high efficiency, a product of epicyclic gearbox design, remains approximately constant down to one-half of the rated power. The gears are compact and the gear tooth contact speeds are low so as to produce low mesh losses. The synchronous generator requires power due to windage, field excitation, resistance, and internal friction losses. The efficiency of the generator remains nearly constant for a power reduction to one-half of the rated value. The assessory loss includes power used by electric motors which drive hydraulic pumps, instrumentation, lubrication pumps, cooling fans, exterior and interior lighting, and maintenance functions. The assessory loss corresponds to the mean power consumed. This power is supplied by either the wind turbine generator or the utility grid. The line resistance and transformer losses are constant for partial power operation.

Figure 3-6 shows design estimates for the rotor and system power coefficients of the Mod-2 as a function of wind speed. The rotor power coefficient shown here represents an attempt to account for all aerodynamic effects including orientational losses associated with operational tolerances on yaw control. The peak value of 0.415 occurs at a design wind speed of 8.94 m/s (20 mph) which was chosen to maximize the annual energy capture associated with a reference wind speed probability distribution function. The design wind speed is primarily a function of the rated power and only secondarily of the wind speed probability distribution function. The maximum system power coefficient (C_{pm}) of 0.382 occurs at a wind speed (V_m) of 9.83 m/s (22 mph) (Douglas, 1979, Boeing, 1978).[*]

Figure 3-6 shows the reduction in power coefficient associated with drive train and electrical losses and spillage. Spillage refers to the loss of the power in the wind which is spilled by blade pitch control in order to obtain the constant rated power at wind speeds between the rated and cut-out values. Figure 3-7 shows the schedule for the variation of blade pitch with wind speed for the Mod-2. The solid and dashed lines show the operational and maximum power schedules, respectively. Between the cut-in and rated wind speeds, the blade pitch is varied in two steps.

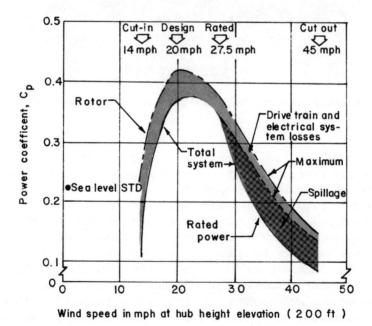

Fig. 3-6. The rotor and system power coefficients for the Mod-2. Adapted from Douglas (1979).

---

[*]The maximum system power coefficient was obtained from Douglas (1979) and the wind speed V_m was read from Figure 3-6 (Boeing, 1978).

POWER OUTPUT FUNCTION MODELS

Quadratic Model for Fixed RPM Machines

Justus et al. (1976a) have developed a simple method for the construction of power output functions. The power output function is defined by the equations:

$$P(V) = 0 \text{ for } V < V_o,$$
$$P(V) = A + BV + CV^2 \text{ for } V_o \leqslant V \leqslant V_r,$$
$$P(V) = P_r \text{ for } V_r \leqslant V \leqslant V_c \tag{3-11}$$

and

$$P(V) = 0 \text{ for } V > V_c$$

where A, B, and C are constants. The constants are evaluated by solving the equations:

$$A + BV_o + CV_o{}^2 = 0,$$
$$A + BV_r + CV_r{}^2 = P_r, \tag{3-12}$$

and

$$A + BV_x + CV_x{}^2 = (V_x/V_r)^3 P_r$$

where

$$V_x = (V_o + V_r)/2. \tag{3-13}$$

The constants A, B, and C are therefore functions of the characteristic wind speeds V_o, V_r, and V_c, and the rated power. The quadratic equation for the power over the wind speed range $V_o < V < V_r$ represents a reasonable approximation for fixed rpm machines. Power output functions constructed using this method are shown in Figure 3-1 for the DOE Mod-0A, Mod-1, and Mod-2 machines as the curves labeled 2. The parameters A, B, and C were calculated using the values of the characteristic wind speeds given in Table 3-1.

The Effect of Air Density. The values of the characteristic wind speeds which are specified for a given air density must be modified in order to construct a power output function which is appropriate for a different value of the air density. The cut-in and rated wind speeds depend upon the power in the wind. In order to achieve equal power at a reference site and a particular site, the values of the density (ρ_0) and cut-in or rated wind speed (V_o) for the reference site must be related to the values of the density (ρ) and cut-in or rated wind speed (V) for the particular site by the equation.

$$\rho_0 V_o{}^3 = \rho V^3. \tag{3-14}$$

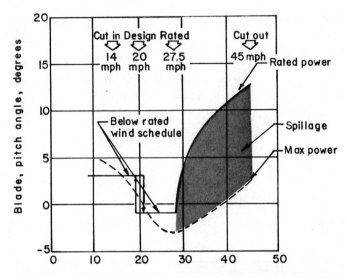

Wind speed in mph at hub height elevation (200ft)

Fig. 3-7. The schedule for the variation of blade pitch with wind speed for the Mod-2. The solid and dashed lines show the operational and maximum power schedules, respectively. Adapted from Boeing Engineering and Construction Company (1978).

The cut-in and rated wind speeds therefore require a cube root correction

$$V = V_o \, (\rho_0/\rho)^{1/3} \tag{3-15}$$

The value of the cut-out wind speed is governed by wind pressure. For equal pressures, the equation

$$\rho_0 \, V_o{}^2 = \rho \, V^2 \tag{3-16}$$

must be satisfied and a square root correction is required

$$V = V_o \, (\rho_0/\rho)^{1/2} \tag{3-17}$$

Manufacturers usually reference the characteristic wind speeds to the standard sea level density (1.225 kg/m^3).

Cubic Model for Fixed RPM Machines

Justus and Mikhail (1978a) have developed a power law model for fixed rpm machines which incorporates additional parameters: the system maximum power coefficient (C_{pm}) and corresponding design wind speed (V_m), and the system power coefficient at rated wind speed (C_{pr}). The power law is defined by the equations:

$$\begin{aligned} P(V) &= 0 \text{ for } V < V_o, \\ P(V) &= [C_{pm}P_r/(C_{pr}V_r{}^3)] \ [(1-A'+B') \, V^3 \\ &\quad + (2A'-3B') \, V_m V^2 + (3B'-A') \, V_m{}^2 V - B'V_m{}^3] \\ &\quad \text{for } V_o \leqslant V \leqslant V_r, \end{aligned} \tag{3-18}$$

$$P(V) = P_r \text{ for } V_r \leqslant V \leqslant V_c,$$

and

$$P(V) = 0 \text{ for } V > V_c$$

where the constants A' and B' are determined by the conditions

$$P(V_o) = 0$$

and

$$P(V_m)/P_r = C_{pm} V_m^3/(C_{pr}V_r^3). \tag{3-19}$$

The power output curve for the Mod-2 based on values of C_{pm}, C_{pr}, and V_m of 0.382, 0.334, and 9.83 m/s (22.0 mph), the characteristic wind speeds of Table 3-1, and equations 3-18 is shown in Figure 3-1. The curve based on the quadratic model and a line which corresponds to linear variation between the cut-in and rated wind speeds are shown for comparison. The curve for the cubic power law model lies closer to the line corresponding to linear variation than the curve for the quadratic power law model.

Quadratic Model for Variable RPM Machines

When operating conditions permit the rotor rpm to be varied, the rpm can be adjusted so as to hold the power coefficient at its maximum value, C_{pm}. Justus and Mikhail (1978a) have also developed a power law model for such variable rpm machines. The power output function is defined by the equations:

$$
\begin{aligned}
P(V) &= 0 \text{ for } V < V_o, \\
P(V)/P_r &= a + b\,(V/V_r) + c(V/V_r)^2 \text{ for } V_o \leqslant V \leqslant V_o + \triangle, \\
P(V)/P_r &= (V/V_r)^3 \text{ for } V_o + \triangle \leqslant V \leqslant V_r, \\
P(V) &= P_r \text{ for } V_r \leqslant V \leqslant V_c,
\end{aligned}
\tag{3-20}
$$

and

$$P(V) = 0 \text{ for } V > V_c$$

where $V_o \leqslant V < V_o + \triangle$ is the speed range near the cut-in speed in which power train losses yield $C_p\,(V) < C_{pm}$. The coefficients a, b, and c are defined by the conditions:

$$
\begin{aligned}
P(V_o) &= 0, \\
P(V_o + \triangle) &= [(V_o + \triangle)/V_r]^3,
\end{aligned}
\tag{3-21}
$$

and

$$b + 2c\,[(V_o + \triangle)/V_r] = 3\,[(V_o + \triangle)/V_r]^2$$

where the conditions require that $P(V)$ approaches zero as V approaches V_o and that $P(V)$ is continuous with and has continuous slope with the V^3 variation which applies in the range $V_o + \triangle < V < V_r$.

POWER OUTPUT CALCULATIONS

Effect of Wind Speed Probability Distribution

Figure 3-8 shows the relationship between estimates of the annual energy output for the Mod-0A, Mod-1, and Mod-2 wind machines and the annual mean wind speed referenced to the appropriate hub height elevation for each machine. The outputs were calculated on the basis of the quadratic power output functions shown in Figure 3-1, Rayleigh and Weibull wind speed probability distribution functions, and equation 3-1. Results are presented for different wind speed probability distribution functions to illustrate effects associated with the shapes of distributions which yield the same mean wind speed. For each machine, the output is shown for Rayleigh reference wind speed distributions and two sets of Weibull reference distributions which were described in Chapter 2: the set based on Putnam's Figure 37, and

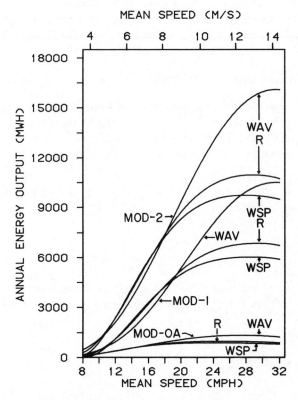

Fig. 3-8. Estimate of the annual energy output for the Mod-0A, Mod-1, and Mod-2 versus mean wind speed at hub height elevation. The results were obtained using quadratic power functions based on the parameters of Table 3-1. The results corresponding to the Weibull Smith-Putnam, Rayleigh, and Weibull average variability wind speed reference distributions are labeled WSP, R, and WAV, respectively.

the set employed by Justus et al. (1978) to describe average variability. The variation of the Weibull shape parameter with mean wind speed is different for each set of functions. In the former set, the shape parameter decreases as the mean wind speed increases; the reverse is true for the latter set. For the Rayleigh and Weibull Smith-Putnam distributions, the outputs of the Mod machines increase with the mean wind speed according to a linear approximation with good accuracy over a range of about 3.13 m/s (7 mph). The linear range is significantly larger for the Weibull average variability reference distributions. Over the linear range, an increase in the mean wind speed from 14 to 16.2 mph at hub height, i.e., an increase of 1 m/s, will produce additional output in the amounts of 29, 43, and 57 percent for the Mod-0A, Mod-2, and Mod-1, respectively. The order 0A, 2, and 1 corresponds to increasing values of the rated wind speed; thus, at 6.26 m/s (14 mph), the increase in mean wind speed has the greatest impact upon the performance of the Mod-1, a machine designed for applications in high wind regions.

Figure 3-9 shows the capacity factor for the Mod-0A, Mod-1, and Mod-2 versus the mean wind speed at hub height. Results for the quadratic power output law

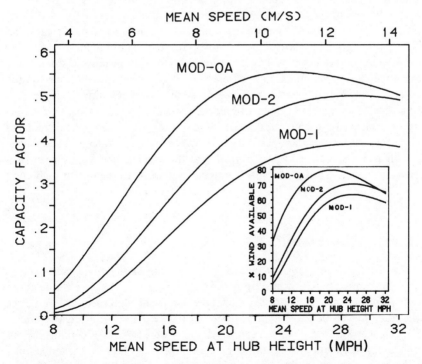

Fig. 3-9. Capacity factor for the Mod-0A, Mod-1, and Mod-2 versus mean wind speed at hub height elevation. The values were calculated using the quadratic output law and Rayleigh wind speed distributions. The insert shows the percentage of time that the wind speed lies in between the cut-in and cut-out values.

and Rayleigh wind speed distributions are shown. The curves are shifted to the right in the series Mod-0A, Mod-2, and Mod-1 due to increasing values of the rated wind speed. The insert shows the percentage of the time that the wind speed lies between the cut-in and cut-out values.

Figure 3-10 provides a description of the performance of the Mod-2 in terms of the variation of capacity factor with mean wind speed at hub height. Results are shown for the quadratic power law model and the wind speed reference distributions employed to obtain the results shown in Figure 3-8. For mean wind speeds up to approximately 8.05 m/s (18 mph) at hub height, the results for the three reference distributions are in good agreement. For wind speeds in excess of this value, the capacity factor given by the Smith-Putnam reference distributions is significantly smaller than that for the Rayleigh distributions, and the capacity factor given by the Weibull average variability distributions is significantly larger than that for the Rayleigh distributions.

Figure 3-11 shows the capacity factor for the Mod-2 versus the Weibull shape parameter k. The capacity factors were calculated by using the quadratic power function of Figure 3-1. Each curve shows the variation of capacity factor with k for a specified value of the mean wind speed. For given values of k and $\langle V \rangle$, the scale factor c was evaluated by using equation 2-13. For low and high mean wind speeds, the capacity factor is especially sensitive to the variation of k. The insert of Figure 3-10 shows the capacity factor for the Mod-2 versus mean wind speed based on the quadratic power function and reference wind speed distributions for the Justus et al. (1978) low, average, and high variability cases. Significant differences between the values appear at low and especially high mean wind speeds. The differences are small over the practical intermediate wind speed range. The differences at low mean wind speeds are academic due to economic considerations involving siting. Figures 3-8, 3-10, and 3-11 suggest that the shape of the wind speed distribution may have a significant impact on wind system performance for high wind sites.

Effect of Power Output Model

The values of calculated capacity factors will depend upon the model employed to develop the approximate power output function. Figure 3-12 shows the relationship between the capacity factor for the Mod-2 and mean wind speed at hub height for the three power output functions shown in Figure 3-1 and Rayleigh reference wind speed distributions. At a mean wind speed of 8.94 m/s (20 mph) at hub height, the linear function result is 1.2 percent larger than that for the cubic function; the quadratic power function result is 4.5 percent smaller than that for the cubic function. The quadratic function provides the most conservative estimate at all values of the mean wind speed. The three results converge at high mean wind speeds.

Relationship Between Vertical Wind Shear and Performance

The performance of a wind turbine generator is often described in terms of a figure which shows either the annual energy output or the capacity factor versus the annual mean wind speed corresponding to a height of 30 ft above ground level. The out-

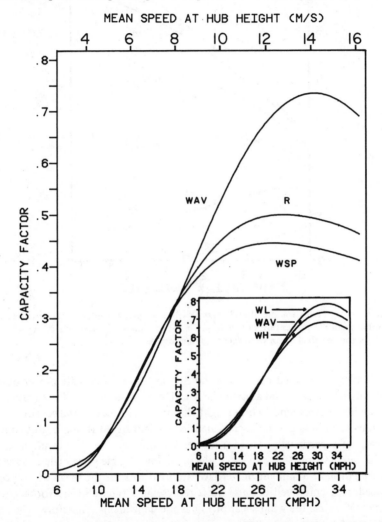

Fig. 3-10. Capacity factor for the Mod-2 versus mean wind speed at hub height elevation. The results were obtained using the quadratic power law function. The results corresponding to the Weibull Smith-Putnam, Rayleigh, and Weibull average variability wind speed reference distributions are labeled *WSP*, *R*, and *WAV*, respectively. The insert shows the capacity factor versus mean wind speed for the quadratic power law function and the Weibull low, average, and high variability wind speed reference distributions; these results are designated by *WL*, *WAV*, and *WH*, respectively.

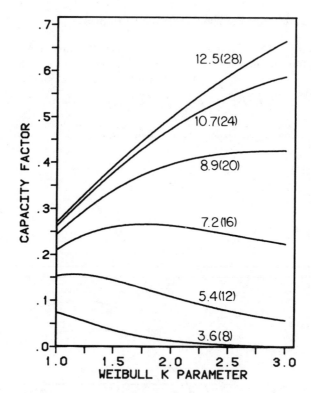

Fig. 3-11. Capacity factor for the Mod-2 versus Weibull shape parameter k. The results correspond to the quadratic power output function. Each curve shows the variation of capacity factor with k for the specified mean wind speeds in m/s (mph).

puts are usually computed using equation 3-1, i.e., an approximate power output function for the machine and a particular set of wind speed probability distribution functions which correspond to hub height elevation. The mean wind speeds for the distributions at hub height are then scaled to a height of 30 ft above ground level on the basis of a particular model for the vertical wind profile. For example, equations 2-48 and 2-49 or 2-52 and 2-53 which employ variable power law exponents, or a simple constant exponent power law, are used to scale the mean wind speeds. This procedure is employed because 30 ft represents a typical height for measured wind data. Of course the wind shear profile is site dependent. Therefore, when this method is used to describe wind system performance, it is important to specify the wind shear profile which is employed to scale the mean wind speeds. For example, Figure 3-13 shows the capacity factor for the Mod-2 based on the quadratic power output function and the Rayleigh wind speed distribution versus the mean wind

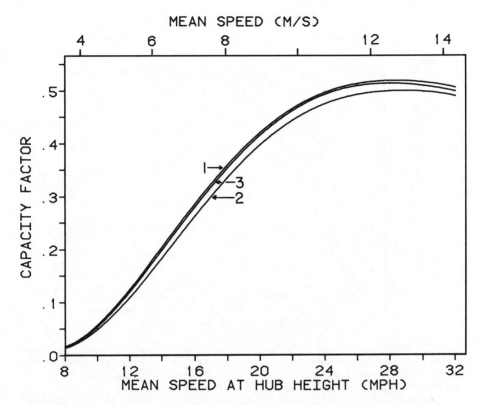

Fig. 3-12. Capacity factor for the Mod-2 versus mean wind speed at hub height. The results correspond to the Rayleigh reference wind speed distributions and the three power output functions shown in Figure 3-1. The results for linear, quadratic, and cubic power functions are labeled 1, 2, and 3, respectively.

speed which corresponds to 30-ft elevation above ground level. Three schemes were employed to scale mean wind speeds from hub height to 30 ft. Results are shown for the variable exponent power law based on surface roughness exponents of 0.15 and 0.37. Results are also shown for the constant one-seventh power law. The large variation of the output at a given mean wind speed with the surface roughness exponent emphasizes the importance of vertical wind shear in relation to wind system performance. For a mean wind speed of 6.26 m/s (14 mph) at 30 ft, the values of the capacity factor for the surface roughness exponents 0.15 and 0.37, and the constant one-seventh power law are approximately 0.27, 0.40, and 0.33, respectively. Due to the site dependent nature of the wind shear profile, a description of performance in terms of mean wind speeds at hub height elevation is preferable to one in terms of mean wind speeds at 30 ft.

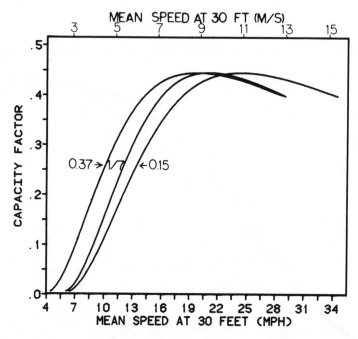

Fig. 3-13. Estimate of the capacity factor for the Mod-2 versus mean wind speed at 30-ft elevation. Three schemes were used to scale mean wind speeds from hub height to 30 ft. Results are given for the variable exponent power law based on surface roughness exponents of 0.15 and 0.37. Results are also shown for the constant one-seventh power law.

Additional Factors Which Influence Performance Estimates

Anemometer Error. Justus (1978a) has reviewed the treatment of anemometer error. If $\overline{V}$ is the true horizontal wind speed in the direction of the wind azimuth line and $\overline{V}_{obs}$ is the observed value, the relationship between these values can be described by

$$\overline{V} = \overline{V}_{obs} (1 - e_u)(1 + e_v)(1 - e_w)(1 - e_{dp}) \qquad (3\text{-}22)$$

where the "u error" (e_u), "v error" (e_v), and "w error" (e_w), are associated with the turbulence response characteristics of the sensor, and the data processing error, e_{dp}, is due to totalizing the wind along the instantaneous direction rather than the resultant direction over the averaging interval. The bar indicates that the wind speeds correspond to a short-term averaging period, e.g., one minute. The u, v, and w designations correspond to the direction of the mean wind, the perpendicular to the mean wind in the horizontal plane, and the vertical. Propeller and vane anemometers are affected by the v error; cup anemometers are not. Usually, the integrated number of turns of a propeller or cup anemometer are taken over a short-term averaging period and the direction is averaged over the same period. The effective speed for the specified interval will be overestimated if it is assumed that the speed represents the mean flow in the average direction. This effect is called data processing error. The factor $(1 - e_v)$ provides a good estimate of the data processing error. For a cup anemometer, an approximate description of error is given by

$$\overline{V} = \overline{V}_{obs} \left[1 - 0.5 \left(\sigma_u/\overline{V}_{obs}\right)^2\right] \left[1 - 0.5 \left(\sigma_w/\overline{V}_{obs}\right)^2\right]$$
$$x \left[1 - 0.5 \left(\sigma_v/\overline{V}_{obs}\right)^2\right] \qquad (3\text{-}23)$$

where σ_u, σ_v, and σ_w are the longitudinal, horizontal, and vertical rms gust magnitudes, respectively. A typical value for the ratios $\sigma_u/\overline{V}_{obs}$, $\sigma_w/\overline{V}_{obs}$, and $\sigma_v/\overline{V}_{obs}$ is 0.2; for this case, equation 3-23 reduces to $\overline{V} = 0.94 \, \overline{V}_{obs}$.

Wind Gusts. If one-minute average wind speeds $(\overline{V})$ are to be used in performance studies involving a wind turbine with a response time of several seconds, the effect of the short period gusts about the one-minute average on the one-minute average output must be accounted for. Since $\langle v \rangle$ (the average wind speed about the one-minute average) is zero, the expressions for $\langle V \rangle$, $\langle V^2 \rangle$, and $\langle V^3 \rangle$ for the one-minute averaging period are

$$\langle V \rangle = \langle \overline{V} + v \rangle = \overline{V}, \qquad (3\text{-}24)$$

$$\langle V^2 \rangle = \langle (\overline{V} + v)^2 \rangle = \overline{V}^2 + \langle v^2 \rangle = \overline{V}^2 \left[1 + (\sigma_u/\overline{V})^2\right], \qquad (3\text{-}25)$$

and

$$\langle V^3 \rangle = \langle (\overline{V} + v)^3 \rangle = \overline{V}^3 + 3 \langle v^2 \rangle \overline{V} + \langle v^3 \rangle \approx \overline{V}^3 \left[1 + 3 (\sigma_u/\overline{V})^2\right] \qquad (3\text{-}26)$$

where the term $\langle v^3 \rangle$ may usually be neglected. Equations 3-25 and 3-26 may be used in place of the terms V^2 and V^3 of equations 3-11, 3-18, and 3-20 to account for the effects of turbulence. The effects of turbulence add to the power whereas anemometer corrections subtract from it. Under typical operating conditions, accurate one-minute average outputs may be obtained by neglecting both errors. Putnam's Figure 72 shows the output of the Smith-Putnam test unit versus time. It is clear that gusts can have a significant impact upon the performance of a large wind turbine.

Kirchhoff (1979) studied the dynamic response of a 25-kW wind turbine generator. He measured the power spectral density of the horizontal gustiness of the wind and the transfer function between this gustiness and the generator power output. He showed that the system responded like a low pass filter with a cut-off frequency of approximately 0.03 Hz. The wind turbine was able to capture about 70 percent of the power available in the turbulence.

Frost (1977) has described three methods of introducing turbulence into analyses of rotor response to wind fluctuations: the spectral approach, the discrete gust approach, and the turbulence simulation technique. Shepherd (1978) has given a review of the effect of turbulence and gusts on wind turbine design.

Wind Shear. The horizontal component of wind velocity varies with height due to vertical wind shear. Justus et al. (1976b) have developed a simple model based on the power law profile with constant exponent which gives the ratio P/P_0 of the shear averaged power flux (P) to the uniform flow power flux (P_0) through the turbine disk at hub height. Figure 3-14 shows the ratio P/P_0 as a function of the power law exponent n and the relative rotor radius R/Z_h where R is the rotor radius and Z_h is the hub height elevation. For $n \leqslant 0.35$, the effect of wind shear is small. However for larger n and large values of R/Z_h, the effect is significant.

Air Density. Power output laws developed on the basis of equations 3-11, 3-18, and 3-20 may be adjusted for changes in air density by scaling the characteristic wind speeds which are used to formulate the output function; see equations 3-15 and 3-17. Alternatively, the unadjusted power output function $P(V)$ may be multiplied by the density ratio ρ/ρ_o, and modified values of the cut-in, rated, and cut-out wind speeds may be determined, e.g., the modified cut-in wind speed is the wind speed for the modified function which gives $P(V) = 0$.

Combined Effects. Power output functions having the forms of equations 3-11 and 3-18 which include corrections for anemometer error, turbulence, wind shear, and density may be expressed as

$$P(\overline{V}) = (P/P_0)\,(\rho/\rho_0)\Big\{A + B\overline{V} + C\overline{V}^2\,[1 + (\sigma_u/\overline{V})^2]\Big\} \qquad (3\text{-}27)$$

and

$$P(\overline{V}) = [C_{pm}P_r/(C_{pr}V_r^{\,3})]\,(P/P_0)\,(\rho/\rho_0)$$
$$\times \Big\{[1 - (A' + B')]\,\overline{V}^3\,[1 + 3\,(\sigma_u/\overline{V})^2]$$
$$+ (2A' - 3B')\,V_m\overline{V}^2\,[1 + (\sigma_u/\overline{V})^2]$$
$$+ (3B' - A')\,V_m^{\,2}V - B'V_m^{\,3}\Big\} \qquad (3\text{-}28)$$

respectively, *where $\overline{V}$ is the one minute average wind speed obtained by applying an appropriate anemometer correction and a vertical wind shear correction which scales from anemometer to hub height, and the constants A, B,* and *C (or A', B',* and *C')* are based on characteristic wind speeds which correspond to the standard sea level density ρ_0. Equations 3-27 and 3-28 give power outputs for wind speeds between the cut-in and cut-out values. In general, these equations will not give zero at V_0 or P_r at V_r. The conditions $P(V) = 0$, and $P(V_r) = P_r$ are used to define new values of the cut-in and rated wind speeds. Negative values of $P(\overline{V})$ obtained from the modified function are set equal to zero, and values of $P(\overline{V}) > P_r$ are set equal to P_r. If the values of A, B, and C (or A', B', and C') are calculated on the basis of modified characteristic wind speeds for air density ρ (given by equations 3-15 and 3-17), then the factor (ρ/ρ_0) is not required in equations 3-27 and 3-28. The adjustment which defines modified values of the cut-in and rated wind speeds is also required for this

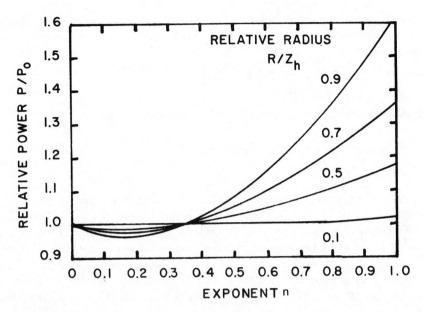

Fig. 3-14. Circular integrated relative power P/P_0 for different values of the relative radius R/Z_h versus the vertical power law exponent. Adapted from C.G. Justus, *Wind and Wind System Performance*. (1978a).

correction procedure. The cut-out wind speeds must be modified to satisfy the pressure condition

$$\rho \overline{V_c}^2 \, [1+(\sigma_u/\overline{V_c})^2] = \rho_0 \, \overline{V}_{c,0}^2 \, [1+(\sigma_u/\overline{V}_{c,0})^2] \tag{3-29}$$

where $V_{c,0}$ is the cut-out wind speed for the reference density ρ_0. Equations 3-20 for variable rpm machines may be modified in a similar fashion to take account of the combined effects on performance.

Linear Regression Models

Justus and Mikhail (1978a) and Justus (1978a) have developed linear regression models which describe wind system performance. They performed calculations for a variety of wind turbine designs using equation 3-1 and various reference wind speed probability distributions. They analyzed the results in terms of a multiple linear regression equation of the form

$$C_f = \beta_0 + \beta_1 (V_0/V_r) + \beta_2 (C_{pr}/C_{pm}) + \beta_3 (\langle V \rangle/V_r)^2 \\ + \beta_4 (\langle V \rangle/V_r)^3 \tag{3-30}$$

where β_0 through β_4 are the regression constants. Standard statistical tests of the regression results demonstrated that this equation gave a good fit over a wide range of representative characteristic wind speeds and reference wind speed distributions. Over the restricted range of mean wind speeds $0.5 \leqslant (\langle V \rangle / V_r) \leqslant 1$, the simpler linear regression form

$$C_f = \gamma_0 + \gamma_1 (V_0 / V_r) + \gamma_2 (C_{pr} / C_{pm})$$
$$+ \gamma_3 (\langle V \rangle / V_r) \tag{3-31}$$

was shown to reproduce the calculated results with good accuracy.

WIND TURBINE ARRAY PERFORMANCE

Definitions

A wind turbine generator farm or cluster is comprised of a number of individual wind turbine generator units. Due to their spatial proximity, e.g., < 10 km, the power output from each unit may be considered to be in phase with the power from all of the units, i.e., each unit has the same power versus time profile. Figure 3-5 shows an artist's drawing of a farm of Mod-2 wind turbines. A wind turbine generator array is comprised of a set of units or farms which are dispersed over an area which is large enough to ensure that wind diversity will affect the array power.

The Significance of Array Studies

An understanding of the performance of wind turbine generator arrays is important for economic and operational assessments of electrical generation. A time series description of array performance is required for (1) economic cost modeling analysis, (2) the evaluation of capacity credit (the amount of conventional generating capacity which can be displaced by wind power capacity), (3) loss of load probability analysis (the determination of the effect of wind power generation on the probability that a given generating system will fail to meet the demand for electrical power), (4) the evaluation of the optimum mixture of generation units of different types, (5) the evaluation of the maximum penetration of wind power capacity in a given system, and (6) the evaluation of operating reserve requirements.

Array Power Availability

The performance characteristics of large arrays of wind turbines have been studied by Justus and coworkers (Justus and Hargraves, 1977, Justus, 1978a,b, Justus and Mikhail, 1978b,c). In one study, array simulations were developed on the basis of wind data from 28 and 25 National Weather Service airport sites in the New England-Middle Atlantic and Central U.S. Regions, respectively. One-minute average wind speeds adjusted to hub height elevation on the basis of equations 2-48 and 2-49 were used to calculate outputs for wind turbine designs corresponding to rated powers of 500, 1125, and 1500 kW. The power output functions for these machines were constructed using the quadratic model. At a given time the mean array output $\langle P \rangle$ was calculated by averaging over the set of National Weather Service sites which constituted the array. The mean output is given by

$$\langle P \rangle = \sum_{i=1}^{n} P(V_i)/n \qquad (3\text{-}32)$$

where $P(V_i)$ is the power for a given machine, V_i is the wind speed at site i, and n is the number of sites in the array. Of course the output per generator is the same for n machines at n sites as it is for n sites with x machines at each site, i.e., n farms consisting of x machines. Statistical analyses were performed for each month of a five-year time series. Results were also obtained as a five-year average by averaging corresponding monthly data over the five yearly sets of data.

Figure 3-15 shows the five-year average single site and array power output frequency distribution for the Central U.S. region. This figure illustrates the improvement in power availability which may be achieved by dispersing wind turbines over a large area. For the wind turbine rated at 1125 kW, an array power level of 200 kW per generator has an availability of 90 percent, i.e., 10 percent cumulative probability. The power level will exceed 200 kW 90 percent of the time for the array, whereas a power level of 200 kW per generator for the single site case has only 62 percent availability, i.e., 38 percent cumulative probability. For a per generator power level of 100 kW, the array gives 99 percent availability versus 71 percent for the single site case. The two dashed curves show the frequency distributions for zero wind speed correlation and 100 percent correlation, i.e., for the cases of a

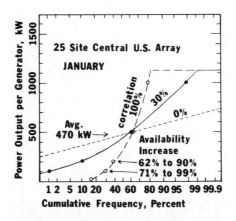

Fig. 3-15. Power output frequency distribution for 1125-kW wind turbine. The average power output is 470 kW for the array and individual site cases. One-hundred percent correlation indicates individual site, 30 percent correlation is the observed average correlation of the actual array, and 0 percent correlation corresponds to a theoretical uncorrelated array. From Justus (1978b).

theoretical uncorrelated array and one machine (or one farm), respectively. The actual spatial correlation for the Central U.S. array was 30 percent.

Power Availability with Storage. Justus (1978b) performed an approximate analysis of the availability of array power with storage. The statistics of array power return times were evaluated at power levels of 100 and 200 kW per generator. The array power return time $t_R(P)$ is defined as the time required for the array power output to return above the level P after once going below that value. The array power return time is sometimes referred to as a run duration. The array power is expressed as power per generator; thus, the power of an array of n generators is nP. Return times and the monthly distribution of return times for different power levels were evaluated. The results for corresponding months during a five-year period were combined. If the storage system is designed to provide P kW per generator for T hours, the total energy stored in kWh for n generators is nPT. If an array power return time $t_R(P)$ has a 90 percent cumulative probability then 90 percent of the time when the array output power falls below nP, the stored energy (nPT) will be sufficient to maintain the power at nP until the array power returns to the nP level. The treatment assumes that the storage device was fully charged at the beginning of the discharge period. A more accurate analysis of power availability with storage requires a time series simulation of storage status since a period during which stored energy is required could begin with a partially charged storage device.

Figure 3-16 shows the observed frequency distribution of return times at a power level of 200 kW for the Central U.S. array in July. For the 500-kW wind turbine, the power availability of the array will be 90 percent when the storage system can

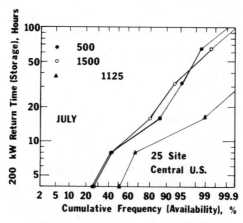

Fig. 3-16. Frequency distribution (availability) for various return times (amount of storage) for 500-, 1125-, and 1500-kW wind turbines in the Central U.S. array for July. From Justus (1978b).

supply n times 200 kW of power for 20 hours, i.e., 4000 kWh per generator. For an availability of 95 percent, approximately 32 hours of storage is required.

Table 3-2 gives the availability at power levels of 100 and 200 kW per generator without storage ($T = 0$) and with storage based on wind data for seven coastal sites in the New England region (a subset of the larger New England-Middle Atlantic array). Values are presented for individual site and array cases for January. The results corresponding to zero storage at 10 kW indicate that the "improved" statistics of arrays yields power at a low level with very high availability. For the array case with storage ($T \neq 0$), Table 3-2 gives the storage times required to produce power levels of 100 and 200 kW per generator with 90, 95, and 99 percent availability. For the 500-kW wind turbine, the availability at the 200 kW per generator level for the individual site case without storage is 56 percent. An availability of 90 percent for the array case at the same power level is achieved with 29 hours of storage. The data of Table 3-2 can also be interpreted in terms of the probability of a power lull (once begun) lasting as long as time T if no storage is available, i.e., 10, 5, and 1 percent lull probabilities correspond to 90, 95, and 99 percent power availability (R) values.

Table 3-2. Availability of power levels of 10, 100, and 200 kW per generator with and without storage for seven site coastal array or individual site cases. T is the storage time in hours and R is the probability of available power in percent.

	Rated Power of Generator (kW)	500		1500	
		T (h)	R (%)	T (h)	R (%)
10 kW/Gen.	Individual Site	0	78	0	60
	Array	0	98	0	91
100 kW/Gen.	Individual Site	0	64	0	54
	↑	0	83	0	75
	Array	20	90	22	90
		27	95	28	95
	↓	47	99	44	99
200 kW/Gen.	Individual Site	0	56	0	49
	↑	0	65	0	62
	Array	29	90	28	90
		39	95	38	95
	↓	66	99	64	99

From Justus (1978b).

Seasonal Variation of Array Performance. The seasonal variations of the monthly mean power output for 500-, 1125-, and 1500-kW machines in the Central U.S. array are shown in Figure 3-17. The output from the 1500-kW machine shows the greatest sensitivity to seasonal variations since this machine requires high winds for effective operation. The annual average and seasonal maximum (winter) and minimum (summer) power outputs for the simulated arrays in the New England and Central U.S. regions are shown in Table 3-3. The ranges are based on five years of data. The annual average capacity factors for the 500, 1125, and 1500 kW machines are approximately 0.40–0.50, 0.30–0.40, and 0.15–0.25. The decreasing values of the capacity factor in the series follows the increasing trend in values of the rated wind speed: 9.4, 10.4, and 13.1 m/s for the 500-, 1125-, and 1500-kW machines, respectively. The values of the power output for the 1500 kW machine in New England range from a summer low of about one-half of the annual average to a winter high of about 1.4 times the annual average. The seasonal variations for the 500- and 1125-kW units are less severe. The seasonal ranges amount to approximately 0.75 to 1.2 times the annual average.

The simulation studies showed that the availability of power may be significantly improved by dispersing wind turbines in large arrays. Much higher availability levels may be achieved by adding storage capacity to the electrical network. However, a tentative analysis of the diurnal and seasonal variation of wind power levels indicated that significant peak load displacement could be achieved by wind turbine generators, i.e., due to high afternoon wind power levels which match demand from air conditioners and steady high wind power levels during the winter.

Array Power Output Model. Figure 3-18 shows wind speed probability distributions for Pacific Coast sites (Justus and Mikhail, 1978c). Distributions are shown

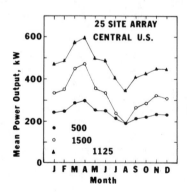

Fig. 3-17. Seasonal variation of monthly mean power output for 500-, 1125-, and 1500-kW machines for the Central U.S. array. The means are based on five years of data. From Justus (1978b).

Table 3-3. Annual average and seasonal maximum and minimum power outputs for three wind turbines. The range of values includes variations from moderate winds (inland arrays) to good winds (coastal array) in New England and good winds in the Central U.S.

Region	Wind Turbine Power Rating (kW)	Seasonal Maximum Power (kW)	Annual Average Power (kW)	Seasonal Minimum Power (kW)
New England	500	240–290	190–240	140–190
	1500	350–490	240–340	130–190
Central U.S.	500	310–340	240	140–160
	1500	540–610	320–330	110–150
	1125	630–690	460–470	250–290

From Justus (1978b).

based on observed wind speeds and the simulation method for the array and individual site cases. Although the mean values of the wind speed, wind power, and machine output are the same for the array and individual site cases, the distributions for these quantities are significantly different for the two cases. The array distributions are generally narrower than those for individual sites. For an array, low and high values of the wind speed, wind power, and machine output are less probable and values of the wind speed, wind power, and machine output near the average are more probable than for the individual site case.

Figure 3-18 also shows the distribution for an array model developed by Justus and Hargraves (1977). Given the observed mean wind speed at hub height $\langle V \rangle_1$ and standard deviation σ_1 for a single "representative" site, the statistical behavior of the winds across an array of arbitrary size may be estimated. For an array consisting of n sites characterized by a wind speed average spatial correlation $\langle \rho \rangle$, the mean array wind speed at hub height $\langle V \rangle_n$ and standard deviation σ_n are given by

$$\langle V \rangle_n = \langle V \rangle_1 \tag{3-33}$$

and

$$\sigma_n = \sigma_1 \left\{ [1+(n-1)\langle \rho \rangle]/n \right\}^{1/2} \tag{3-34}$$

where equation 3-34 represents a generalization to the case of non-zero correlation, i.e., $\sigma_n = \sigma_1/n^{1/2}$ for independent ($\langle \rho \rangle = 0$) sites. The values of $\langle V \rangle_n$ and σ_n may then be used to obtain the Weibull wind speed probability distribution for the array.

Justus and Mikhail (1978b) have improved this method. The value of σ_n is calculated instead by using the equation

$$\sigma_n = 1.11 \, \sigma_0 \left\{ [1+(n-1)\langle \rho \rangle]/n \right\}^{1/2}. \tag{3-35}$$

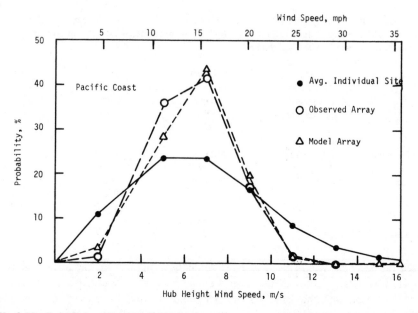

Fig. 3-18. Graphical comparison between the wind speed distribution observed at the average single site, observed for the array, and calculated using the model parameters for the Pacific Coast region array. The model parameters are: $\langle V \rangle_0 = \langle V \rangle_n = 6.59$ m/s, $\sigma_0 = 3.02$ m/s, $\langle \rho \rangle = 0.19$, and $n = 14$. From Justus and Mikhail (1978c).

where σ_0 is interpreted as the root mean square value of the standard deviations σ_i for the individual sites

$$\sigma_0 = [\sum_{i=1}^{n} \sigma_i^2 / n]^{1/2}. \tag{3-36}$$

The regression constant 1.11 accounts for 97 percent of the variance among observed and theoretical values of σ_n / σ_0, i.e., where the theoretical values are given by equation 3-35. Using equations 2-15 and 3-35, the Weibull parameters for the array distribution are given by

$$k_n = (\sigma_n / \langle V \rangle_n)^{-1.086}$$

$$= (1.11 \, \sigma_0 / \langle V \rangle_0)^{-1.086} \left\{ [1 + (n-1) \langle \rho \rangle] / n \right\}^{-0.543} \tag{3-37}$$

and

$$c_n = \langle V \rangle_0 / (1 + 1/k_n) \tag{3-38}$$

where $\langle V \rangle_0$ is the average array wind speed. The quantities σ_0 and $\langle V \rangle_0$ define a representative site.

The average spatial cross-correlation $\langle \rho \rangle$ is calculated from the hourly or three hourly wind speed time series for the n sites which constitute the array. The cross correlation between site i and site j is

$$\rho_{ij} = \langle (v_i - \langle v \rangle_i)(v_j - \langle v \rangle_j) \rangle /$$
$$[\langle (v_i - \langle v \rangle_i)^2 \rangle \langle (v_j - \langle v \rangle_j)^2 \rangle]^{1/2} \tag{3-39}$$

where $v_i(t)$ and $v_j(t)$ are the time series wind speeds for sites i and j, respectively, and $\langle v \rangle_i$ and $\langle v \rangle_j$ are the corresponding monthly mean wind speeds. The average cross-correlation is given by the expression

$$\langle \rho \rangle = \sum_{ij} \sum \rho_{ij} / N \tag{3-40}$$

where $N = n(n - 1)/2$ is the number of pairs of sites. Figure 3-18 shows good agreement between the results for the observed and model array cases.

If the average cross-correlation is unknown, its value may be estimated on the basis of the mean array size and the results from available regional studies of cross-correlation. If sites i and j are separated by a distance r_{ij}, the average distance between site pairs (the mean array size) is

$$\langle r \rangle = \sum_{ij} \sum r_{ij} / N. \tag{3-41}$$

Figure 3-19 shows the relationship between average cross-correlation and average distance for the Pacific Coast array. This correlation profile is thought to be anomalously low due in part to the effects of complex terrain.

Given the parameters k_n and c_n which define the Weibull wind speed distribution for the array, the mean array power for a given machine and other array statistical information may be readily calculated. For simplicity, the array power output $P(V_n)$ can be taken to be a linear function of the hub height array wind speed V_n

$$P(V_n)/P_r = a + b \ (V_n/V_r). \tag{3-42}$$

Equation 3-42 may be recognized as equation 3-31 with

$$a = \gamma_0 + \gamma_1(V_0/V_r) + \gamma_2(C_{pr}/C_{pm}), \tag{3-43}$$
$$b = \gamma_3, \tag{3-44}$$

and

$$V_n = \langle V \rangle.$$

The array capacity factor is

$$C_{fn} = \langle P \rangle_n / P_r = \int_0^\infty (P(V_n)/P_r) f_w(V_n) \, dV_n \tag{3-46}$$

where $\langle P \rangle_n$ is the mean array power and $f_w(V_n)$ is the Weibull probability distribution function for the array.

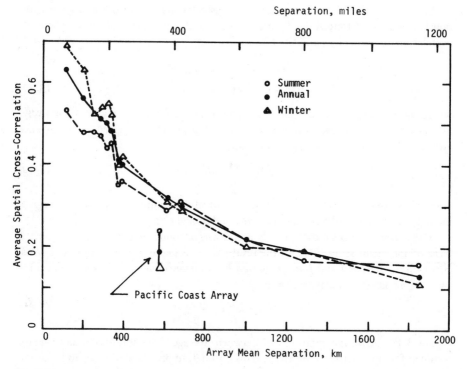

Fig. 3-19. Average spatial cross correlation versus array mean separation for the Pacific Coast array. From Justus and Mikhail (1978c).

The probability that the array power output is between the power limits P_i and P_j is

$$\text{prob}\,(P_i \leqslant P \leqslant P_j) = \exp\,[-(V_i/c_n)^{kn}] - \exp\,[-(V_j/c_n)^{kn}] \qquad (3\text{-}47)$$

where V_i and V_j are the array wind speeds corresponding to P_i and P_j, namely

$$V_i = [(P_i/P_r)-a]\,V_r/b \qquad (3\text{-}48)$$

and

$$V_j = [(P_j/P_r)-a]\,V_r/b. \qquad (3\text{-}49)$$

Figure 3-20 shows the probability in percent that the array capacity factor lies in intervals of width $0.1\,\langle P \rangle_n/P_r$ versus $\langle P \rangle_n/P_r$ for three model arrays. Model array results are shown for three values of the average spatial correlation: 0.06, 0.13, and 0.20. The value 0.13 corresponds to a model which describes a Continental array with good accuracy (Justus and Mikhail, 1978c). Decreasing values of the spatial correlation yield narrower probability distributions, i.e., increased probability for

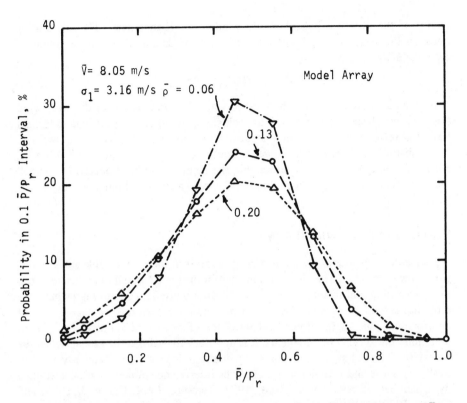

Fig. 3-20. Probability per 0.1 $\langle P\rangle_n/P_r$ interval versus $\langle P\rangle_n/P_r$ for three model arrays. From Justus and Mikhail (1978c).

power levels near the mean array value and decreased probabilities for low and high power levels.

Time Series Model for Array. Justus (1978a) has developed a model which allows the simulation of output on an hour-by-hour basis. The Weibull distribution probability that the wind speed V_1 at a representative array site is less than an arbitrary value V is

$$P(V > V_1) = \exp\left[-(V_1/c_1)^{k_1}\right] \qquad (3\text{-}50)$$

where k_1 and c_1 are the Weibull parameters for the representative site. The corresponding probability for the array wind speed V_n is

$$p(V > V_n) = \exp\left[-(V_n/c_n)^{k_n}\right]. \qquad (3\text{-}51)$$

The simulation model assumes that the array wind speed at a given time is the speed which would have a probability in the array distribution which is equal to the probability of the observed individual site wind speed at that time. The relationship between the array and representative site wind speeds may therefore be found by equating 3-50 and 3-51.

$$V_n = c_n \, (V_1/c_1)^{k_1/k_n} \qquad (3\text{-}52)$$

A time series for V_1 may thus be used to generate a time series for the array wind speed V_n which may then be used to evaluate array power outputs and additional statistical information. For example, Justus (1978a) used this method to evaluate the probability of array wind power output changes over various time intervals. Such information is useful for evaluating the ability of utility "operating reserve" units to "smooth" wind power fluctuations on an hour-by-hour and day-by-day basis.

PERFORMANCE OF WIND FARMS

It is advantageous to group wind turbine generators in farms in suitable areas of high wind power density. The close proximity of units lowers installation, interconnection, operation, and maintenance costs. Given a limited area with high winds, the machines should be "packed" so as to minimize the performance degradation of interior units due to the downwind influence of one or more units. In order to determine the spacing requirements, it is necessary to evaluate the rate at which the energy of the wind stream flowing through a farm is renewed horizontally and vertically by shear and turbulence effects. The interference problem has been studied by a number of researchers (Reed, 1974, Templin, 1974, Crafoord, 1975, and Garate, 1977). Different working assumptions were employed in each study and differences involving the relationship between farm efficiency and unit separation were significant.

The General Electric Wind Energy Mission Analysis researchers (Garate, 1977) employed a model which used momentum theory to determine far wake distances and widths as a function of speed deficiency at the core of the wake. A model which describes the dissipation blade tip vortices was used to determine the length of the wake which is stabilized by vortex flow. This length was used as a minimum boundary of the wake influence and no attempt was made to ascertain the effects of vortex flow on the delay of turbulent mixing and speed recovery. The worst wake condition occurs at the rated wind speed. For wind speeds above the rated value for a given turbine, power is spilled; thus, the momentum through the rotor increases and the speed defect decreases. The analytical model assumes that the rotor achieves maximum power extraction. The calculations were performed for a horizontal-axis machine rated at 1500 kW with a nominal rotor diameter of 200 ft. Other significant machine characteristics which were employed in the model study were a tip speed ratio of 10, a rotor speed of 40 rpm, and a rotor operating lift coefficient of 1.0.

The vortex flow model yielded a minimum separation distance of 3000 ft, i.e., 15 rotor diameters. The momentum flow model was used to evaluate the group efficiency of a given number of staggered machines under conditions of steady wind flow. The results are shown in Figure 3-21. The results emphasize the importance of minimizing the number of rows of wind turbines aligned in a given direction, i.e.,

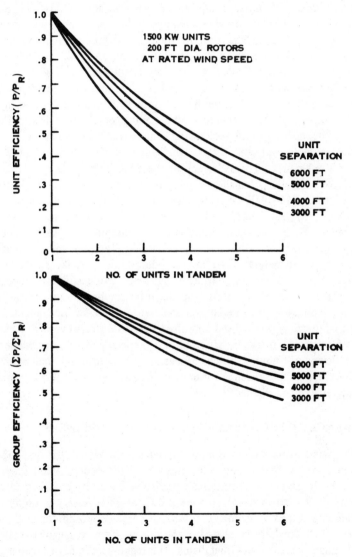

Fig. 3-21. Wake interference effects on unit and farm efficiency. From Garate (1977).

for a separation distance of 3000 ft, the unit efficiency is 80 percent for a unit in the second row and 45 percent for a unit in the third row.

The General Electric researchers analyzed the relationship between area density and farm efficiency. The assumptions employed were (1) no more than three units can be aligned in any given direction, (2) the spacing between units is 3000 ft, (3) the units consist of horizontal-axis machines (1500 kW/200 ft), and (4) the wind rose is symmetrical. Of course, if a given site has prevailing winds, the geometric matrix should be designed to provide maximum farm efficiencies. For example, if the prevailing winds always come from one direction, the most efficient farm would consist of a line of units perpendicular to that direction. A simple analysis based on these assumptions showed that square and rectangular matrices give the highest power/area densities whereas semicircular or triangular matrices provide the highest farm efficiencies. The maximum number of units which can be located in a given area without exceeding the three-unit limit for any given direction is 13. The units form a rectangle. The 13-unit rectangular matrix was analyzed in terms of farm efficiency as a function of wind direction. The results are shown in Figure 3-22 in the form of a polar plot. The efficiency of this farm for the case of an omnidirectional wind rose may be obtained by integrating the polar plot; the result is 90 percent for a 13 x 1.5 = 19.5-MW farm contained in an area of 2.5 square miles.

Crafoord (1975) and Templin (1974) used different representations for the boundary layer. Their wind farm models incorporated increased surface roughness and friction speed factors. Both researchers assumed that the wake is smeared out within a few rotor diameters. The effect of reduced boundary layer profiles on the performance of successive rows of wind turbines was studied. Crafoord computed the effect of spacing, machine rating, and the total number of units on the relative power output compared to a single unobstructed unit. Like the General Electric researchers, Crafoord computed the unit efficiency for different rows in a quadratic array. His results were markedly different. For a spacing of 15 rotor diameters, he found an efficiency of approximately 95 percent for a unit in the tenth row of the array. The difference in efficiency for units in the size range 50 kW-5 MW was small for a spacing of 15 rotor diameters.

PERFORMANCE OF THE NASA/DOE 200-kW WIND TURBINE

The Mod-0A wind turbine is based on the design for the Mod-0, the first large wind turbine developed by NASA under the auspices of the Federal Wind Energy Program. In order to gain early operational experience in a utility environment, design changes were limited to those which were shown to be necessary to obtain a functionally adequate 200-kW machine. Clayton, New Mexico was chosen as the site for the first Mod-0A. This city is located in the northeastern corner of the state near the Texas Panhandle area of the Great Plains. This region represents a large geographic area with excellent wind potential. The Mod-0A at the Clayton site is shown in Figure 3-23. Glascow and Robbins (1979) have described the utility operational experience with this machine.

The first rotation took place on November 30, 1977. The fully automated machine was turned over to the Town of Clayton Light and Water Plant on March 6,

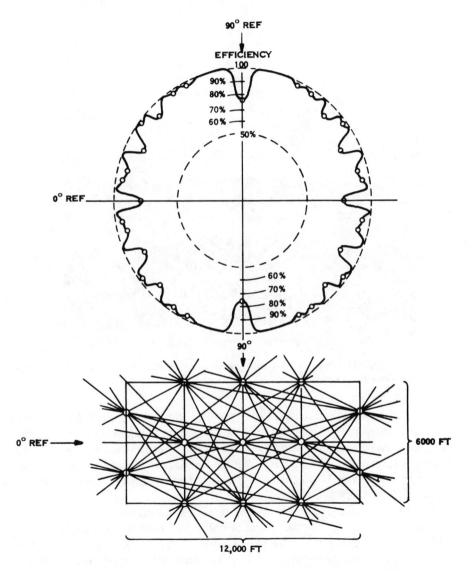

Fig. 3-22. Farm efficiency analysis for a thirteen unit rectangular arrangement. From Garate (1977).

Fig. 3-23. NASA/DOE 200-kW wind turbine generator at Clayton, New Mexico. Courtesy of NASA Lewis Research Center.

1978. The general characteristics of the utility site and the operational history through January 1, 1979 are given in Table 3-4. During this period of utility operation, the machine was shut down for three periods which totaled 50 days. Rotor blade changes were made during two shutdown periods and the third shutdown was caused by a generator bearing failure. Through January 1, 1979, the wind turbine produced 260,240 kWh of electrical energy.

Figures 3-24 and 3-25 show a running total of the wind turbine's electrical output and operating time for the period of utility operation through December 31, 1978. The average power for this period based on the time during which the machine was producing power was 91.8 kW. The wind turbine operated 44.2 percent of the time throughout the period and the capacity factor was 0.202, i.e., .442 x 91.8/200. For both the capacity factor and operating time calculations, the shut-

Table 3-4. Information for the utility application of the Mod-OA at Clayton, New Mexico.

General Characteristics

Population	3,000
Annual energy consumption (1978)	15,100 MWh
Peak demand	3.8 MW
Average daytime demand	2.8 MW
Mean annual wind speed @ 9.1 m	5.82 m (13.0 mph)
Hub height 30 m	7.2 m (16.1 mph)

Operational History

November 30, 1977	First rotation
January 19, 1978	First 100 hrs of operation
March 6, 1978	Turnover to utility
May 24, 1978	1000 hrs of operation, 94,000 kWh
June 2, 1978	Shutdown to inspect and replace blades
June 28, 1978	Returned to service
September 11, 1978	Shutdown to install modified blades
September 25, 1978	Returned to service
October 12, 1978	2000 hrs of operation 181,000 kWh
December 1, 1978	Shutdown—generator bearing failure
December 11, 1978	Returned to service
January 1, 1979	2810 hrs of operation 260,240 kWh

From Glascow and Robbins (1979).

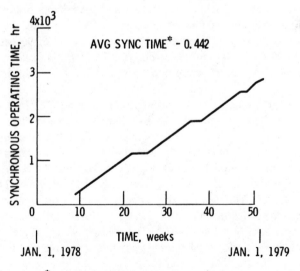

*OMITTING 3 SHUTDOWN PERIODS SHOWN (50 days)

Fig. 3-24. Mod-OA 200-kW wind turbine output for the period March 6 through December 31, 1978. From Glascow and Robbins (1979).

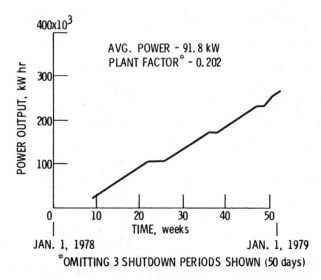

Fig. 3-25. Mod-0A 200-kW wind turbine synchronous operating time. From Glascow and
Robbins (1979).

down periods were omitted. Figure 3-26 shows the weekly wind turbine availability for the period of utility operation. The availability is expressed as the ratio
of the number of hours of synchronous operation to the number of hours the wind
speed falls within the cut-in and cut-out values times 100. As expected, the weekly
availability was low initially and approached the project goal of 90 percent as the
year progressed. The various symbols in Figure 3-26 identify weeks when certain
repairs were necessary or unique conditions existed. The circles indicate periods of
normal operation.

  The measured and predicted values of the power output at different wind speeds
were found to be in good agreement. These values are shown in Figure 3-2. However, the output produced during monthly periods was found to be significantly less
than that predicted on the basis of data from the on-site meteorological tower and
the standard performance model. For example, the mean wind speed at hub height
elevation for the period March 1 through December 31, 1978 was 7.24 m/s (16.2
mph) (Glascow, 1979). The value of the capacity factor based upon the Rayleigh
wind speed frequency distribution and the quadratic power output function defined
by the parameters of Table 3-1 is 0.417.[*] The sea level values of the cut-in, rated,
and cut-out wind speeds were used to construct the power output function. The

---

[*]The values of the cut-in and rated wind speeds given by Neustadter (1979) (9.5 and 22.4 mph,
respectively) were used for this calculation. Glascow and Robbins (1979) report values of 10.8
and 21.4 mph.

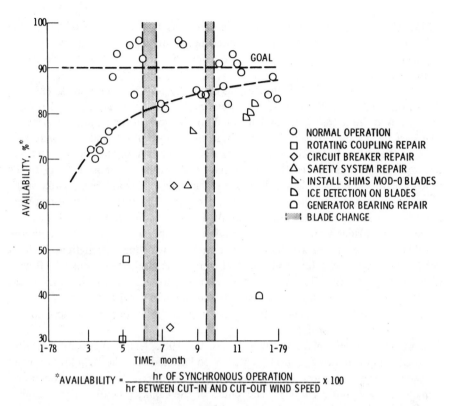

Fig. 3-26. Availability of the Mod-0A 200-kW wind turbine, March 6 through December 31, 1978. From Glascow and Robbins (1979).

density adjustment for the higher elevation of the Clayton site would yield a lower value for the capacity factor. Furthermore, the mean wind speed for the entire seven-month period, used to obtain the calculated capacity factor, may have been significantly higher than the value for the period during which the machine was in operation. However, it is improbable that these effects can account for the entire difference between the observed and calculated values of the capacity factor. NASA engineers believe that the performance losses are caused primarily by the time required for startup and shutdown operations. The predicted energy will, of course, always be larger than the actual energy since momentary excursions in the wind speed below the cut-in or above the cut-out values will initiate a shutdown cycle even when the wind speed returns to an operational value. Investigations are under way to modify the microprocessor logic so as to maximize energy capture without reducing the 30-year design goal for the operational life of the machine.

Additional factors which compromise the accuracy of the simple performance model were mentioned earlier. The large difference between the operational and estimated values of the capacity factor emphasizes the importance of developing an accurate dynamical performance model, i.e., a model which employs time series wind velocity data in conjunction with operational yaw and startup/shutdown strategies.

In March and May of 1978, inspections of the Mod-0A revealed that fasteners on the rotor blades near the root section were being lost. In June, the wind turbine was removed from the tower and lowered to the ground for detailed inspection. Cracks were discovered in the aluminum skins of both blades and they were sent back to the Lewis Research Center for repair. The shank of the root section was removed and it was learned that the damage was restricted almost entirely to the inboard 2 m of the 19-m blades, the area in which the load from the wing-like structure was transferred to the shank or root section. Structural modifications were made and included the addition of a set of doublers to the exterior surfaces of the blades. The blades were reinstalled and the machine was returned to operation. The blade problems were attributed to two causes: inadequate strength at the joint between the wing structure and the steel root shank, and blade loads which were higher than predicted. The high loads occur during normal operation at wind speeds near the cut-out value, during safety system shutdowns when the blades are feathered quickly to stop the rotor, and at times when the yaw brake does not supply normal restraint during yaw corrections.

During the period July 3 through August 29, 1980, less than two months after the dedication ceremony, a Mod-0A with wooden blades at the Kahuku Point site on Oahu, Hawaii, produced 178 MWh during 1,078 hours of synchronous operation (*Wind Energy Report*, 1980b). The values of the capacity factor and availability for this period are approximately 0.65 and 79 percent, respectively.

4
The Development of
Large-Scale Wind Turbines

Many of the large-scale wind turbine prototypes being developed today resemble the Smith-Putnam test unit. The majority of today's engineering teams have judged that a two-bladed, horizontal-axis design presents the best opportunity for commercial success in the near term. In this chapter, a brief review is presented of current efforts to develop machines with a rated power which equals or exceeds 200 kW. Actual design and development projects, and not concepts or possibilities, are described.

UNITED STATES MOD PROGRAM

The U.S. Wind Energy Program was initiated in 1973 by the National Science Foundation as part of the solar energy program. In 1974, a five-year wind energy program plan was developed as part of the Solar Energy Plan of the Project Independence Blueprint. (Savino, 1973). The program included the 1973 workshop recommendation to design, build, and test a nominal 100-kW, 125-ft diameter rotor wind turbine. This wind turbine was designated Mod-0 and the NASA Lewis Research Center was chosen to develop the project. In 1975, the management responsibility for the Federal Wind Energy Program was transferred to the Energy Resource and Development Administration (ERDA) and more recently to the DOE. NASA Lewis has continued to manage the phase of the Federal program which involves the development of the technology necessary for the design, fabrication, and operation of large, horizontal-axis wind turbines. The Mod-0 project was followed by other projects involving utility applications of four 200-kW wind turbines similar to Mod-0 and designated Mod-0A, one 2-MW, 200-ft diameter wind turbine designated Mod-1, and three 2.5-MW, 300-ft diameter wind turbines designated Mod-2. Figure 4-1 indicates the locations which were chosen for these test units. Additional DOE wind turbine candidate sites are also shown. Other projects, including two advanced multimegawatt projects (Mod-5) and two advanced 200- to 500-kW projects (Mod-6H and Mod-6V), are aimed at achieving lower energy costs. The General Electric Company and the Boeing Engineering and Construction Company have been selected for the Mod-5 projects. The Rockwell International Company has been selected for the Mod-6H project. NASA Lewis is also responsible for a comprehensive technology

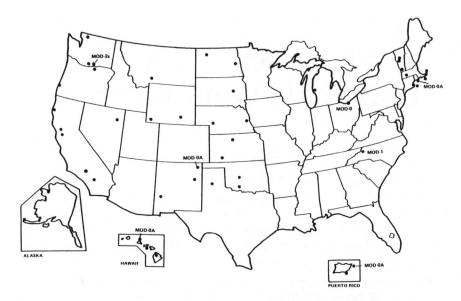

Fig. 4-1. Installations and candidate wind energy sites. From DOE 1980.

support program. Thomas and Donovan (1978), Robbins and Thomas (1979), and Thomas and Robbins (1979) have summarized the Mod program.

Mod-0

The Mod-0 is installed at NASA's Plum Brook facility near Sandusky, Ohio. Since 1975, this machine has served as a test bed for evaluating design concepts and validating analytical methods and computer codes which are being used to design advanced machines. A schematic drawing of the nacelle is shown in Figure 4-2. The design and operating characteristics of the Mod-0 have been described by Thomas and Richards (1978), Glascow and Birchenough (1978), and Thomas and Donovan (1978). Additional technical references are cited in these reports.

The Mod-0 has two aluminum blades. The rotor is normally operated downwind from a (100-ft) steel open-truss tower. The test unit has also been operated with the nacelle in an upwind configuration to assess effects on system structural loads and machine control requirements. The cut-in, rated, and cut-out wind speeds are 4.47 m/s (10 mph), 8.05 m/s (18 mph), and 13.4 m/s (30 mph), respectively. In the wind speed range from 8.05 to 13.4 m/s, the machine produces 100 kW of electrical power. The rotor speed is maintained at constant rpm by pitching (rotating) the blades about their lengthwise (spanwise) axes to control the rotor torque at different wind speeds. This type of speed control is referred to as full-span pitch control. The

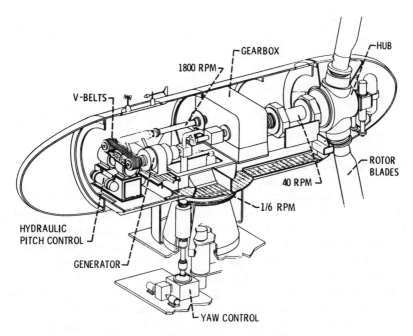

Fig. 4-2. Schematic of the Mod-0 wind turbine nacelle. From Thomas and Donovon (1978).

nominal rotational speed of the Mod-0 is 40 rpm; however, a belt drive feature in the drive train system permits operation at different speeds for test purposes. Power is transmitted from the rotor through a speed increasing gearbox to a synchronous generator which produces 60 Hz/480 V, three-phase power at 1800 rpm. For wind speeds in the range from 4.47 m/s to 8.05 m/s, the rotational speed is maintained at 40 rpm by controlling the electrical load. A microprocessor controls the automatic operation of the machine during startup, synchronization, and shutdown.

A yaw drive system maintains the proper orientation of the nacelle and windstream. The alignment is maintained by redundant yaw motors through a double reduction self-locking worm drive which engages a bull gear attached to the nacelle. Operational studies were also conducted under free yaw conditions.

Glascow and Birchenough (1978) cited results based on the Mod-0 downwind rotor test experience:

1. Power level can be controlled effectively with a closed loop integral and proportional control system, operating on output power, if adequate damping is present in the drive train.
2. The wind direction as seen at hub height is highly variable and the error band should be set quite wide (approximately ±25° for Mod-0) for wind turbines with positive yaw drive systems.

3. Frictional damping in yaw must be provided for the Mod-0 wind turbine to prevent large rotor and yaw drive loads from occurring during yawing operations. A damping force must also be provided when the nacelle is not yawing, otherwise the nacelle must be locked to the tower.
4. Mod-0 tests indicate that a potential exists for a free yaw downwind rotor machine. However, steps must be taken to increase the stabilizing force with rotor offset or coning, and some positive restraining force or damping in yaw may be required.

The fluid coupling shown in Figure 4-2 provides the damping which was required to maintain effective power control in gusty wind conditions. The yaw drive operates at 1°/s and the yaw controller senses the directional error measured by a wind vane mounted on the nacelle. The wind vane signal is modified by a filter with a 30-second time constant. The controller's ±25° deadband must be exceeded for several seconds before the yaw correction is initiated. When activated, the yaw motors drive the nacelle until the filtered yaw error signal is within ±18°. Figure 4-3 shows the behavior of the wind speed, alternator power, and nacelle yaw error during three minutes of operation. Initial tests with a dual yaw drive system indicated that elastic yaw restraint produced high yaw response with attendant excessive rotor loads in gusty winds above 8 m/s. A yaw brake consisting of a two-stage pressure brake system was used to lock the nacelle to the tower during normal operation and to provide damping during yaw maneuvers. This system was also used for the Mod-0A machines. The free yaw tests indicated that the turbine was at least neutrally stable in yaw

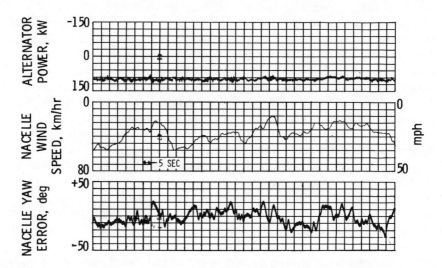

Fig. 4-3. Power control for the Mod-0 with fluid coupling. From Glascow and Birchenough (1978).

while operating at 40 rpm in synchronization with the grid and no drag from the brake. The machine did oscillate in yaw at a low frequency. Application of the yaw brake stabilized the nacelle in yaw and produced acceptable operation.

The Mod-0 is currently being run in an automatic, unattended mode in synchronization with the Ohio Edison network.

Mod-0A

NASA Lewis designed and manufactured the 200-kW Mod-0A wind turbine for applications in utility power systems. The Mod-0A program is designed to obtain early operation and performance data while gaining experience in a user environment. Robbins and Sholes (1978) and NASA Lewis Research Center (1978) have described the Mod-0A machine and program. The key issues which will be addressed in the program are (1) impact of the variable power output on the utility grid, (2) compatibility with utility requirements (power voltage and frequency control), (3) demonstration of unattended, fail-safe operation, (4) reliability of the wind turbine system, (5) required maintenance, and (6) initial public reaction and acceptance.

The first Mod-0A was installed at Clayton, New Mexico in November, 1977. The operation and performance experience with this machine was described in Chapter 3. Machines have also been installed in Culebra, Puerto Rico (July, 1978), Block Island, Rhode Island (May, 1979), and Oahu, Hawaii (1980). The Westinghouse Electric Corporation plays an important role as a contractor in the Mod-0A program.

The Mod-0A is basically an uprated version of the Mod-0. Figure 3-25 shows this machine at the Clayton, New Mexico site. The primary design specifications and operating parameters are summarized in Table 4-1.

The blade is made of aluminum and consists of a main load carrying spar and ribs covered by a thin sheet metal skin. The cylindrical blade root shank is made of steel. The blades are rigidly attached to the hub, i.e., the blade orientation is limited to changes in pitch. The hub transmits the torque to the low speed shaft and transmits all other blade loads into the bedplate through the shaft bearings. The blades are built by the Lockheed California Company.

The pitch-change assembly consists of a hydraulic supply, a rack and pinion actuator, and gears which rotate the blades in the hub. A schematic drawing of the assembly is shown in Figure 4-4. Hydraulic pressure moves a pair of racks. The racks rotate a pinion and a master gear which in turn rotates the blades through bevel gears bolted to the blade spindle. The hydraulic supply is mounted in front of the nacelle. Hydraulic fluid is brought into the main shaft through rotating seals and transmitted to the rack and pinion actuator mounted on the rotor hub.

The drive train assembly transmits the rotor torque to the generator. A 1:45 fixed ratio gearbox transmits the power to the high-speed shaft.

The safety system is designed to protect the wind turbine from catastrophic failure. Sensors monitor significant parameters such as the rotor speed, generator current, vibration, yaw error, electrical load, pitch system hydraulic fluid level, microprocessor failure, and so forth. When any sensor signal is outside the normal operating range, a shutdown procedure is initiated. For example, if rotor overspeed is detected, the blades are feathered. A pneumatic backup system will feather the blades if the hydraulic system fails. If these systems both fail, a redundant rotational speed sensor actuates an emergency brake to stop the rotor.

Table 4-1. 200-kilowatt wind turbine design specifications.

Rotor

Number of blades. 2
Diameter, ft 125
Speed, rpm. 40
Direction of
rotation . . . . . . counterclockwise (looking upwind)
Location relative to tower . . . . downwind
Type of hub . . . . rigid
Method of power regulation. . . . . . . .variable pitch
Cone angle, deg 7
Tilt angle, deg. 0

Blade

Length, ft 59.9
Material . . . . .aluminum
Weight, lb/blade. 2300
Airfoil .NACA 23000
Twist, deg 26.5
Solidity, percent .3
Tip chord, ft. 1.5
Root chord, ft. 4
Chord taper . . . . linear

Tower

Type . . . . pipe truss
Height, ft. 93
Ground clearance, ft 37
Hub height, ft 100
Access . . . . .hoist

Transmission

Type . . . . three-stage conventional
Ratio. 45:1
Rating, hp 460

Generator

Type . . . . . .synchronous ac
Rating, kVA. 250
Power factor. 0.8
Voltage,
V . . . . 480 (three phase)
Speed, rpm. 1800
Frequency, Hz. 60

Orientation Drive

Type . . . . .ring gear
Yaw rate, rpm. 1/6
Yaw drive . . .electric motors

Control System

Supervisory . .microprocessor
Pitch actuator . . . . hydraulic

Performance

Rated power, kW 200
Wind speed at 30 ft, mph
(at hub):
Cut-in 6.9 (9.5)
Rated 18.3 (22.4)
Cut-out 34.2 (40)
Maximum
design 125 (150)

Weight (klb)

Rotor
(including blades). 12.2
Above tower. 44.9
Tower 44.0
Total 88.9

System Life

All components, yr 30

From Robbins and Sholes (1978).

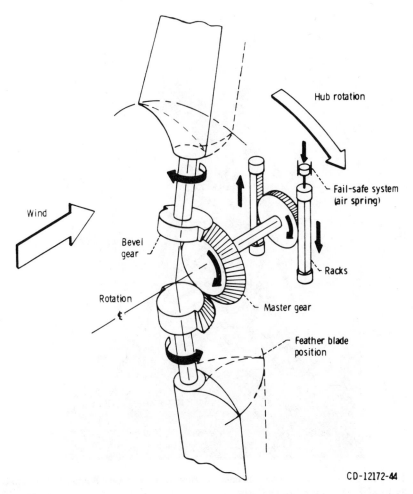

Fig. 4-4. Blade pitch-change system. From NASA Lewis Research Center (1978).

Mod-1

The megawatt class Mod-1 wind turbine is a two-bladed, 200-ft diameter wind turbine with a rated power of 2 MW. The hub height is 140 ft and the rotor is located downwind of the tower. The rotor operates at a fixed speed of 35 rpm. A photograph of the Mod-1 on Howard's Knob near Boone, North Carolina is shown in Figure 4-5.

NASA Lewis's specifications and design requirements were based on the Mod-0 design and operational experience. State-of-the-art technology was used to minimize technical risk. General Electric's Space Division was the prime contractor. The Boeing Engineering Company manufactured the two steel blades. The Mod-1

Fig. 4-5. The Mod-1 wind turbine generator on Howard's Knob near Boone, North Carolina. Courtesy of NASA Lewis Research Center.

was put in operation in July, 1979. The mountaintop site is 1347 m above sea level. After a preliminary test period, the Blue Ridge Electric Membership Corporation, a rural electric cooperative, will operate the machine and use its power. Poor and Hobbs (1979a,b) and the General Electric Company (1979) have described the Mod-1 project.

Table 4-2 gives a summary of the design parameters. A schematic drawing of the Mod-1 nacelle is shown in Figure 4-6. The rotor speed is controlled by full-span pitch control. The blades are attached to the hub barrel via three-row cylindrical roller bearings which allow the pitch angle of the blade to be varied 105° from full feather to maximum power. The blade pitch is controlled by hydraulic actuators which provide a maximum pitch rate of 14°/s. The rotor assembly is supported by a single bearing which has two rows of tapered rollers.

The Mod-1 blade represents a compromise between aerodynamic performance and ease of fabrication. The blade is comprised of a $97\frac{1}{2}$ ft long steel welded monocoque spar and a monolithic foam-filled bonded trailing edge afterbody (Van Bronkhorst, 1979). The blade is tapered in planform and thickness. It uses a NACA 44XX series airfoil with thickness ratio varying from 33 percent at the root to 10 percent

Table 4-2. Summary of design parameters for the Mod-1.

Rated power	2000 kW @ 11.4 m/s (25.5 mph) 1800 kW @ 11.0 m/s (24.6 mph)
Cut-in wind speed	5 m/s (11 mph) max.
Cut-out wind speed	15.9 m/s (35 mph)
Maximum design wind speed	67 m/s (150 mph at shaft centerline — assume no wind shear)
Rotors per tower	1
Location of rotor	Downwind of tower
Direction of rotation	CC (looking upwind)
Blades per rotor	2
Cone angle	9°
Inclination of axis rotation	None
Rotor speed control	Variable blade pitch
Rotor speed	34.7 rpm
Blade diameter	61 m nom. (200 ft)
Airfoil	44XX series
Blade twist	11° linear
Tower	Steel truss
Blade tip to ground clearance	12 m (40 ft)
Hub	Rigid
Transmission	Fixed ratio gear
Generator	60 Hz/synchronous
Yaw rate	0.25°/s
Control system	Electro mechanical/microprocessor

System life:
  Dynamic components
  Static components
	
	30 years (with maintenance)
	30 years (with maintenance)

All wind speeds measured at 9-m (30-ft) elevation.
From General Electric Company (1979).

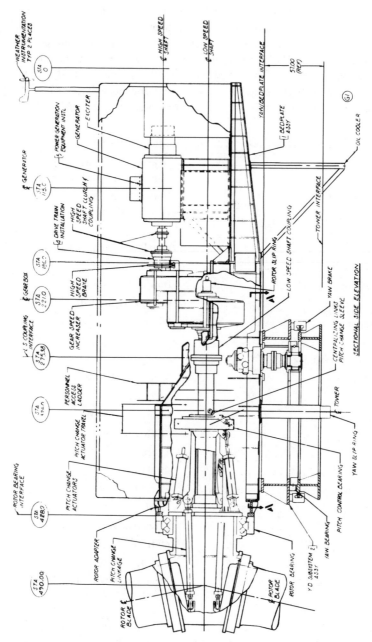

Fig. 4-6. Mod-1 nacelle configuration. From Poor and Hobbs (1979a).

at the tip. The 11° twist varies linearly from root to tip. The Mod-1 blade assembly and blade cross-section are shown in Figure 4-7. The spar is fabricated from A533 Grade B, Class 2 high strength, low-carbon steel. It consists of upper and lower panels; each panel is welded in six sections. The lower panel is stiffened by chordwise intercostals and a spanwise tee-shaped stiffener. The trailing edge is fabricated from urethane foam with 301 stainless steel skins.

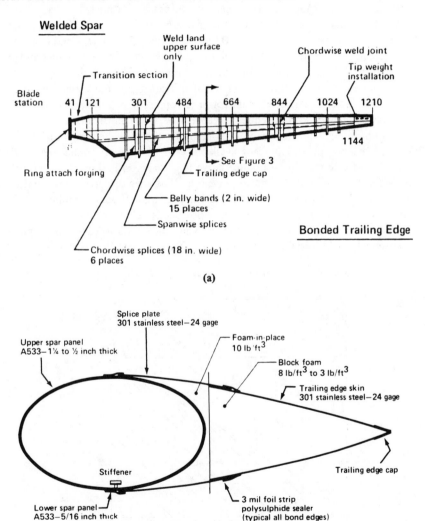

Fig. 4-7. (a) Mod-1 blade assembly, and (b) Mod-1 blade cross section. From Van Bronkhorst (1979).

The drive train consists of the low-speed shaft and couplings, a three-stage gear-box and the high-speed shaft which drives the generator. A dry disk slip is incorporated in the high-speed shaft for protection against torque overloads. A disk brake is provided to stop the rotor and hold it in parked position. The high-speed shaft drives the GE synchronous 4160-V AC generator at 1800 rpm. Voltage control is provided by a shaft-mounted, brushless exciter which is controlled by a solid state regulator and power stabilizer. The generator output is transmitted by cables and a slip ring at the yaw bearing down the tower to a control enclosure. Power is transmitted to the utility interface via a 2000-kilovolt ampere (kVA) step-up transformer.

The welded steel bedplate supports all of the tower-mounted equipment and provides a load path between the rotor and yaw structure. The yaw drive system consists of lower and upper structures, a cross-roller bearing, and dual hydraulic motors and hydraulic brakes. The steel truss tower is composed of tubular members. Tubular members were chosen to reduce tower shadow loads on the blades.

The Mod-1 design rpm was chosen to maximize the annual energy output (6500 MWh) at sea level for 100 percent availability in an 18-mph annual mean wind regime. The steady state performance curve (C_p versus λ) is shown in Figure 3-4. Figure 4-8 shows the Mod-1 operating envelope. The envelope gives the operational modes and limits for variations in wind speed and direction (Poor and Hobbs, 1979b). The nonoperating mode is indicated for wind speeds below the cut-in value of 11 mph. A five-minute average wind speed and yaw angle deviation above 11 mph and 5°, respectively, will initial a yaw maneuver. Normal operation is obtained if the

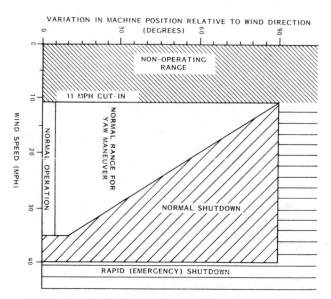

Fig. 4-8. Mod-1 operating envelope. From Poor and Hobbs (1979b).

five-minute average yaw angle deviation is within the 5° envelope. A normal shut-down is initiated when the five-minute average wind speed exceeds 35 mph or exceeds the wind speed-yaw envelope. An emergency shutdown operation is initiated when the instantaneous wind speed or yaw angle deviation exceeds 40 mph or 90°, respectively.

Although the Mod-1 configuration was largely determined by the NASA Lewis design specifications, some options were left open for design trade-offs between per-formance, structural design requirements, and cost. Table 4-3 gives Poor and Hobbs (1979b) brief description of the trade-off procedure and results for these options.

Spera (1977) has described and compared seven computer codes which have been used to calculate dynamic loads in horizontal-axis wind turbines. Comparisons involving Mod-0 test results and computer predictions have been used to verify the codes and formulate design criteria for advanced machines. Poor and Hobbs (1979b) have described the application of the General Electric Turbine System Synthesis (GETSS) code in the Mod-1 design study. Factors which influence machine design are proper placement of resonant frequencies, accurate prediction of operational loads and deflections, and a design approach which minimizes sensitivity to dynam-ic loading. Resonances and load amplification are avoided by adjusting compo-nent masses and stiffnesses during the design phase to assure that modal frequencies do not coincide with particular integer multiples of the rotor speed. For example, there is a significant increase in loads when unfavorable coupling occurs between blade-flapping and tower-bending modes. In the Mod-1 design study, it was conclud-

Table 4-3. Configuration option trade-offs for the Mod-1.

Blade airfoil:	Performance versus manufacturability cost. Airfoil selection driven by manufacturability. Selected 44XX series.
Blade twist:	Performance versus blade loading. Blade twist driven by structural design requirements. Selected 11°.
Rotor speed:	Maximum energy capture versus torque cost. Oper-ating speed driven by maximum energy capture for a given rotor diameter, rated power, and wind duration curve. Selected 35 rpm.
Rotor cone angle:	Balance blade thrust versus centrifugal loads. Cone angle selected to minimize blade root stress. Selected 9°.
Rotor axis inclination:	Blade clearance versus yaw moments. Rotor coning more effective. Selected 0° axis inclination.
Hub (rigid or teetered):	Blade-hub load reductions versus cost. Rigid hub less costly. Selected rigid hub.

From Poor and Hobbs (1979b).

ed that given a blade flap frequency in the range of 2.15 to 2.7 P (2.15 to 2.7 times per rotation), unfavorable coupling could be avoided by ensuring that the tower bending frequency exceeded 2.8 P. As a consequence, the preliminary design requirement of 2.2 P was raised to 2.8 P.

Spera et al. (1979) have given a preliminary analysis of load data for the Mod-1. A total of 10 vibrational modes with frequencies in the range of 0.3 to 10 Hz were identified in the modal study of the Mod-1. All of the important measured frequencies were within 20 percent of calculated values. This accuracy is sufficient to avoid operating resonances. Resonances at the rated rotor speed can be avoided or minimized through modal analysis and design changes. Resonances at lower speeds during startup or shutdown cannot be avoided. Modal analysis may be used to assure that two such resonances do not occur at the same time by requiring blade frequencies and tower frequencies to be separated.

A large sample of cyclic load data is required to assess fatigue design procedures. The data should encompass all wind speeds and directions in the operating range, gustiness conditions, stopping transients, and so forth. In a preliminary assessment of the Mod-1 performance, Spera et al. (1979) analyzed a small data sample consisting of strain gauge readings taken at four blade locations during 2800 rotor revolutions. During the measurement period, wind speeds ranged from 5 to 14 m/s and the machine power reached 1.6 MW. Figure 4-9 shows the chordwise cyclic load factor for inboard blade sections versus the cumulative probability of occurrence of this factor. This fatigue load factor equals the cyclic chordwise load (one-half of the load range during each rotor revolution) divided by the nominal calculated cyclic load. The GETSS code was used to calculate the load at 14.4 m/s. The design spectrum line corresponds to the log normal distribution with the median value equal to the nominal GETSS case and an estimate for the standard deviation based on

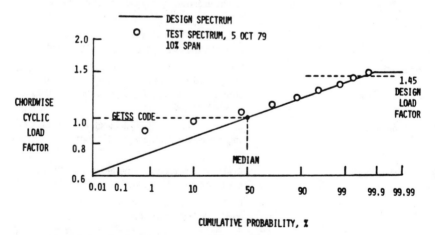

Fig. 4-9. Comparison of measured and design blade fatigue load spectra for inboard blade sections in the chordwise direction. From Spera et al. (1979).

early tests of the Mod-0. The Mod-1 blades were designed for infinite life at chord-wise loads which were 1.45 times the nominal GETSS case. Figure 4-9 shows that the design loads were estimated to exceed observed loads 99 percent of the time. Figure 4-10a shows cyclic flatwise load spectra for inboard stations at 10 and 39 percent span. The log standard deviation based on early Mod-0 data is higher than it was for the chordwise loads, and the design load factor is 1.90. The test data are in good agreement with the design spectrum. Figure 4-10b shows the cyclic load spectra at 75 percent span. Measured load factors are significantly larger than the

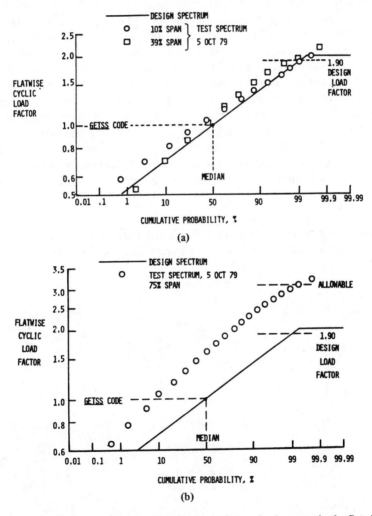

Fig. 4-10. Comparison of measured and design blade fatigue load spectra in the flatwise direc-tion: (a) inboard blade sections at 10 and 39 percent span, and (b) outboard blade sections at 75 percent span. From Spera et al. (1979).

design values for equal probabilities of occurrence; however, the measured loads appear to be within allowable fatigue loads.

The General Electric Company improved the Mod-1 design in a trade-off study managed by NASA Lewis (Poor and Hobbs, 1979a). The goal was a design for a lighter and less expensive machine with the same operational characteristics as the Mod-1. This machine was designated Mod-1A. The objectives were to reduce the weight from 297,000 kg to 182,000 kg or less, reduce the second unit cost from $2,900/kW to $1,000/kW, and reduce the cost of energy from 18¢/kWH to 5¢/kWH (1978 dollars). Schematic drawings of the three design concepts which were considered for trade-off studies are shown in Figure 4-11. System number three was selected. Its significant characteristics are a teetered hub, two downwind blades with partial-span pitch control, an integral gearbox structure, and a soft tower.

The teetered hub concept gave the lowest loads for a two-blade system. Hydraulic partial-span torque control is incorporated in the outer 15 percent of the span. Structural weight is eliminated by the integral gearbox concept. The gearbox-bedplate incorporates the rotor and yaw support structure into the gearbox casing. The conical shell tower has a lateral bending frequency of 1.2 P. The comparison of the silhouettes of the Mod-1 and Mod-1A design in Figure 4-12 shows the significant reduction in size. Figure 4-13 illustrates the difference in weights for major components. The Mod-1A weight amounts to approximately 49 percent of the Mod-1 weight.

Mod-2

The Mod-2 project was initiated in 1976. The Boeing Engineering and Construction Company was awarded a four-year contract by DOE-NASA in August, 1977.

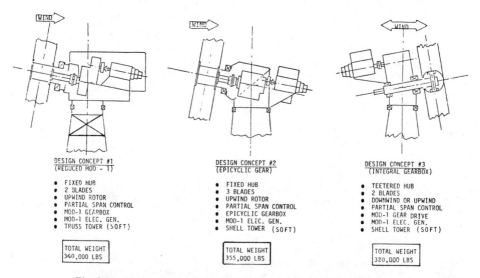

Fig. 4-11. Three candidate systems. From Poor and Hobbs (1979a).

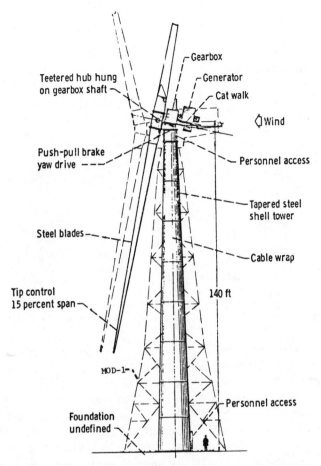

Fig. 4-12. Mod-1A outline and comparison with Mod-1. From Poor and Hobbs (1979a).

The contract calls for the design, fabrication, construction, installation, testing, and checkout of a 2.5-MW horizontal-axis machine. The current contract calls for the development of three machines and long-lead procurement for a fourth unit. The Mod-2 has been designed by using the technology base produced by the research and development efforts on the Mod-0, Mod-0A, and Mod-1 machines, and is referred to as a second generation system. The primary objective for the end hardware is direct and efficient commercial application. The project has been structured to achieve this goal through a substantial effort in concept selection, comparatively few design specifications and requirements, and encouragement of commercial application practice. Figure 4-14 shows two of the three machines which have been installed at Goodnoe Hills, a mountain site located about 21 km east of Goldendale, Washington, near the Columbia River. This site was chosen through a candidate site

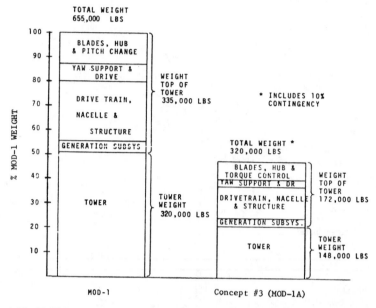

Fig. 4-13. Weight comparison for the Mod-1 and Mod-1A. From Poor and Hobbs (1979a).

competition in which the DOE invited utility companies to propose sites for demonstration projects. The machines were dedicated in May, 1981. The machines will be operated by the Bonneville Power Administration and connected to the Northwest power grid through lines owned by the local utility — the Klickitat County Public Utility.   Douglas (1979), Lowe and Engle (1979), and Boeing Engineering and Construction (1978) have described the Mod-2 project.

NASA Lewis's firm design requirements were the following: horizontal axis, minimum rotor diameter of 300 ft, 14-mph mean annual wind speed at 30-ft elevation, 30-year service life, and unattended remote site operation. Required sensitivity studies conducted during the concept study phase produced significant improvements in the original design. The driver in the trade-off studies was the cost of electricity. The major characteristics of the Mod-2 system and the nacelle arrangement are shown in Figure 4-15. This machine has a 300-ft diameter, tip-controlled, teetered, upwind rotor.   The rotor axis (hub) is located 200 ft above ground level. The all-steel rotor is supported by the low-speed shaft through an elastormeric bearing which permits teetering. Rotor torque is transmitted by an attenuating quill shaft to a step-up planetary gearbox. The gearbox drives a 2500-kW synchronous generator at 1800 rpm. Teeter and rotor brakes are used to eliminate motion when the machine is not operating. The drive train, generator, electronic control system, pitch and yaw hydraulic system, and other support equipment are housed in the nacelle. The nacelle is aligned with the wind by a single hydraulic motor which

Fig. 4-14. Two Mod-2 wind turbines installed at the Goodnoe Hills test site. Courtesy of Bonneville Power Administration (1981).

drives a planetary reduction gear. The nacelle is supported by a shell-type tower with a conical base.

Davison (1979) has described the Mod-2 rotor. The specifications and requirements are given in Table 4-4 and the rotor configuration is shown in Figure 4-16. The rotor is of hollow steel shell construction with steel spar members. The material is A-633 steel. The plate thickness varies from 0.60 inches at the blade root to 0.18 inches at the tip. The rotor is composed of three sections. The view in Figure 4-16 is taken from the nacelle side. The hub, middle, and tip sections are 60, 75, and 45 ft long, respectively. The weights of the components are tip – 10,400 pounds each, middle – 40,500 pounds each, and hub – 67,800 pounds. The total weight of the rotor is 169,600 pounds. The maximum and minimum (tip) chords are 136 and

56.6 inches, respectively. The maximum airfoil thickness varies from 6.78 inches at the tip to 57 inches at the hub. The outer 30 percent of the blade is rotatable through 100°. When the controllable tip is in the feathered position, the tip section trailing edge is downwind. See the small view labeled "high-wind attitude" in Figure 4-16.

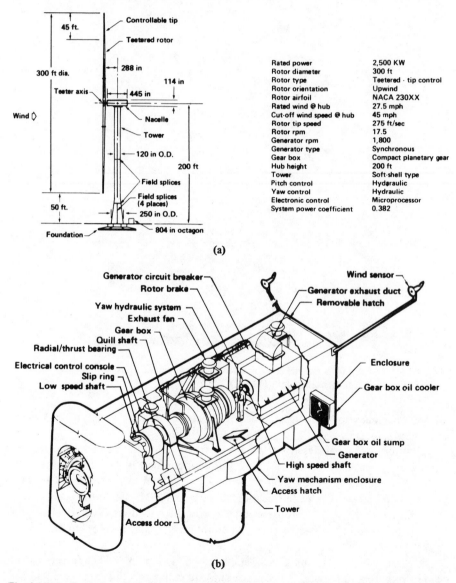

Rated power	2,500 KW
Rotor diameter	300 ft
Rotor type	Teetered - tip control
Rotor orientation	Upwind
Rotor airfoil	NACA 230XX
Rated wind @ hub	27.5 mph
Cut-off wind speed @ hub	45 mph
Rotor tip speed	275 ft/sec
Rotor rpm	17.5
Generator rpm	1,800
Generator type	Synchronous
Gear box	Compact planetary gear
Hub height	200 ft
Tower	Soft-shell type
Pitch control	Hydraulic
Yaw control	Hydraulic
Electronic control	Microprocessor
System power coefficient	0.382

(a)

(b)

Fig. 4-15(a). General configuration and features of the Mod-2. (b) Mod-2 nacelle arrangement. From Douglas (1979).

Table 4-4. Specifications and requirements for the Mod-2 rotor.

- External configuration requirements
  - Rotor diameter $\geqslant$ 300 ft
  - Airfoil contour = NACA 230XX
  - Twist about 50% chord = –2.5° to +4°
  - Controllable tip = 30% semi-span
  - Pitch control = +5° to –97°
  - Teeter = ±5°

- Environmental requirements
  - Design rotational speed = 17.5 rpm
  - Cut-off wind speed @ hub = 45 mph
  - Steady winds plus gusts
  - Lightning, temperature, precipitation, projectile
  - Nonoperative: snow, ice, extreme winds
  - Handling and transportation

- Internal design requirements
  - Weldable, low-cost steel construction
  - Commercial tolerances
  - Limit operating loads
  - Fatigue loads, 30-year life
  - Operating fault loads: overspeed, inadvertant
    feathering and braking

From Davison (1979).

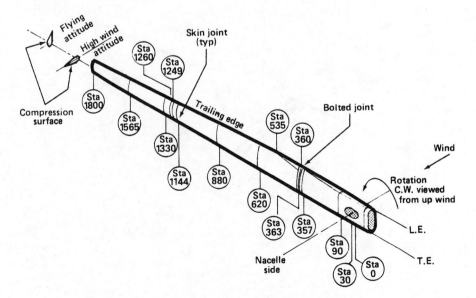

Fig. 4-16. Mod-2 rotor configuration. From Davison (1979).

A bolted field joint is present at station 360 to permit interchangeability of blade elements in case of damage and to facilitate shipping. The hole at station zero accommodates the teeter assembly and the stub of the low-speed shaft.

The spindle assembly is shown in Figure 4-17. The angle of attack is controlled by a tip actuator which pushes the stub fitting on the trailing edge of the tip. The spindle rotates on two lubricated roller bearings located in the outboard end of the middle section. If a major system failure occurs, the actuators drive the blade tips at rates of 4 to 8°/s by using energy stored in separate hydraulic accumulators.

The hub section of the blade is shown in Figure 4-18. The cut-away drawing shows the teeter assembly. The teeter assembly minimizes 1-P flapwise loads. It is crucial to the relief of the high-frequency cyclic loading to the low-speed shaft. Trade studies showed that use of a teetered hub rather than a rigid hub yields reduced weights for the rotor assembly, nacelle, and tower, and consequently reduced material and fabrication costs. The blade is supported by two elastomeric bearings which transmit the dead weight, bending, shear, and torsion into the teeter trunnion. The elastomeric bearings consist of concentric layers of bonded sheet steel and rubber. The trunnion is welded to the stub of the low-speed shaft. A teeter stop limits the travel to ±5°, and a teeter brake holds the blade and prevents flapping motion when the rotor is parked.

The blade area constitutes 3 percent of the rotor disk area. The five rotor sections are assembled in the field. The rotor is lifted into place for connection to the low-speed shaft by a 240-ton ringer crane. Figure 4-19 illustrates this construction method.

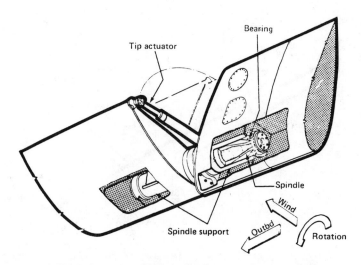

Fig. 4-17. Mod-2 spindle assembly. From Davison (1979).

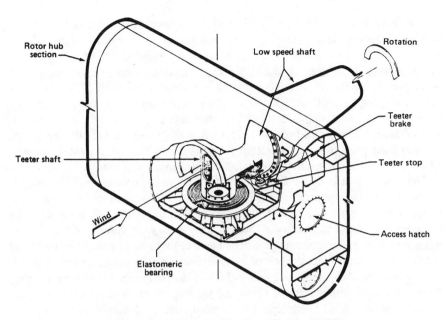

Fig. 4-18. Mod-2 hub assembly. From Davison (1979).

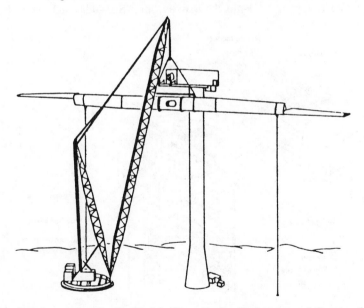

Fig. 4-19. Erection method for the Mod-2. From Boeing Engineering and Construction Company (1978).

The drive train subassembly consists of a low-speed shaft, quillshaft, gearbox, high-speed shaft, couplings, parking brake, and generator. The rotor torque is transmitted to the gearbox through a soft quillshaft. The quillshaft reduces the 2-P rotor fatigue effects at the gearbox and improves the quality of the generator output. A three-stage epicyclic gearbox increases the rotational speed from 17.5 rpm to the generator speed of 1800 rpm. Electrical power is produced by a four-pole synchronous generator containing an integral brushless exciter. It is a three-phase, 60-Hz, 4160-V generator rated to provide 3125 kVA at 0.8 power factor. Proper voltage is maintained prior to synchronization with the utility through excitation control. After synchronization, excitation control provides constant power factor output.

The yaw system is shown in Figure 4-20. The yaw system connects the nacelle to the tower. On command from the control system, the yaw system rotates the nacelle into the wind at a rate of $1/4°/s$, or holds it in position. Rotor and nacelle loads are transferred to the tower through the yaw bearing which is of the crossed-roller type with an integral ring gear. The raceway diameter is approximately 3.0 m (12 ft). Commands from the central system actuate the hydraulic motor. The motor advances itself along the ring gear and drives the outer race of the roller bearing. Wind direction sensors mounted on top of the nacelle transmit signals to the yaw control system. Wind excursions are averaged over a 30-second interval and the command for a yaw maneuver is initiated when an excursion exceeds $20°$. A hydraulic brake is applied to damp the yaw motion. Six additional brakes hold the

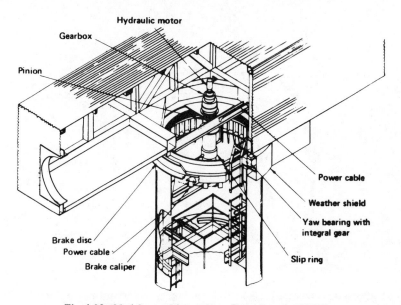

Fig. 4-20. Mod-2 yaw drive system. From Douglas (1979).

nacelle from inadvertent yawing. The yaw brake calipers are spring activated and hydraulically released. This fail-safe feature allows the brakes to be applied if the hydraulic system fails.

The upper portion of the steel shell tower is 3.0 m (10 ft) in diameter and the tower flares out to a diameter of 6.4 m (21 ft) at the base. The tower is soft in the sense that its frequency is lower than the rotor frequency. A stiff tower has a higher frequency than the rotor. The soft tower has a much lower weight than a stiff open-truss tower and the shell-type construction is less expensive to fabricate on a cost-per-weight basis. The soft tower permits greater flexibility in design work since rotor stiffness and weight are not serious restraints. Table 4-5 gives a weight summary for the Mod-2.

Table 4-5. Weight summary for the Mod-2.

Element	Weight (Pounds)
Blade	100,336
Hub	67,722
Pitch control	1,509
Rotor subassembly	169,567
Low speed shaft and bearings	22,865
Quill shaft and coupling	9,483
Gearbox	39,000
High speed shaft and coupling	600
Rotor brake system	280
Lubrication system	6,664
Generator	17,000
Drive train	95,892
Nacelle structure	33,680
Yaw drive	17,742
Rotor support structure	7,152
Environmental control and fire prevention	870
Cabling and electrical facilities	645
Instruments and controls	690
Generator accessory unit	2,500
Nacelle	63,279
Tower	246,536
Cable installation	4,130
Cable transition	500
Lightning protection	300
Tower subassembly	251,466
Total above foundation	580,204

From Boeing Engineering and Construction Company (1978).

Mod-X Design

In 1978, NASA Lewis initiated the Mod-X conceptual design study (NASA Lewis Research Center, 1979). The goal was the design of a 200-kW wind turbine which would produce electricity at a site with a mean annual wind speed of 6.26 m/s (14 mph) referenced to 9.14 m (30 ft) above ground level at the lowest possible cost. The design effort benefited from experience gained through the design and construction of the Mod series machines and a 100-kW unit designed and built by Dr. Ulrich Hütter of West Germany in the late 1950s (Hütter, 1964, 1973). A schematic drawing of the Hütter machine is shown in Figure 4-21.

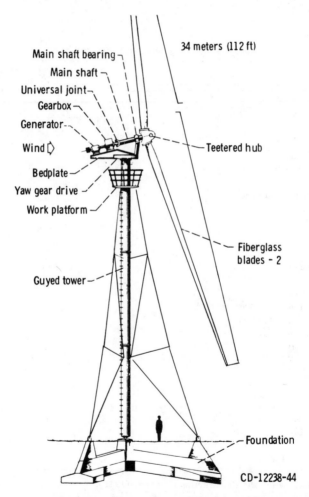

Fig. 4-21. Schematic of the Hütter 100-kW wind turbine. From NASA Lewis Research Center (1979).

The base-line Mod-X configuration resulted from an evaluation and trade-off study of major components and subsystems. Machine features that were shown to be potentially cost-effective and technically feasible were incorporated in the base-line design and changed only if shown to be too costly or impractical during the evaluation process. Weight goals were established on the basis of the major components of the Mod-0A and Hütter machines; the weights of these machines are 90,000 and 29,000 pounds, respectively. An examination of the Hütter design indicated that the low weight was achieved through a combination of the load alleviation, simplification, and compactness achieved by using a teetered hub, a feature which was also identified in the Mod-1A and Mod-2 design studies. A weight goal of 35,000 pounds was selected for the Mod-X. Emphasis was given to reducing the weights of the four major subsystems of the Mod-0A which account for 82 percent of the total cost. These subsystems include the rotor, drive train, yaw drive, and tower.

The major characteristics of the Mod-X are given in Table 4-6. The selected Mod-X base-line configuration is shown in Figure 4-22. A schematic drawing of the Mod-X pod assembly is shown in Figure 4-23. The Mod-X pod assembly uses the gearbox as the main load-bearing component. The rotor and the pitch change mechanism are supported on the low-speed shaft and the gearbox rests on a small bedplate which is attached to the tower by a pair of hinge pins. The hinges permit the bed-

Table 4-6. Major characteristics of the Mod-X.

Rated power	200 kW
No. blades	2
Blade type	Wood, steel spar and rib, steel, transverse fiberglass tape, other
Blade configuration	Downwind
Hub type	Teetered
Hub height	30.5 m (100 ft)
Speed and power control	Pitchable blades (partial or full span), hydraulically operated
Rotor support	Low speed shaft of gearbox
Gearbox	3-stage, parallel shaft
Generator	Synchronous or induction depending upon application
Yaw drive	Passive
Control safety	Microprocessor
Tower	Cantilever rotating cylinder
Foundation	Factory precast concrete vaults, dirt-filled
Design wind regime	6.26 m/s (14 mph) at 9.1 m (30 ft)
Applications	Isolated (synchronous generator)
	Utility grids (induction generator)

From NASA Lewis Research Center (1979).

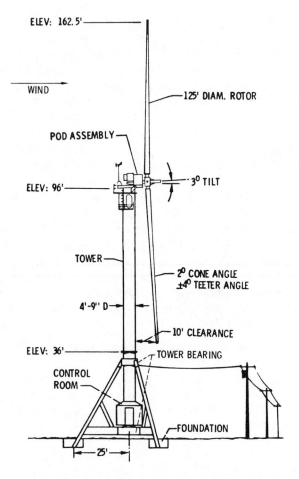

ELEV: 162.5'

WIND

125' DIAM. ROTOR

POD ASSEMBLY

ELEV: 96'

3° TILT

TOWER

2° CONE ANGLE
±4° TEETER ANGLE

4'-9" D

10' CLEARANCE

ELEV: 36'

TOWER BEARING

CONTROL
ROOM

FOUNDATION

25'

Fig. 4-22. Selected Mod-X base-line configuration. From NASA Lewis Research Center (1979).

plate to tilt during the installation of the pod. The bedplate is tilted 3° from the horizontal during operation. The pod assembly components are weatherproof.

The rotor has two blades that are coned 2°. The four blade concepts which are considered are shown in Figure 4-24. It was judged that the blade design and material cannot be selected on the basis of available information. More blade design and cost studies are needed and the final blade concept may be different from those considered in the Mod-X study. The Mod-X teetered hub design borrows features from both the Hütter and Mod-1A designs. Figure 4-25 shows the teetered hub and a schematic drawing of the blade pitch-change assembly. It was judged that more detailed design and cost studies for the blades and pitch-change mechanism are needed in order to select the full or partial pitch control configuration for the Mod-X. The

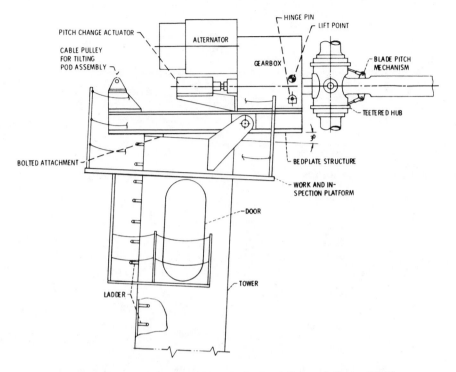

Fig. 4-23. Mod-X pod assembly. From NASA Lewis Research Center (1979).

hydraulic pitch-change mechanism selected for the Mod-X is shown in Figure 4-25. This mechanism is similar to the one used in the Hütter 100-kW machine. In the Hütter system, a rotating hydraulic seal is required since the actuator rotates with the low-speed shaft. In the Mod-X, this seal is eliminated by fixing the hydraulic actuator to the pod.

Synchronous and induction generators were selected for utility and isolated applications, respectively.

The tower design for the Mod-X is unique. The main section of the tower consists of a cantilevered cylinder which encloses an access ladder and electrical and hydraulic conduits. The pod assembly is rigidly attached to the tower. The tower is mounted on bearings and can rotate about its centerline. The yaw control moments are provided by aerodynamic drag on the blades. The base of the tower is used to house the control and hydraulic equipment. A tripod tower base was selected to support the tower bearings and resist overturning forces. The Mod-X foundation concept consists of three earth-filled concrete vaults.

The estimated weights and costs of the Mod-X design concept are summarized in Table 4-7. Costs are given in 1978 dollars for the second unit and the one-hundredth

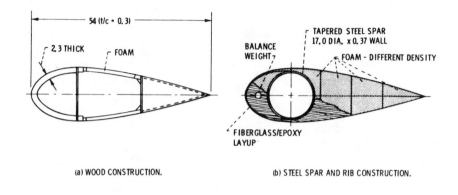

(a) WOOD CONSTRUCTION.

(b) STEEL SPAR AND RIB CONSTRUCTION.

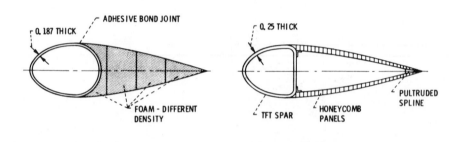

(c) STEEL CONSTRUCTION.

(d) TRANSVERSE FIBERGLASS TAPE WOUND CONSTRUCTION.

Fig. 4-24. Mod-X blade concepts. From NASA Lewis Research Center (1979).

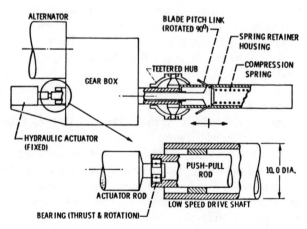

Fig. 4-25. Schematic of the Mod-X blade pitch assembly. From NASA Lewis Research Center (1979).

Table 4-7. Cost and weight summary.
($ 1978)

	Weight (lb)	2nd Unit Cost ($)	Production Quantity Cost ($)
Rotor			
Hub	2,800	19,600	8,120
Blades	4,700	70,000	30,000
Pitch change	1,700	12,000	8,700
Gearbox	15,000	36,000	25,300
Electrical			
Generator	2,500	6,500	5,000
Switchgear and wiring	1,600	7,600	5,400
Capacitors	1,300	500	500
Sliprings	100	900	810
Structure			
Tower	41,500	62,250	31,540
Bedplate	1,500	2,250	1,140
Bearings	- - - - - -	6,000	5,400
Foundations	- - - - - -	4,000	4,000
Installation	- - - - - -	8,250	8,250
Control	200	9,000	6,700
Safety	20	2,000	1,000
Shop ass'y and test	- - - - - -	18,000	4,000
Shipping	- - - - - -	3,000	3,000
Installation and checkout	- - - - - -	4,500	4,500
Subtotal	- - - - - -	272,350	153,360
15% G & A	- - - - - -	40,850	23,000
15% profit	- - - - - -	46,980	26,450
Total	72,920	360,180	202,810

From NASA Lewis Research Center (1979).

unit. The cost reduction for the 100-unit production run amounts to 44 percent. Table 4-8 provides a comparison of the weight and cost of second Mod-X and Mod-0A machines. The costs include 15 percent general and administrative (G&A) and 15 percent profit. The table entries show that significant weight and cost reductions were obtained for all of the major components. The weight and cost reductions amount to 19 percent and 79 percent, respectively. Cost per kW figures for the Mod-0A and Mod-X are $8690/kW and $1800/kW, respectively. The figures for the cost of energy will be given in Chapter 5.

Table 4-8. Mod-X/Mod-0A weight and cost comparison.
($1978)

	Mod-0A			Mod-X		
	Weight (lb)	Cost	Percent of Total	Weight (lb)	Cost	Percent of Total
Rotor—blades and hub and pitch control system	13,800	$814K	47	9,200	$134K	38
Gearbox—generator— hub support	26,800	260	15	17,500	56	15
Yaw bearing: drive and brake system	3,900	77	4	------	8	2
Tower	45,000	90	5	43,000	85	24
Foundation	------	60	4	------	5	1
Electronic control system	------	70	4	200	12	3
Safety system	------	21	1	20	3	1
Electrical system	------	70	4	3,000	12	3
Shipping	------	18	1	------	4	1
Installation	------	258	15	------	41	12
	90,000	$1,738K	100%	72,920	$360K	100%

From NASA Lewis Research Center (1979).

TVIND WIND TURBINE

The most unusual effort to design and build a MW-scale wind turbine was initiated in 1975 by teachers and students from the Tvind College complex near Ulfborg, Denmark. Assistance was provided by volunteer experts from scientific and technological institutes in Copenhagen who served as part-time volunteers. This effort was initiated for two reasons. The wind turbine, named Tvindkraft, was built to provide inexpensive electricity and heat for a complex of three schools. The Tvind project was also meant to convince governmental authorities that Denmark's consistently strong winds could be harnessed to provide at least some of the generating capacity which would be provided by proposed nuclear power plants. The project has been financed by the pooled paychecks of the Tvind schools' 80 teachers. The Tvind Community is located on the northwestern coastal plateau of Jutland about 10 km from the coast in the direction of the prevailing westerly winds. The Tvind project has been described in the *Consulting Engineer* (1976), and by Dornberg (1979), and Pederson (1979).

The Tvind wind turbine has a three-bladed, 54-m diameter, downwind rotor which is mounted on a 53-m, slipformed concrete tower. The rotor drives a 2-MW alternator. The major characteristics are given in Table 4-9. The Tvind wind turbine is shown in Figure 4-26.

Table 4-9. Main characteristics of the Tvind wind turbine.

Rotor location	Downwind
Rotor diameter and area	54 m and 2290 m^2
Hub height	53 m
Number of blades	3
Maximum rotational speed	42 rpm
Blade construction	Fiberglass spar
Regulation	Pitch control
Generator	2 MW
Performance	Self-starting at appr. 5 m/s, 2 MW at appr. 14.8 m/s
Transmission	Conventional, ratio appr. 1:18.3
Tower	Reinforced concrete
Weight of one blade	5200 kg
Weight of nacelle	100,000 kg

From Pederson (1979).

Early plans called for factory construction of fiberglass blades, but manufacturers were reluctant to undertake this task. The Tvind group obtained experience with fiberglass technology by building three fishing boats for the Continuation School and then built the blades themselves. The variable pitch blades are made of epoxy-reinforced fiberglass and polyester foam, except for the integral steel mounting rings which connect them to the hub. Strength was added to the bond between the glass-reinforced plastic (GRP) blade and its mounting ring by wrapping the continuous glass strand around the mounting bolts which link the end plate and the mounting ring. The blades are 27 m long, have a maximum width of 2.1 m, and weigh approximately 5200 kg each. They were manufactured inside a large tent borrowed from the Danish army. Figure 4-27 shows the Tvind team carrying the first blade from the construction tent. The blade profiles are based on the NACA 23000 series. The profile at the root is 23021 and at the tip is 23012. The twist of the blade varies from -3° at the tip to +40° at the root.

The design rotational speed of the Tvind machine is 40 rpm. The pitch-change assembly is hydraulically operated. The design calls for the blade pitch to be held constant at wind speeds up to 15 m/s, feathered to maintain the rated output for wind speeds between 15 and 20 m/s, and feathered for cut-out at a wind speed of 20 m/s. A unique fail-safe device is provided for overspeed protection. If the rotor speed exceeds 42 rpm, parachutes installed in the blade tips are released. The parachutes are held in place by trap doors secured by magnets. The magnets are released by centrifugal force.

Fig. 4-26. The Tvind wind turbine. Courtesy of the Tvind Schools.

The yaw drive mechanism is hydraulically operated. The yaw rate of the 100,000-kg nacelle is one-third of a revolution per hour. The low-speed shaft from the rotor hub to the gearbox is a salvaged ship propeller shaft. The step-up gearbox was originally designed as a 19:1 step-down gearbox for a mine winder. The mine did not take delivery and the Tvind group obtained this 18,000-kg unit for a small fraction of its original price. The high-speed shaft drives a three-phase, 800-rpm AC generator rated at 2 MW and 3300 V. The generator output is transmitted through high-tension cables in the tower to transformers at ground level. The power is transformed to 380 V, a rectifier-inverter converts this to DC, and a thyristor-type inverter finally yields AC constant frequency power at 50 Hz.

The machine is expected to produce 4000 MWh per year. The Tvind community's electrical demand amounts to 1 MW. Tvindkraft's power will be used for the electrical and water-heating needs of the community and excess power may be sold to the electricity supply company. Many technical problems were solved during the construction and test phase of the project. At the present time, larger parachutes for the fail-safe system are being tested. It is expected that the machine will begin running without limitation in 1981. The Tvind project is a remarkable example of initiative and cooperation.

Fig. 4-27. Students, teachers, and the mill team remove the first blade from the construction tent. Courtesy of the Tvind Schools.

DANISH NATIONAL PROGRAM

In 1976, the Danish government invited utilities to manage a program for the development of large wind turbine generators. The costs were to be shared in a 3:1 ratio between the government and utilities. The Danish Wind Energy Program was initiated in 1977. The program is administered and coordinated by the Research Association of the Danish Electricity Supply Undertakings (DEFU). Engineers from Electric Service Amalgamation (ELSAM), ELKRAFT, and DEFU are in charge of the management of individual projects. The Technical University of Denmark and the Risø National Laboratory are participating in the implementation of the projects. The program has been described by Pederson (1979), Grastrup (1979), and DEFU (1979).

Early program activities included measurements on the 20-year-old Gedser wind turbine that has been restored with a view to limited test operation, the design and construction of two new wind turbines, and site evaluation and selection for the new plants.

The Gedser wind turbine was designed and built in 1956-1957. After a test period, the turbine was used in normal operation from 1959 to 1967. During this period, it produced approximately 2242 MWh of electrical energy. In 1964, the year of highest annual production, the output was 367.14 MWh. The capacity factors based on the full period and best year are approximately 0.14 and 0.21, respectively. The main characteristics of the Gedser wind turbine are given in Table 4-10. The U.S. DOE helped support the Gedser restoration and measurement project. The cooperation consisted of a mutual exchange of measured results and analyses for the Gedser and U.S. Mod series wind turbines. The U.S. interest in the Gedser machine was prompted by design features which were different from those of the Mod machines and the favorable long-term experience with the Danish machine. The Gedser wind turbine has a three-bladed rotor with fixed pitch and the rotor is located on the upwind side of a concrete tower. A schematic drawing of the Gedser machine is shown in Figure 4-28. The machine was operated from November, 1977 to April, 1979. The measurement program provided information concerning the dynamic behavior of the rotor, the power curve for this stall-regulated machine, the power quality obtainable from an induction generator, and measuring techniques (Pederson, 1979, DEFU, 1979).

The main effort of the Danish program has been the design and construction of two 630-kW wind turbines, Model A and Model B. The technical specifications for these machines are shown in Table 4-11 and schematic drawings are shown in Figure 4-29. The machines have been erected near the provincial town of Nibe. The principal difference in these machines is that Model A is stall-regulated while Model B is fully pitch-regulated. The hubs for both machines are rigid (non-flapping) and made of welded, heavy steel plates.

Both rotors have identical airfoils, NACA 4412-4433 trapezoidal shape, and 11° twist. The blades have a 12-m outer section made entirely of fiberglass laminate

Table 4-10. Main characteristics of the Gedser wind turbine.

Rotor location	Upwind
Rotor diameter	24 m
Number of blades	3
Blade tip speed	38 m/s
Rotational speed	30 rpm
Rotor area	450 m^2
Blade construction	Steel main spar, wooden webs, aluminum skin Heavily stayed Braking flaps in blade tips
Regulation	Stall regulated, no pitch control
Generator	Asynchronous 200 kW, 750 rpm
Transmission	Double chain 1:25
Tower	Stiffened concrete cylinder
Hub height	24 m
Performance	Self-starting at 5 m/s, 200 kW at 15 m/s Typical annual production – 350,000 kWh/y
Weight of one blade	1650 kg

From Pederson (1979).

(polyester, E-glass) and an 8-m inner section comprising fiberglass shells attached to a steel spar. Model A has a hub with the 3 inner-most 8-m blade sections stayed to it. A shaft running through these sections carries the outer 12-m blade sections and permits the pitch to be shifted to four discrete values: a starting position of 15°, running positions of 1° for wind speeds below 10 m/s and –4° for wind speeds above 10 m/s, and a brake position of –20°. Model B has bearings carrying cantilevered blades with 90° of varying pitch. Schematic drawings of the Model A and B blades are shown in Figure 4-30. Figure 4-31 shows the design pitch regulation strategy and constant power curves for Models A and B. Figure 4-32 shows the design power output curves for these machines.

The yaw drive and pitch regulation mechanisms are hydraulically operated. Yaw restraint is provided by four disk brakes which can resist a yaw moment of 300 kg/nm. The tower is a reinforced concrete conical structure. The calculated values of the two lowest natural frequencies are 1.38 Hz and 7.34 Hz (2.4 and 12.8 P, respectively). During startup, the machines will pass through resonance with the 3-P excitation.

Tests of the Models A and B began in 1979 and 1980, respectively. In the next phase of the Danish program, activities will include measurements on and evaluation of Models A and B, study of the possibilities for large-scale wind generation, environmental assessments, and theoretical studies of relevant aerodynamic issues.

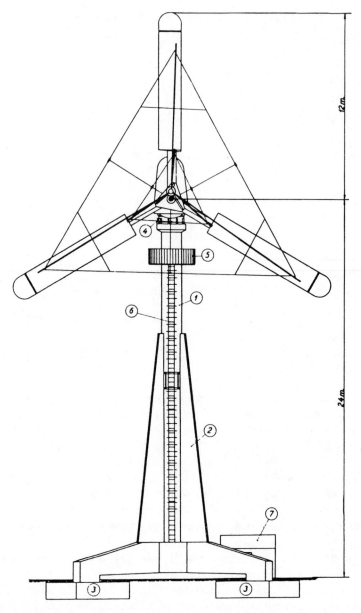

Fig. 4-28. Schematic drawing of the Gedser 200-kW wind turbine. Courtesy of H. Grastrup, ELSAM.

Table 4-11. Technical specifications for Models A and B.

	Mod A	Mod B
Rotor diameter	40	40
Hub height, m	45	45
Tower height, m	appr. 41	appr. 41
Rotor location	upwind	upwind
Number of blades	3	3
System life, yr	25	25
Wind speed:		
Cut-in, m/s	5	5
Rated, m/s	appr. 13	appr. 13
Cut-out, m/s	25	25
Weight of 1 blade, kg		3.370
Rotor speed, rad/s	appr. 3.5	appr. 3.5
Rotor cone angle, deg	6	6
Rotor tilt angle, deg	6	6
Yaw rate, deg/s	0.4	0.4
Pitch regulation		
Range, deg	+15 to –20	+90 to –1
Maximum speed, deg/s	6	8
Normal speed, deg/s	1	6
Generator:		
Type	Asynch., 4-pole	Asynch., 4-pole
Installed power	appr. 630 kVA	appr. 630 kVA
Weight, kg	appr. 4000	appr. 4000
Transmission:		
Type	conventional	conventional
Ratio	appr. 10000	appr. 10000
Weight, kg		

From Pederson (1979).

WTG MACHINE

WTG Energy Systems, Inc. of Angola, New York has designed, built, and tested a three-bladed, upwind, fixed pitch, horizontal-axis wind turbine which produces its rated power of 200 kW at a wind speed of approximately 13.4 m/s (30 mph). The design for the WTG machine is based on the Gedser wind turbine. The machine, designated MP1-200, is installed on Cuttyhunk Island, Massachusetts, as part of the

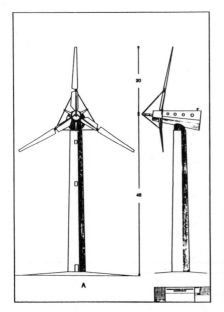

Fig. 4-29. Nibe 630 kW demonstration wind turbines: (a) Model A, and (b) Model B. From Grastrup (1979).

island's independent diesel-powered utility system. A photograph of the MP1-200 at the Cuttyhunk site is shown in Figure 4-33. A schematic drawing is shown in Figure 4-34. The general characteristics, control system, and blades of this machine have been described by Spaulding (1979) and Rose (1979). A similar machine, designated MP2-200, is being installed on Cape Breton Island, Nova Scotia.

The MP1-200 produces supplemental energy for the Town of Gosnold. Most of the town's electrical energy is produced by six diesel generator sets ranging in capacity from 30 to 175 kW. The total capacity amounts to 455 kW. Startup, shutdown, and synchronization operations are manually controlled. The wind turbine must be able to maintain synchronization with diesel plants and operate isochronously when it is the only generator supplying the town's load. The pitch of the MP1-200 rotor is adjusted to attain its peak power coefficient at a wind speed of 8.05 m/s (18 mph). At higher speeds, the rotor stalls in a smooth and predictable fashion. The power produced by the synchronous generator increases up to its peak value and levels off. The speed of the fixed pitch rotor, and hence the generator rpm, is maintained by an off-the-shelf controller. The controller is based on an Intel 8080A microprocessor chip which uses a modified assembly language developed by the manufacturer. In isolated networks where wind generators can have significant impact, a responsive, independent, speed control system is required to track the higher frequency components present in gusty winds which have speeds within the operational range.

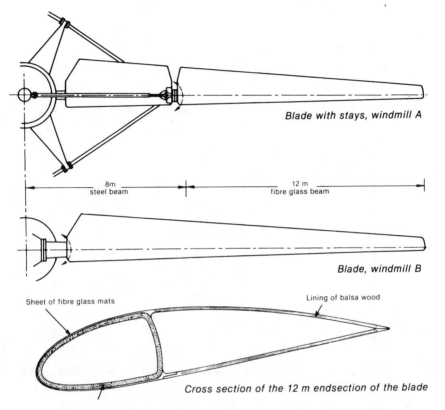

Blade with stays, windmill A

8m
steel beam

12 m
fibre glass beam

Blade, windmill B

Sheet of fibre glass mats

Lining of balsa wood

Cross section of the 12 m endsection of the blade

Fig. 4-30. Blades of the Models A and B wind turbines. From DEFU (1979).

Nova Scotia Power has purchased an MP2-200 wind turbine for use in a wind-hydro pumpback project (*Wind Energy Report*, 1979a). The turbine has been installed within 1000 ft of an existing pumping plant near the Wreck Cove hydroelectric plant. When the wind speed falls in the operational range and the reservoir is empty, the output will be fed to the Nova Scotia grid. When wind and water are available, output from the wind turbine will operate the pumps. When there is water and no wind, power from the grid will operate the pumps. WTG Energy Systems, Inc., estimates that the turbine will produce 600 MWh/year at the Wreck Cove site. This amount corresponds to a capacity factor of 0.34.

SOUTHERN CALIFORNIA EDISON PROJECT

The Southern California Edison Company (SCE) has initiated a program to construct and test a 3-MW wind turbine at a site in the San Gorgonio Pass near Palm Springs, California (Scheffler, 1979a,b, 1980). Southern California Edison has become especially interested in wind energy conversion for two reasons. Over

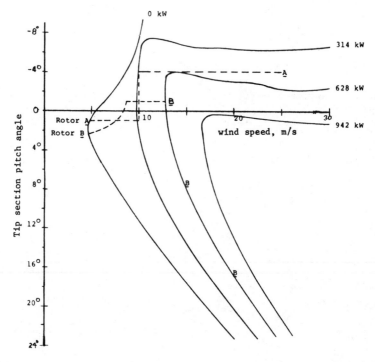

Fig. 4-31. Constant power curves and pitch regulation schedules for the Models A and B wind turbines. From Pederson (1979).

two-thirds of SCE's generating capacity consists of oil-fired systems. The high costs of low-sulfur fuel oil and embargo possibilities provide ample reason for reducing this dependence on oil. Second, the SCE wind resource shows great promise. A site in the San Gorgonio Pass near their Devers substation was selected in the first competition for Federal wind turbine candidate sites. Although this site has not been chosen for a Federal demonstration project, measurements at the DOE meteorological tower indicate mean annual wind speeds and power densities of 8.40 m/s (18.8 mph) and 951 W/m^2, respectively, at 200-ft elevation. Southern California Edison evaluated wind turbine designs of both DOE and private suppliers and one of the private designs studied was that of Charles Schachle's Wind Power Products, Inc. Schachle's development program began in 1970. His efforts culminated in the construction of a 140-kW prototype system with a blade span of 21.9 m (72 ft) which has been operational since May 1977 at Moses Lake, Washington. Southern California Edison decided to purchase and test a Schachle machine with a rated power of 3 MW.

The purpose of the project is to demonstrate the technical and economic feasibility of this machine as an intermediate step toward the widespread utilization of

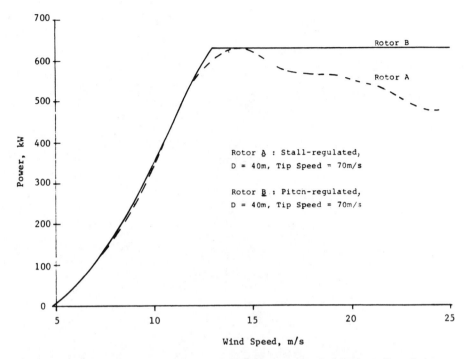

Fig. 4-32. Design power output curves for the Models A and B wind turbines. From Pederson (1979).

commercial hardware. Rights to the Wind Power Products, Inc. machine were acquired by the Bendix Company in 1979. Bendix Wind Power Products, Inc. has assumed responsibility for the installation of the 3-MW machine at the SCE site (*Wind Energy Report*, 1979b).

The machine has a three-bladed upwind rotor with a diameter of 50.3 m (165 ft). It is designed to produce 3 MW of electrical power at a wind speed of 17.9 m/s (40 mph). A simulation of the Schachle-Bendix machine is shown in Figure 4-35. The rotor is aligned with the wind flow through the rotation of a tubular steel tower which is mounted on a concrete and steel base. The blades are made of laminated wood and fiberglass. The machine uses a hydraulic pump-motor link between the rotor hub and the generator which is located at ground level. This permits a less complex gearbox design and allows the rotor blades to operate more efficiently at speeds which vary in proportion to the speed of the wind. The fluid drive system is expected to achieve power output which is approximately double that of constant speed machines at high wind speeds. The hydraulic system is designed to operate at a fixed pitch in the interval from the cut-in to rated wind speed. Above the rated wind speed, hub pitch control is used to maintain a constant maximum rotor speed.

Fig. 4-33. WTG MP1–200 wind turbine on Cuttyhunk Island. Courtesy of WTG Energy Systems, Inc.

The machine was rated at 17.9 m/s (40 mph) to maximize the annual energy output at the SCE test site. An analysis of the wind data collected near this high-wind site indicated that a machine rated at 17.9 m/s would produce approximately twice the energy of a machine rated at 11.2 m/s (25 mph). The value of the additional energy is expected to more than offset the increased cost of the higher rated machine. The machine is expected to produce an annual output of approximately 6000 MWh at the test site. This amount corresponds to a capacity factor of 0.23.

The wind turbine and data recording system are capable of unattended operation. Southern California Edison personnel at the Devers substation, one-quarter mile from the site, will monitor the operation and performance of the test unit. The Schachle-Bendix machine was interconnected in December 1980. Southern California Edison has plans to test additional large machines as they become available.

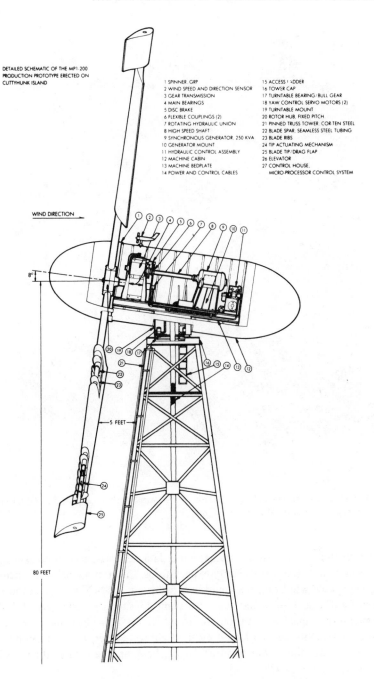

DETAILED SCHEMATIC OF THE MP1-200
PRODUCTION PROTOTYPE ERECTED ON
CUTTYHUNK ISLAND

1 SPINNER, GRP
2 WIND SPEED AND DIRECTION SENSOR
3 GEAR TRANSMISSION
4 MAIN BEARINGS
5 DISC BRAKE
6 FLEXIBLE COUPLINGS (2)
7 ROTATING HYDRAULIC UNION
8 HIGH SPEED SHAFT
9 SYNCHRONOUS GENERATOR, 250 KVA
10 GENERATOR MOUNT
11 HYDRAULIC CONTROL ASSEMBLY
12 MACHINE CABIN
13 MACHINE BEDPLATE
14 POWER AND CONTROL CABLES

15 ACCESS LADDER
16 TOWER CAP
17 TURNTABLE BEARING/BULL GEAR
18 YAW CONTROL SERVO MOTORS (2)
19 TURNTABLE MOUNT
20 ROTOR HUB, FIXED PITCH
21 PINNED TRUSS TOWER, COR-TEN STEEL
22 BLADE SPAR, SEAMLESS STEEL TUBING
23 BLADE RIBS
24 TIP ACTUATING MECHANISM
25 BLADE TIP/DRAG FLAP
26 ELEVATOR
27 CONTROL HOUSE,
   MICRO-PROCESSOR CONTROL SYSTEM

WIND DIRECTION

8°

5 FEET

80 FEET

Fig. 4-34. Schematic of the MP1-200. From Rose (1979).

Fig. 4-35. Simulation of the Schachle-Bendix 3-MW wind turbine at the San Gorgonio Pass site. Courtesy of Southern California Edison.

If the results of the test programs, wind data analyses, economic assessments, and environmental assessments verify expectations, the commercial implementation of megawatt scale wind turbines could be a reality on the SCE system in the 1980s. Southern California Edison has tentative plans to site approximately 100 MW of wind generation capacity at various locations in the San Gorgonio Pass between 1986 and 1991. Studies conducted by the California Energy Commission and the utility suggest that the power potential of the Pass may be greater than 1000 MW (*Wind Energy Report*, 1978).

UNITED KINGDOM DESIGN STUDY

A design feasibility and cost study of large-scale wind turbines was supported by the United Kingdom (UK) Department of Energy on a cost-sharing basis. The study group, known as the Wind Power Group, consists of Hawker Siddeley Dynamics (now British Aerospace Dynamics), Cleveland Bridge and Engineering Company, Electrical Research Association, and Taylor Woodrow Construction. A design was sought which would minimize the cost of energy at high wind hill sites in the UK while using a universal turbine design. The Wind Power Group (1979) has described this study effort.

The result was a base-line design for a two-bladed, 60-m diameter, horizontal-axis wind turbine which produces 3.7 MW at a wind speed of approximately 22 m/s (49 mph). The fixed pitch rotor operates upwind of the tower at a nominal fixed speed of 34.1 rpm and the hub is located 45 m above ground level. The major characteristics of the base-line design are given in Table 4-12. Two different tower-nacelle designs were developed. One design incorporates a circular reinforced concrete tower and a steel framed nacelle assembly. The turbine gearbox and generator are fully cantilevered from the tower. The nacelle's steel frame is covered with glass-reinforced plastic. The nacelle is supported from the tower by the orientational gear. A schematic drawing of this design is shown in Figure 4-36. The tower comprises a hollow, reinforced concrete cylinder of 8-m outside diameter with a 0.35-m thick wall. The second design incorporates an open lattice steel tower and a steel-framed nacelle which supports the turbine, gearbox, and generator. The tower comprises circular hollow steel sections. A schematic drawing of this design is shown in Figure 4-37.

The blades are made of welded mild steel to BS 4360 Grade 50. The rotor is composed of five main sections and is assembled with bolted flange joints. The upwind rotor configuration was chosen to minimize cyclic stress caused by tower shadow effects. The rigid steel construction is expected to yield a maximum blade deflection of 1.4 m. The turbine may not be self-starting. Provision is made to use the generator as a motor to start the fixed pitch rotor. Between the cut-in and cut-out wind speeds of 7 and 27 m/s, the power output and speed are governed by matching turbine/generator torque. The power output curve is shown in Figure 4-38. Spoilers are deployed and the generator is disconnected when the wind speed exceeds the cut-out value of 27 m/s. The spoilers are deployed by releasing air pressure which holds them in the retracted position. When the rotor speed slows to 10 percent of full value, a friction brake is applied. When stopped, the blades are turned to the horizontal position and the yaw drive system turns them parallel to the wind.

Coupling between the turbine shaft and the gearbox is provided by a standard double engagement gear. Speedup is provided by a two-stage, divided drive, helical

Table 4-12. Major characteristics of the base-line United Kingdom wind turbine.

Rotor diameter	60 m (197 ft)		
Number of blades	2		
Design rotational speed	34.1 rpm		
Hub type	Rigid		
Tower height	45 m (148 ft)		
Overspeed protection	Spoilers and friction brakes		
Generator			
Type	Induction		
Rating	4.1 MVA		
Power factor	0.9		
Voltage	3.3 kV		
Speed	750 rpm		
Frequency	50 Hz		
Blade characteristics			
Radius	10%	50%	100%
Chord, m	3.795	2.515	0.915
Angle	20.00°	4.43°	–1.25°
Airfoil section, NACA:	4421	4414	4412
Weights			
Blades and hub	40,000 kg		
Shaft, housing, and bearings	19,500 kg		
Transmission	43,300 kg		
Moment of inertia (about axis of rotation)	6.12×10^6 kgm^2		
Characteristic wind speeds			
Cut-in	7 m/s (15.7 mph)		
Rated	22 m/s (49.2 mph)		
Cut-out	27 m/s (60.5 mph)		
Survival	60 m/s (134 mph)		
System life	20 years		

From Wind Power Group (1979).

spur coaxial gear transmission rated at 3.9 MW with a nominal 35:750 ratio. The high-speed shaft is coupled to an eight-pole, cage-type induction generator rated at 3.7 MW.

The turbines are expected to produce approximately 9600 and 11,000 MWh/y of electrical energy at sites with annual mean wind speeds of 9 and 10 m/s, respectively, referenced to 10-m elevation above ground level. These outputs correspond to capacity factors of 0.30 and 0.34, respectively. The Wind Power Group has estimated that 211 60-m base-line machines located at hill sites on the West Coast of

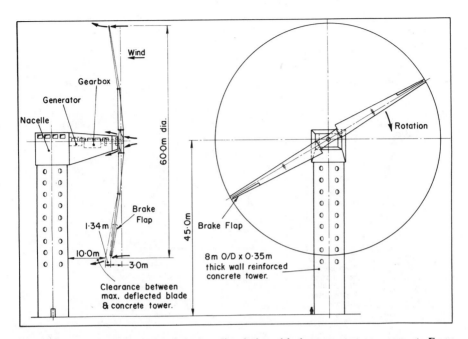

Fig. 4-36. General arrangement of the base-line design with the concrete tower concept. From "Development of Large Wind Turbine Generators," a study supported by the U.K. Department of Energy through a shared cost contract with Hawker Siddely Dynamics Ltd, Cleveland Bridge & Engineering Co Ltd, Electrical Research Association Ltd, and Taylor Woodrow Construction Ltd.

Scotland having annual mean wind speeds in the range from 8.6 to 12.0 m/s could produce approximately 2220 GWh of electrical energy.

The two-blade, fixed-pitch configuration for the high-wind UK turbine was selected on a tentative basis. The designers judged that this configuration would yield a minimum cost of energy despite the lower power coefficients and larger oscillatory loads which occur with two blades rather than three or more, and the higher generator and transmission costs required by a fixed-pitch design. Component trade-off studies and wind tunnel tests were conducted to confirm the selection of the configuration prior to prototype construction. The decision to begin construction was made in January 1981.

PROGRAM OF THE FEDERAL REPUBLIC OF GERMANY

The Wind Energy R&D Program of the Federal Republic of Germany is part of the government's Program for Energy Research and Technologies. Theoretical studies, wind data analyses, and experience with the 100-kW Hütter wind turbine indicated

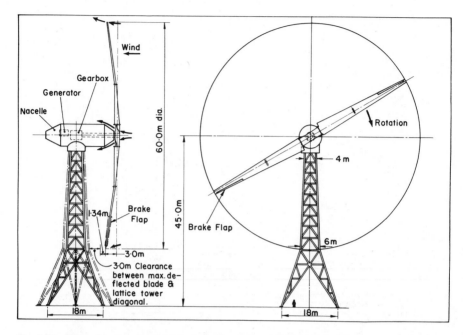

Fig. 4-37. General arrangement of the base-line design with the steel lattice concept. See credit for Figure 4-36.

that large-scale machines could supply significant generation to the national grid. In 1976, the Growian (*Gr*osse *Wi*ndenergie *An*lage) program was initiated. The program is designed to develop and test large-scale wind turbines from the 100-kW scale up to the 5-MW scale while also assessing the technical, economic, and institutional requirements for their widespread use. The German program has been described by Neumann and Windheim (1978) and Windheim and Neumann (1979).

The West German Ministry for Research and Technology awarded a contract to Maschinenfabrik Augsburg-Nürnberg Aktiengesellschaft (MAN-Neue Technologie) to design the Growian I wind turbine. Körber (MAN-Neue Techologie, 1979a,b) has summarized the design work. The Growian I is a two-bladed, horizontal-axis machine with a rated power of 3 MW. The major characteristics of this machine are given in Table 4-13. An artist's drawing of the Growian I is shown in Figure 4-39 and a schematic drawing of the nacelle arrangement is shown in Figure 4-40.

The two-bladed rotor has a diameter of 100.4 m and operates downwind from a slender cylindrical tower made of either reinforced concrete or steel. The hub is located approximately 100 m above ground level. The 46-m steel spar rotor blade is assembled in three segments. The steel spar extends from the root to the tip and molded fiberglass segments give the blade its airfoil form. The blades are mounted on a teetered (pendulum) hub. They rotate about their bladewise axes on anti-friction bearings. Blade pitch is adjusted by a motor driven linkage. Overspeed protection is provided by a device activated by centrifugal force which turns the blades to their feathered position.

A planetary gear system of ratio 1:81 is used to step up the rotor speed from approximately 18.5 rpm to the generator speed of 1500 rpm. The drive of the planetary gear is bolted to the rotor shaft. The gear system consists of two plane-

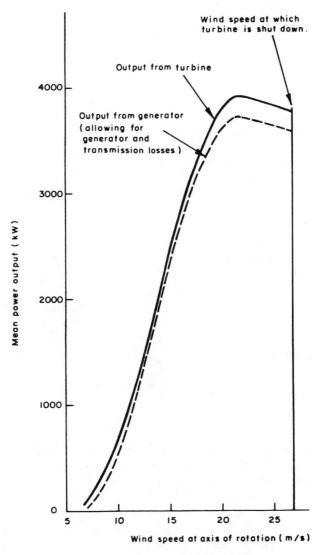

Fig. 4-38. Power-wind speed characteristic for the base-line design. See credit for Figure 4-36.

Table 4-13. Major characteristics of the Growian I.

Two-blade rotor with teeter (pendulum) hub	
Downwind operation	
Blade of steel-spar design with glass fiber airfoil	
Single guyed tower	
Controlled yaw operation	
Rotor diameter	100.4 m
Rotor speed	18.5 rpm ± 15%
Hub height	100 m
Blade profile	FX-77-W
Blade thickness (root/tip)	4.25 m/1.3 m
Maximum power coefficient	0.45
Tower height	96.6 m
Tower diameter	3.5 m
Gearing	1:81
Generator	3 MW/1500 rpm ± 15%, 6.3 kV/50 Hz
Yaw drive system	Electromechanical, rate: 0.5°/s
Mean annual energy output	12 GWh[a]
Power to area ratio at rated power	380 W/m^2
Characteristic wind speeds	
Cut-in	6.3 m/s (14.1 mph)
Rated	12.0 m/s (26.8 mph)
Cut-out	24.0 m/s (53.7 mph)
Survival	60.0 m/s (134 mph)
Mass of tower head with rotor	240 t

[a]This output estimate is based on an annual mean wind speed of 6 m/s at 10 m above ground level, the value for a coastal site on the North German Plain.

From Windheim and Neumann (1979) and Körber (1979).

tary stages and one spur gear stage. A disk brake installed at the high-speed side of the gearbox is able to arrest the rotor at full speed in an emergency. A universal shaft connects the gearbox and generator.

An asynchronous generator, whose rotor is energized with alternating current via slip rings, permits operation at variable speeds. The frequency and the static voltage are held constant when the wind turbine is connected to the supply network. The speed of rotation and power output are regulated by controlling the blade pitch and the generator moment. The generator deviates elastically in the subsynchronous or hypersynchronous modes of operation during fluctuations in the rotor speed until synchronous operation is reestablished by an overriding speed regulator. The performance characteristics of the wind turbine are shown in Figure 4-41. The wind turbine normally operates in the static control range. The ex-

Fig. 4-39. Simulation of the Growian I.   Courtesy of the Wind Energy R&D Program of the Federal Republic of Germany.

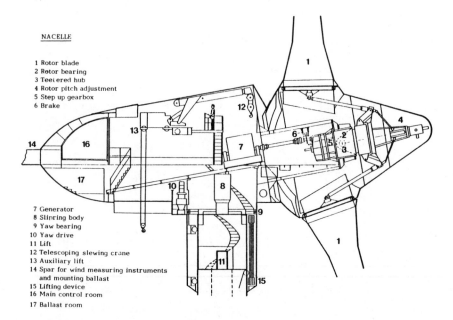

NACELLE

1 Rotor blade
2 Rotor bearing
3 Teei ered hub
4 Rotor pitch adjustment
5 Step up gearbox
6 Brake

7 Generator
8 Slinring body
9 Yaw bearing
10 Yaw drive
11 Lift
12 Telescoping slewing crane
13 Auxiliary lift
14 Spar for wind measuring instruments
   and mounting ballast
15 Lifting device
16 Main control room
17 Ballast room

Fig. 4-40. Configuration of the Growian I nacelle. Courtesy of the Wind Energy R&D Program of the Federal Republic of Germany.

tended limits of the dynamic control range are employed to control brief fluc-tuations in the rotor speed.

The yaw drive system is operated by geared motors whose pinions engage a gear ring integral with the tower. The central system senses signals from wind measuring instruments mounted on a 20-m spar which projects from the windward side of the nacelle.

The erection concept is illustrated in Figure 4-42.

The Growian-Bau and Betriebsgesellschaft mbH, a company which consists of the three German electrical supply utilities HEW, RWE, and Schleswag, initiated the contruction phase of the Growian I in 1979. The project is scheduled for comple-tion in 1982. The wind turbine will be operated at the Kaiser-Wilhelm-Koog which is located at the mouth of the river Elbe near the North Sea.

In 1978, Messerschmitt-Bölkow-Blohm GmbH (MBB) initiated a study of ad-vanced concepts for large wind turbines. This study produced the Growian II design. The Growian II is a one-bladed, 145 m diameter, horizontal-axis machine which is expected to produce 5 MW at a wind speed of 11.3 m/s (25.3 mph). The hub is located approximately 120 m above ground level. The major characteristics of this machine are given in Table 4-14 and a simulation is shown in Figure 4-43. The Growian II project is being developed under the auspices of KFA Julich GmbH. A one-third scale demonstration unit with a rated power of 300 kW is to be built and tested.

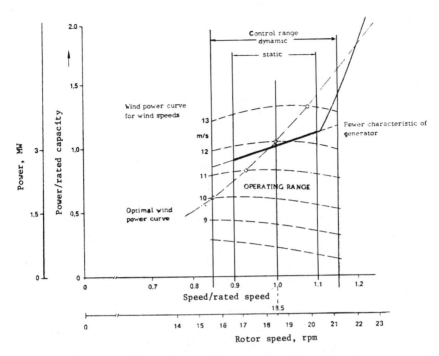

Fig. 4-41. Performance characteristics of the Growian I. Courtesy of the Wind Energy R&D Program of the Federal Republic of Germany.

The preparation of the manufacturing documentation for a full-scale prototype was completed in 1980.

SWEDISH PROGRAM

The Swedish wind energy program was initiated in 1975. The National Swedish Board for Energy Source Development is responsible for the implementation of the program. The program is designed to provide a factual basis for a parliamentary decision in 1985 concerning the development of large wind generation capacity. Brandeis (1979) has given an overview of the Swedish program and Hugosson (1979a,b) has described the specifications, siting, and selection for large wind turbine prototypes.

A two-bladed, horizontal-axis experimental unit was designed and built by Saab Scania. The major characteristics of this research unit are given in Table 4-15. The turbine is presently under test at Kalkugnen, 200 km north of Stockholm on the Baltic coast. The unit began providing test data for the aluminum blade/rigid hub combination late in 1977. This combination accumulated 846 hours of operation before the hub was changed in May, 1978. Over 2000 hours of operation were then

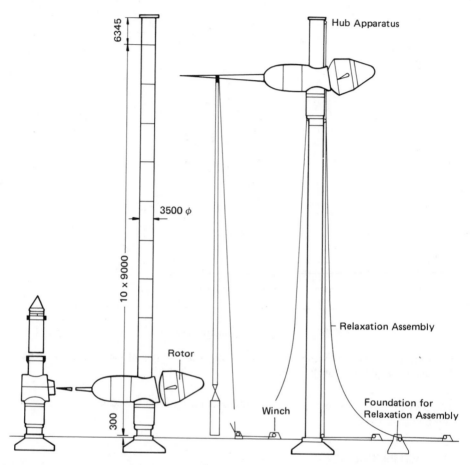

Fig. 4-42. Erection concept for the Growian I. Courtesy of the Wind Energy R&D Program of the Federal Republic of Germany.

obtained with the aluminum blade/flapping-hub combination. The GRP (glass-reinforced plastic) blade/flapping-hub combination is now under test. Finally, the CRP/GRP (CRP is compounded reinforced plastic) blade/flapping-hub combination will be tested.

The Swedish researchers concluded that theoretical studies concerning cost, technique, and availability for MW-scale systems could be verified best through the design and construction of prototype systems. Technical specifications were developed for the design of large wind turbine systems and in 1978, a Request for Proposals was mailed to interested companies (Hugosson, 1979a). The specifications for the prototype systems permitted the proposers to choose two or three blades, a diameter within the range 70 to 90 m, and a rated power within the range 2 to 4 MW. Mandatory requirements included blade pitch control, remote control and monitoring, and access to the nacelle during operations.

Hugosson (1979b) has described the prototype designs which were selected. One prototype is contracted to Karlskronavarvet AB (KKRV), part of the Swedish State Shipyard Group, together with Hamilton Standard (HS), a division of United

Table 4-14. **Major characteristics of the Growian II.**

System concept	Single blade, downwind rotor
Installed capacity	5 MW
Projected life span	⩾ 20 years
Hub height	appr. 120 m
Rotor diameter	145 m
Blade profile	Wortmann FX 77-W
Blade thickness (root tip)	7m/1.3 m
Rotor speed	16-18 rpm
Rotor concept	Single blade with counterweight Flapping hinge and pitch control
Pitch control	Electromechanical
Yaw control	Electronically regulated
Gear system	Two stage planetary gearing with high stage spur gear
Generator	Double fed asynchronous machine
Generator speed	1500 ± 200 rpm
Tower	Cylindrical reinforced concrete tower, 3.5 m in diameter, three stays
Characteristic wind speeds referenced to hub height elevation	
Cut-in	6.6 m/s (14.8 mph)
Rated	11.25 m/s (25.2 mph)
Cut-out	20 m/s (45 mph)
Weights Rotor (blade, counterweight, hub, pitch control)	appr. 100 t
Nacelle (gearbox, shaft, generator, structure, yaw mount, and ballast)	appr. 300 t
Tower (structure and cables)	appr. 850 t

From: Messerschmitt-Bölkow-Blohm GmbH (1979).

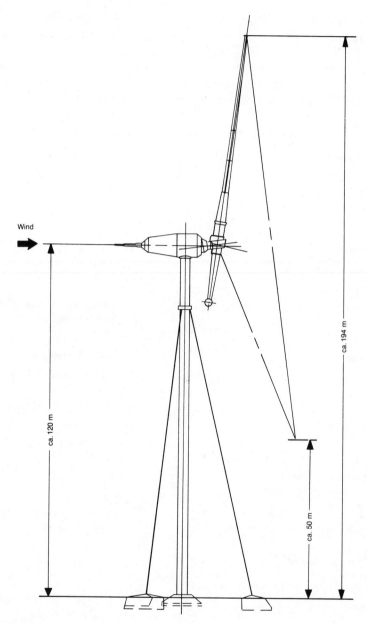

Fig. 4-43. Simulation of the Growian II. Courtesy of the Wind Energy R&D Program of the Federal Republic of Germany.

Table 4-15. Major characteristics of the Swedish research unit.

Tower	Concrete, diameter 2 m
Hub height	25 m
Hub type	(a) Rigid; (b) Flapping
Turbine diameter	(a) 18 m; (b) 24 m
Turbine rpm	77
Rotor blades	(a) Aluminum; (b) GRP: (c) CRP + GRP
Rated power	63 kW (75 kW)
Generator	380 V, asynchronous
Grid voltage	10 kV

From Hugosson (1979a).

Technologies Corporation. The other prototype is contracted to Karlstads Mekaniska Werkstad AB (KMW) together with ERNO, a division of the German VFW-Fokker group. The contracts of $10 million each call for turnkey delivery at specified sites.

Sweden's two largest utilities, Statens Vattenfallsverk and Sydkraft AB have contracted to play important roles during the installation and operation phase of the projects. The energy produced by the prototypes will be purchased by the utilities. The KKRV-HS prototype will be sited at Maglarp in the province of Skåne within the Sydkraft operating area, and the KMW-ERNO unit on the island of Gotland within the Vattenfall operating area. The main characteristics of the two prototypes are summarized in Table 4-16. The project schedule is given in the lower part of the table. Figures 4-44 and 4-45 show artist's drawings of the KKRV-HS and KMW-ERNO machines, respectively.

Both prototype units are two-bladed horizontal-axis machines. The 3-MW KKRV-HS prototype features include teetered hub, downwind rotor, full-span pitch control, soft system dynamics, steel shell tower, synchronous generator, and free yaw operation. The 2-MW KMW-ERNO prototype features include rigid hub, upwind rotor, full-span pitch control, stiff system dynamics, reinforced concrete tower, induction generator, and controlled yaw operation. The two designs utilize a wide range of technological options.

HAMILTON STANDARD PROGRAM

Hamilton Standard, a Division of United Technologies, is developing a 3-MW wind turbine in conjunction with Karlskronavaret AB of Sweden. As described earlier, this machine is one of two prototypes which were selected for development by the

Table 4-16. Characteristics of the KKRV-HS and KMW-ERNO prototype units.

Characteristic	KKRV-HS	KMW-ERNO
Hub height, m	80	80
Turbine diameter, m	77.6	75
Number of blades	2	2
Rated power, MW	3	2
Cut-in wind speed, m/s	6	6
Rated wind speed, m/s	14.2	12.5
Cut-out wind speed, m/s	21	21
Annual output, MWh	6000-8000	6000-8000
Rotor characteristics	Downwind GRP-Epoxy Filament wound design Full span pitch control 25 rpm	Upwind Steel box spar with GRP-Epoxy leading and trailing edges Full span pitch control 25 rpm
Hub	Teetered	Rigid
Gearbox	Epicyclic	Parallel plus right angle
Generator	Synchronous	Induction
Tower	Welded steel shell with low supports, diameter 4 m	Slip-form, cable reinforced concrete. Base diameter 9.5 m, tapering off to 4.5 m at the top.
System dynamics	Soft. Uncoupled dynamic modes	Stiff, especially concerning tower
Yaw control	Free in yaw	Hydraulic motors, slow
Grid connection, kV	55	30
Short circuit power, MWA	150	40
Schedule • Contracts signed • Start site wind data collection • Final design reviews • Start building activities at site • Tower erected • Nacelle and turbine installed • Delivery tests • Formal delivery to NE		July-September 1979 November 1979 Spring 1980 Summer 1980 Summer 1981 Autumn 1981 Winter 1981/82 Spring 1982

From Hugosson (1979b).

Fig. 4-44. Karlskronavarvet/Hamilton Standard 3-MW wind turbine at Maglarp. From Hugosson (1979b). Model picture taken at the site by the National Swedish Board for Energy Source Development.

Fig. 4-45. KMW/ERNO 2.4-MW wind turbine at Näsudden. From Hugosson (1979b). Model picture taken at the site by the National Swedish Board for Energy Source Development.

National Swedish Board for Energy Source Development. Hamilton Standard is designing the rotor, manufacturing two 38-m fiberglass blades, and performing a computerized analysis of the entire system. The major characteristics of this machine, designated WTS-3, are given in Table 4-16 under the heading "KKRV-HS." Additional data for this base-line system are given in Table 4-17. Figure 4-46 illustrates the key features of the Hamilton Standard base-line design. The nacelle and rotor system are shown in Figure 4-47.

The blades are of filament-wound fiberglass construction. Blade loads are transferred through a double-lap-type shear connection which eliminates the bending moment and associated peel loads inherent in single-lap types. Figure 4-48 shows how the fiberglass blade spar is connected through the root end retention to the pitch-change bearings. Each blade is attached to the rotor hub through a single plane set of tapered roller-bearings. The bearing is preloaded to keep the rolling element

Table 4-17. Additional characteristics of the WTS-3.

Rotor	
Type	Teetered
Airfoil	230 XX
Tip speed	102 m/s (228 mph)
Pitch control	Hydraulic
Electronic control	Micro processor; fast response
Generator	
Speed	1800 rpm
Frequency	50 Hz
Gearbox	Planetary; dynamically mounted with springs and dashpots
System power coefficient	0.412
Rotor power coefficient	0.448
Maximum survival wind speed	55.9 m/s (125 mph)
Design life	30 years
Annual output	
Mean annual wind speed at 9.14 m (30 ft)	
5.36 m/s (12 mph)	appr. 6,500 MWh
6.26 m/s (14 mph)	appr. 9,500 MWh
7.15 m/s (16 mph)	appr. 12,000 MWh
8.05 m/s (18 mph)	appr. 13,200 MWh
Projected production cost (1979 dollars)	
Installed cost	$1.8 M ($600/kW)
Operating cost	$9000/year (0.5%)
Cost of energy	2.5¢/kWh

From Hamilton Standard (1979a,b).

Key Features

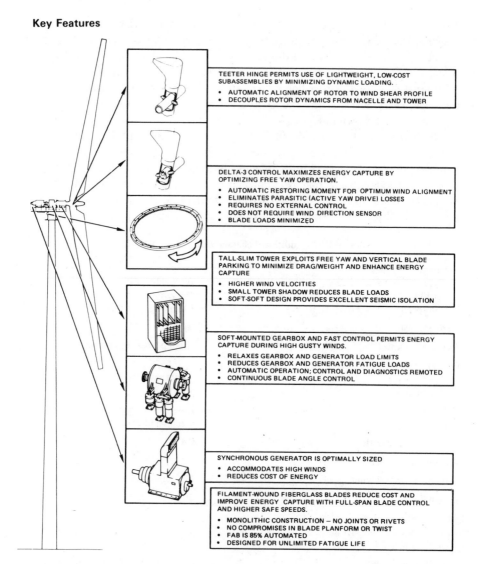

TEETER HINGE PERMITS USE OF LIGHTWEIGHT, LOW-COST
SUBASSEMBLIES BY MINIMIZING DYNAMIC LOADING.

* AUTOMATIC ALIGNMENT OF ROTOR TO WIND SHEAR PROFILE
* DECOUPLES ROTOR DYNAMICS FROM NACELLE AND TOWER

DELTA-3 CONTROL MAXIMIZES ENERGY CAPTURE BY
OPTIMIZING FREE YAW OPERATION.

* AUTOMATIC RESTORING MOMENT FOR OPTIMUM WIND ALIGNMENT
* ELIMINATES PARASITIC (ACTIVE YAW DRIVE) LOSSES
* REQUIRES NO EXTERNAL CONTROL
* DOES NOT REQUIRE WIND DIRECTION SENSOR
* BLADE LOADS MINIMIZED

TALL-SLIM TOWER EXPLOITS FREE YAW AND VERTICAL BLADE
PARKING TO MINIMIZE DRAG/WEIGHT AND ENHANCE ENERGY
CAPTURE

* HIGHER WIND VELOCITIES
* SMALL TOWER SHADOW REDUCES BLADE LOADS
* SOFT-SOFT DESIGN PROVIDES EXCELLENT SEISMIC ISOLATION

SOFT-MOUNTED GEARBOX AND FAST CONTROL PERMITS ENERGY
CAPTURE DURING HIGH GUSTY WINDS.

* RELAXES GEARBOX AND GENERATOR LOAD LIMITS
* REDUCES GEARBOX AND GENERATOR FATIGUE LOADS
* AUTOMATIC OPERATION; CONTROL AND DIAGNOSTICS REMOTED
* CONTINUOUS BLADE ANGLE CONTROL

SYNCHRONOUS GENERATOR IS OPTIMALLY SIZED

* ACCOMMODATES HIGH WINDS
* REDUCES COST OF ENERGY

FILAMENT-WOUND FIBERGLASS BLADES REDUCE COST AND
IMPROVE ENERGY CAPTURE WITH FULL-SPAN BLADE CONTROL
AND HIGHER SAFE SPEEDS.

* MONOLITHIC CONSTRUCTION — NO JOINTS OR RIVETS
* NO COMPROMISES IN BLADE PLANFORM OR TWIST
* FAB IS 85% AUTOMATED
* DESIGNED FOR UNLIMITED FATIGUE LIFE

Fig. 4-46. Key features of the Hamilton Standard base-line design. Courtesy of Hamilton Standard Division of United Technologies Corporation.

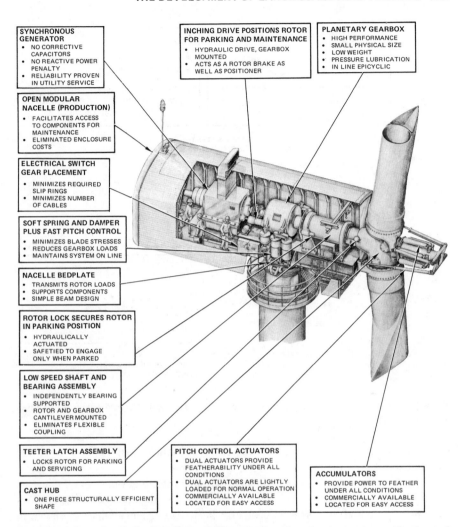

SYNCHRONOUS GENERATOR
- NO CORRECTIVE CAPACITORS
- NO REACTIVE POWER PENALTY
- RELIABILITY PROVEN IN UTILITY SERVICE

INCHING DRIVE POSITIONS ROTOR FOR PARKING AND MAINTENANCE
- HYDRAULIC DRIVE, GEARBOX MOUNTED
- ACTS AS A ROTOR BRAKE AS WELL AS POSITIONER

PLANETARY GEARBOX
- HIGH PERFORMANCE
- SMALL PHYSICAL SIZE
- LOW WEIGHT
- PRESSURE LUBRICATION
- IN LINE EPICYCLIC

OPEN MODULAR NACELLE (PRODUCTION)
- FACILITATES ACCESS TO COMPONENTS FOR MAINTENANCE
- ELIMINATED ENCLOSURE COSTS

ELECTRICAL SWITCH GEAR PLACEMENT
- MINIMIZES REQUIRED SLIP RINGS
- MINIMIZES NUMBER OF CABLES

SOFT SPRING AND DAMPER PLUS FAST PITCH CONTROL
- MINIMIZES BLADE STRESSES
- REDUCES GEARBOX LOADS
- MAINTAINS SYSTEM ON LINE

NACELLE BEDPLATE
- TRANSMITS ROTOR LOADS
- SUPPORTS COMPONENTS
- SIMPLE BEAM DESIGN

ROTOR LOCK SECURES ROTOR IN PARKING POSITION
- HYDRAULICALLY ACTUATED
- SAFETIED TO ENGAGE ONLY WHEN PARKED

LOW SPEED SHAFT AND BEARING ASSEMBLY
- INDEPENDENTLY BEARING SUPPORTED
- ROTOR AND GEARBOX CANTILEVER MOUNTED
- ELIMINATES FLEXIBLE COUPLING

TEETER LATCH ASSEMBLY
- LOCKS ROTOR FOR PARKING AND SERVICING

CAST HUB
- ONE PIECE STRUCTURALLY EFFICIENT SHAPE

PITCH CONTROL ACTUATORS
- DUAL ACTUATORS PROVIDE FEATHERABILITY UNDER ALL CONDITIONS
- DUAL ACTUATORS ARE LIGHTLY LOADED FOR NORMAL OPERATION
- COMMERCIALLY AVAILABLE
- LOCATED FOR EASY ACCESS

ACCUMULATORS
- PROVIDE POWER TO FEATHER UNDER ALL CONDITIONS
- COMMERCIALLY AVAILABLE
- LOCATED FOR EASY ACCESS

Fig. 4-47. Schematic of the base-line nacelle arrangement and rotor system. Courtesy of Hamilton Standard Division of United Technologies Corporation.

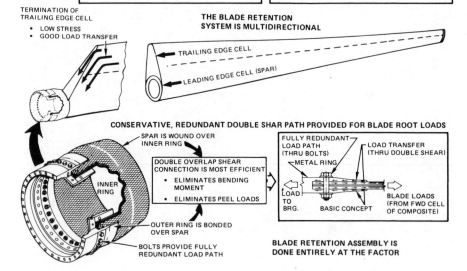

BLADE INCORPORATES SIMPLE EFFECTIVE LIGHTNING CONCEPT

- CONDUCTIVE ONLY IN THE PRESENCE OF A HIGH VOLTAGE CHARGE
- DOES NOT IMPARE T.V. RECEPTION
- PROVIDES PROTECTION TO INNER RING OF BLADE RETENTION

BLADE

- HOLLOW CONTINUOUS STRUCTURE
- NO CHORDWISE OR LONGITUDINAL JOINTS
- AERODYNAMIC DESIGN OPTIMIZED FOR MAXIMUM ANNUAL ENERGY CAPTURE
- NACA 230XX SERIES AIRFOIL

TERMINATION OF TRAILING EDGE CELL
- LOW STRESS
- GOOD LOAD TRANSFER

THE BLADE RETENTION SYSTEM IS MULTIDIRECTIONAL

TRAILING EDGE CELL

LEADING EDGE CELL (SPAR)

CONSERVATIVE, REDUNDANT DOUBLE SHAR PATH PROVIDED FOR BLADE ROOT LOADS

SPAR IS WOUND OVER INNER RING

DOUBLE OVERLAP SHEAR CONNECTION IS MOST EFFICIENT
- ELIMINATES BENDING MOMENT
- ELIMINATES PEEL LOADS

INNER RING

OUTER RING IS BONDED OVER SPAR

BOLTS PROVIDE FULLY REDUNDANT LOAD PATH

FULLY REDUNDANT LOAD PATH (THRU BOLTS)

LOAD TRANSFER (THRU DOUBLE SHEAR)

METAL RING

LOAD TO BRG.

BASIC CONCEPT

BLADE LOADS (FROM FWD CELL OF COMPOSITE)

BLADE RETENTION ASSEMBLY IS DONE ENTIRELY AT THE FACTOR

Fig. 4-48. The base-line blade assembly.   Courtesy of Hamilton Standard Division of United Technologies Corporation.

contact pattern on the race over the full range of moment loadings and to provide constant stiffness to the blade.

The rotor is attached to the main shaft through the teeter hinge and the nacelle is free to yaw in response to the wind force. The free alignment with the wind direction is facilitated by a skewed geometry in the teeter hinge which is known in helicopter technology as the delta-three effect. The teeter axis assembly is shown in Figure 4-49. The bores for the teeter hinge bearings are skewed by 30° to create the delta-three control effect. The bores and the teeter hinge pin are designed for either elastomeric or oil-lubricated antifriction bearings.

The pitch-change assembly is also shown in Figure 4-49. Each blade is controlled by dual hydraulic actuators to satisfy the requirement of complete control system redundancy.

The low-speed shaft, mounted on roller bearings, supports the rotor at the teeter axis, and the gearbox through a rigid coupling. The gearbox is a two-stage, spur

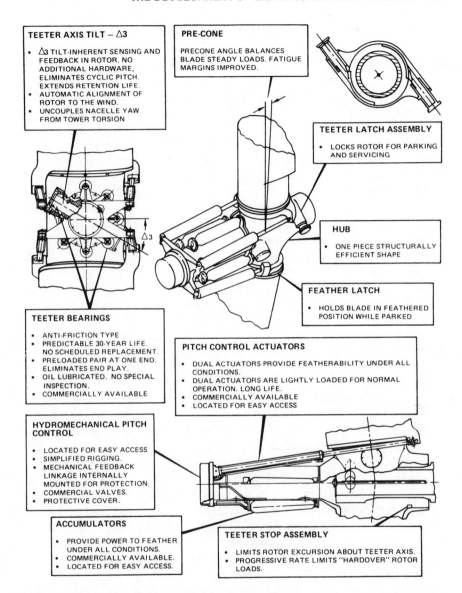

TEETER AXIS TILT – Δ3

- Δ3 TILT-INHERENT SENSING AND FEEDBACK IN ROTOR. NO ADDITIONAL HARDWARE, ELIMINATES CYCLIC PITCH. EXTENDS RETENTION LIFE.
- AUTOMATIC ALIGNMENT OF ROTOR TO THE WIND.
- UNCOUPLES NACELLE YAW FROM TOWER TORSION

PRE-CONE

PRECONE ANGLE BALANCES BLADE STEADY LOADS. FATIGUE MARGINS IMPROVED.

TEETER LATCH ASSEMBLY

- LOCKS ROTOR FOR PARKING AND SERVICING

HUB

- ONE PIECE STRUCTURALLY EFFICIENT SHAPE

FEATHER LATCH

- HOLDS BLADE IN FEATHERED POSITION WHILE PARKED

TEETER BEARINGS

- ANTI-FRICTION TYPE
- PREDICTABLE 30-YEAR LIFE. NO SCHEDULED REPLACEMENT.
- PRELOADED PAIR AT ONE END. ELIMINATES END PLAY.
- OIL LUBRICATED. NO SPECIAL INSPECTION.
- COMMERCIALLY AVAILABLE

PITCH CONTROL ACTUATORS

- DUAL ACTUATORS PROVIDE FEATHERABILITY UNDER ALL CONDITIONS.
- DUAL ACTUATORS ARE LIGHTLY LOADED FOR NORMAL OPERATION. LONG LIFE.
- COMMERCIALLY AVAILABLE
- LOCATED FOR EASY ACCESS

HYDROMECHANICAL PITCH CONTROL

- LOCATED FOR EASY ACCESS
- SIMPLIFIED RIGGING.
- MECHANICAL FEEDBACK LINKAGE INTERNALLY MOUNTED FOR PROTECTION.
- COMMERCIAL VALVES.
- PROTECTIVE COVER.

ACCUMULATORS

- PROVIDE POWER TO FEATHER UNDER ALL CONDITIONS.
- COMMERCIALLY AVAILABLE.
- LOCATED FOR EASY ACCESS.

TEETER STOP ASSEMBLY

- LIMITS ROTOR EXCURSION ABOUT TEETER AXIS.
- PROGRESSIVE RATE LIMITS "HARDOVER" ROTOR LOADS.

Fig. 4-49. The base-line hub and pitch-change assemblies.   Courtesy of Hamilton Standard Division of United Technologies Corporation.

gear epicyclic with a 36:1 gear ratio and 30-year design life. The gearbox output shaft drives the generator through a flexible coupling. Torsional softness in the drive system is obtained by reacting the rotor torque with springs and dampers.

The base-line machine relies on a soft-system design philosophy. The blade motions and deflections are uncoupled from the ground by the teeter hinge, soft drive system, and soft tower. Impulsive or harmonic airloads at the blade are absorbed in the inertia of the blade, blade and hub, or hub and nacelle, depending upon the frequency and the system mode. The soft system approach yields rotor thrustwise motions which are damped by the air. The aerodynamic damping is augmented by damping about a soft torsional mounting of the gearbox. The shaft damping provided in the gearbox mount allows the control system to hold the wind turbine near its performance peak at wind speeds up to the rated value. A responsive blade pitch control system minimizes variations in shaft torque. The blade pitch is managed above the rated wind speed to hold a nonconstant mean torque. The control system considers gust margins and sets a nonconstant limit which is higher than that for the standard rating concept.

Free yaw operation may yield a significant improvement in performance. The wind direction and speed can vary considerably over the area swept by the blades of large machines. Yaw control based on a drive mechanism and a single direction sensor can fail to "square" the disk with the effective wind force. A rotor in the free yaw configuration can be squared by the wind force itself. Hamilton Standard estimates that the abandonment of free yaw control can save 3 percent of parasitic power consumption and regain 3 to 10 percent of the energy capture which is lost due to poor heading control.

Hamilton Standard has been awarded a contract to design, fabricate, install, and test a large wind turbine for the U.S. Bureau of Reclamation. This machine was referred to as a System Verification Unit (SVU) in the Request for Proposals announcement. The NASA Lewis Research Center will provide technical management for the project. The SVU is to determine the technical and economic feasibility of integrating wind generation with the Water and Power Resources Service's Colorado River Storage Project.

The wind turbine will be installed at a site in south-central Wyoming, five miles south of Medicine Bow. The machine, an uprated version of the WTS-3 which has a rated power of 4 MW, was designed to produce energy at a minimum cost at the energetic Medicine Bow site. The mean annual wind speed and power density at the Medicine Bow, Mod-2 candidate site are 9 m/s (20.1 mph) and 975 W/m^2, respectively, referenced to 61-m (200-ft) elevation above ground level. The Hamilton Standard machine, designated the WTS-4, may be the first of 50 large machines which constitute a wind farm at this site. Power produced by these machines would be fed into existing transmission lines which link power plants at the Service's Glen Canyon, Flaming Gorge, and other hydroelectric dams. Some of the demand for power met by releasing water through the hydroelectric plants would be supplied by wind generation. The water saved would then be used to generate additional hydroelectric power during periods when the wind is calm.

The specifications for the WTS-4 are given in Table 4-18. A simulation of the WTS-4 is shown in Figure 4-50. The design power output profile for the WTS-4 is shown in Figure 4-51a. Figure 4-51b shows the design values of the annual energy output versus the mean annual wind speed referenced to 9.1-m (30-ft) elevation above ground level. The erection of the wind turbine at the Medicine Bow site is scheduled for the summer of 1981.

DARRIEUS VERTICAL-AXIS WIND TURBINES

Sandia Laboratories is developing Darrieus vertical-axis wind turbine (VAWT) technology for the U.S. Wind Energy Program. Recently, Braasch (1979) described this effort. Sandia has tested 2-,5-, and 17-m machines. A recent cost study indicates that the Darrieus system may be cost-competitive with horizontal-axis machines in intermediate sizes. Design studies suggest that the cost of energy decreases with VAWT rotor size up to the largest system investigated (1.6 MW).

The Aluminum Company of America (ALCOA) has begun to market a range of sizes of Darrieus VAWTs. Ai (1979) has described the ALCOA program. The Eugene Water and Electric Board and Southern California Edison began testing units with a rated power of 500 kW in 1980 (*Wind Energy Report*, 1979c,d). The ALCOA machines have extruded aluminum blades. The 500-kW unit measures 41 m (134 ft) high by 25 m (82 ft) wide. A larger unit with a rated power of 1 MW is being designed.

DAF Indal, an aluminum fabrication company, will soon market a line of VAWTs ranging in size from 8 to 500 kW (Szostak, 1980).

Table 4-18. WTS-4 specification.

• Rotor diameter	255 ft (77.6 m)
• Hub ℄ height	262 ft (80 m)
• Generator	4.0 MW synchronous AC 60 Hz
• Wind regime	
• Cut-in	7.1 m/s (15.9 mph)
• Rated	16.2 m/s (36.7 mph)
• Cut-out	27 m/s (60.4 mph)
• Downwind rotor	
• Free yaw	
• Soft Tower	
• Low maintenance requirement	
• High on-line availability	

Fig. 4-50. Simulation of the WTS-4 wind turbine. Courtesy of Hamilton Standard Division of United Technologies Corporation.

OFFSHORE WIND POWER SYSTEMS

Heronemus (1972, 1976) has emphasized the advantages of offshore wind turbine arrays. In 1972, he argued that multimegawatt offshore systems could use power to electrolyze water. Hydrogen could then be pumped ashore in pipelines. He envisioned 13,880 six-MW wind turbine units feeding 83 submerged electrolyzer stations. The offshore windpower units occupied sites off the Northeast coast in the vicinity of Georges Bank. He estimated that the offshore system could produce an annual output of 159 billion kWh.

Offshore systems could circumvent significant obstacles to land-based systems including environmental, land use, aesthetic, electromagnetic interference, and other

problems. The disadvantages of ocean-based systems include greater costs for installation, operation, maintenance, and transmission to an onshore distribution system.

The United Kingdom, Netherlands, and Sweden, as well as the U.S., are studying
the feasibility of offshore systems (*Wind Energy Report*, 1979e). The U.S. study

WTS-4 POWER PROFILE (SEA LEVEL)

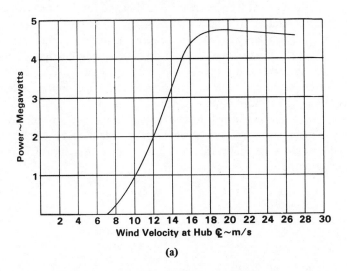

(a)

WTS-4 YEARLY ENERGY OUTPUT

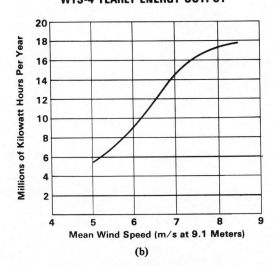

(b)

Fig. 4-51. (a) WTS-4 power output profile at sea level, and (b) WTS-4 annual energy output.
Courtesy of Hamilton Standard Division of United Technologies Corporation.

was conducted by the Advanced Systems Technology Division of the Westinghouse Electric Corporation (Kilar, 1979).

The Westinghouse researchers concluded that (1) offshore systems are technically feasible for much of the U.S. offshore region, (2) for a 500 MW offshore array, the busbar energy costs range from 6.4¢/kWh for the West Coast region to 17.6¢/kWh for the Southeast region (by comparison, a 25-unit array of Mod-2s installed onshore would yield a cost range of 3 to 5¢/kWh, 1978 dollars), (3) the Northeast and Northwest coasts are conducive to offshore development and have busbar energy costs in the range from 7.5 to 10¢/kWh, (4) the cost of electrolysis of water, transmission of hydrogen to shore, and reconversion to electrical energy is nearly six times as large as transmission by submarine cables, and (5) the hydrogen conversion scheme is too far beyond state-of-the-art to permit credible conclusions concerning its technical feasibility (Kilar, 1979, *Wind Energy Report,* 1979e).

Land-based systems have a significant economic advantage over offshore systems. However, the fuel saving capability of wind power systems and the U.S. dependence on unreliable sources of oil provide reason enough to develop effective offshore windpower systems.

5
Economics of Wind Generation

INTRODUCTION

Many studies have been conducted to determine the most economical dimensions of a wind turbine generator. Putnam (1948) determined that the generator rating and propeller diameter for a machine of his general design, which would operate efficiently in a wide range of wind regimes, were in the ranges 1.5 to 2.5 MW and 175 to 225 ft, respectively. Results from a study of horizontal-axis turbines conducted by the General Electric Company (1976) indicate a continuous decrease in the cost of energy with increase in rotor diameter, whereas the results from a study by the Kaman Aerospace Corporation (1976) indicate a specific size for minimum cost, which depends on the wind regime. The results of such studies are sensitive to the assumptions used to determine the cost of the rotor, which represents a large fraction of the total cost, and the reliability of the performance model used to estimate the energy production over the entire size range.

Putnam (1948) estimated the cost and the value of electricity from an array of wind turbines to the Central Vermont Public Service Company. The cost was larger than the value and the utility company could not justify the addition of wind-generating systems. The correct criterion for the economic comparison of different generating devices is total utility system costs. A generating unit should be installed if its cost, when combined with the cost of the generating units which constitute the entire system, will yield a minimum cost of electricity and an adequate level of reliability. In order to apply this criterion, it is necessary to simulate the total utility generating system performance and cost during a period which represents a significant fraction of the operating life of the units under consideration.

Marsh (1979) has developed a total system performance and cost analysis methodology for the assessment of wind generation in utility systems. The methodology was applied to determine the value of wind-generation capacity ($/kW) in several utility systems. Estimates for the costs of mass-produced wind turbine generators and their operation and maintenance requirements may be compared with the value of wind generation to determine the feasibility of wind power applications. Johanson and Goldenblatt (1978, 1979) have developed a similar generic planning methodology for the determination of the value of wind generation in utility systems.

The procedures for the accurate determination of value are complex, time-consuming, and costly. Consequently, simple methods such as the estimation of "busbar energy costs" have often been used to compare the cost of energy from alternate generating devices.

In this chapter, results from a recent study of machine size optimization are presented. A brief review of the simple and comprehensive methods for the determination of the cost and value of wind generation is presented.

MACHINE SIZE OPTIMIZATION TRENDS

During the Mod-2 design study, Boeing (1978) developed analytical models using cost, weight, and performance relationships to estimate the effects of rotor diameter and generator output on the economics of horizontal-axis wind turbines. Figure 5-1a shows the effect of rotor diameter and rated power on the busbar energy cost. The results were obtained for a wind speed frequency distribution with a mean of 6.26 m/s (14 mph) referenced to 30-ft elevation. The Mod-2 parameters (2.5 MW and 300 ft) were chosen to yield a near minimum cost of energy. A broad range of certain combinations of rated power and rotor diameter yield a near-minimum energy cost.

Figure 5-1b shows the effect of the variation of the mean wind speed on the energy costs. The solid line gives the busbar energy cost for the Mod-2. The broken line gives the energy cost which would be obtained from a machine with the optimum rated power and rotor diameter for each of the design wind speed distributions. The

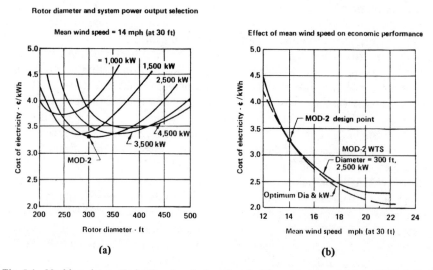

Fig. 5-1. Machine size optimization trends (one-hundredth production unit) in 1977 dollars: (a) rotor diameter and system power output selection, and (b) effect of mean wind speed on economic performance. From Lowe and Engle (1979) and Boeing (1978).

solid curve intersects the broken curve at the Mod-2 design wind speed of 6.26 m/s. At a site with a mean wind speed of 7.60 m/s (17 mph), the Mod-2's energy cost is reduced by 25 percent, a cost which is approximately 2 percent higher than that for the optimum machine for this speed. At a 9.83 m/s (22 mph) site, the numbers are 32 and 7 percent (Lowe, 1978). The difference between the Mod-2 and optimum energy cost over the wind speed range of Figure 5-1b is less than 10 percent.

The optimum dimensions obtained in Boeing's study are in good agreement with the results obtained during Putnam's pioneering effort.

BUSBAR ENERGY COSTS

Computational Method and Parameters

The busbar energy cost (BEC) is evaluated using a levelized fixed charge method. An estimate is obtained for the cost which is required to purchase, install, own, operate, and maintain a given generating unit. The method yields the constant revenue stream with a present value which equals the life cycle cost of the generating system (NASA Lewis Research Center, 1979). The BEC at the terminals of the generating unit's step-up transformer is usually computed by using

$$\text{BEC (¢/kWh)} = \left\{ (IC,\$) (FCR,\%) + (AOM,\$) (LF) (100) \right\} / (AEP,kWh) \qquad (5\text{-}1)$$

where: IC = initial turnkey cost of the generating system; FCR = levelized fixed charge rate which accounts for return on capital, depreciation, allowance for retirement dispersion, income and property taxes, and insurance; AOM = unlevelized annual operation and maintenance (O&M) cost; LF = levelizing factor for O&M; and AEP = anticipated annual energy production (allowance is made for the loss of energy when the wind power is adequate and the generating unit is unavailable).

Estimates for DOE/NASA Machines

Estimates for the busbar energy costs of electricity from the DOE/NASA machines versus the site mean wind speed referenced to 30-ft elevation are shown in Figure 5-2. The results shown as solid lines reflect the machine capital costs for second units and assumes that the machines will operate 90 percent of the time the wind speed falls in the operational range. The second unit costs for the installed prototypes in 1977 dollars are given in Table 5-1. The second unit costs do not include the nonrecurring costs which contribute to the cost of first prototype units.

The fixed-charge rate is determined by the weighted average cost of capital, the general inflation rate, the design life of the unit, the debt/equity ratio of the utility, and other financial parameters. The results shown in Figure 5-2 correspond to a rate of 18 percent, a representative value for investor-owned utilities. The rate assumes a general inflation rate of 6 percent, no allowance for tax preferences, an after-tax

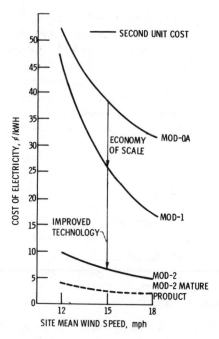

Fig. 5-2. Busbar cost of electricity for the DOE/NASA Mod wind turbines. From Thomas and Robbins (1979).

weighted average cost of capital of 8 percent (10 percent before tax), and a 30-year life cycle.

During the Mod-2 design study, Boeing estimated the O&M cost for production wind turbines operating in a 25-unit cluster to be 1 percent of the capital cost. Since O&M costs will escalate during the life cycle, the O&M cost must be levelized. When the assumed values of the economic factors given above (no real escalation for O&M costs) are used, the levelizing factor is 2.0. The busbar energy costs shown in Figure 5-2 were obtained by assuming that the contribution to the cost from O&M amounts to 2 percent of the capital cost.

Table 5-1. Second unit costs for the DOE/NASA prototype machines.
($1977)

	Mod-0A	Mod-1	Mod-2
Rated Capacity (kW)	200	2000	2500
Cost, $M	1.61	5.40	3.37
$/kW	8050	2700	1350

From Thomas and Robbins (1979).

The Boeing estimate for the annual energy production of the Mod-2 for 90 percent availability at a site with a mean wind speed of 6.26 m/s (14 mph) referenced to 30-ft elevation is 9,750,000 kWh. The assumed economic parameters and equation 5-1 yield a BEC of 6.9¢/kWh for the second Mod-2 (1977 dollars). The dashed curve of Figure 5-2 shows the projected BEC for the one-hundredth production unit. Table 5-2 gives estimates of the installed equipment cost for the one-hundredth unit and the rules which were used to estimate the cost. The total per-unit cost of $1,717,000 is based on the installation of a 25-unit farm. The assumed economic parameters and equation 5-1 yield a BEC of 3.5¢/kWh for a site with a mean wind speed of 6.26 m/s (14 mph) referenced to 30-ft elevation (1977 dollars).

Estimates of the BEC were developed during the Mod-X conceptual design study (NASA Lewis Research Center, 1979). The energy cost goal for the 200-kW baseline design was 1 to 2¢/kWh. While this goal was not achieved, it was shown that the estimated BEC for a site with an annual mean wind speed of 6.26 m/s (14

Table 5-2. Mod-2 cost summary for the 100th production unit.

Turnkey Account	Cost ($K)
Site preparation	162
Transportation	29
Erection	137
Rotor	329
Drive Train	379
Nacelle	184
Tower	271
Initial spares	35
Nonrecurring	35
Total initial cost	1,561
Fee	156
Total turnkey	1,717 ($687/kW)
Annual operations and maintenance	15

Cost-estimating ground rules:

- All costs are in mid-1977 dollars

- Costs are fully burdened and include a 10 percent fee

- Costs of installation and operation are based on a 25-unit farm

- Rate of installation is one unit per month

- Sites are generally flat with few natural obstacles, soil is easily prepared for foundation (land cost is not included)

- Transportation costs are based on rail and truck transport over a distance of 1000 miles

From Boeing (1978).

mph) — the design wind speed — is economical in utility applications where the cost of fuel exceeds 2.30¢/kWh (1978 dollars).

The economic parameters are summarized in Table 5-3. The levelizing factor for O&M expenses (1.886) corresponds to a discount rate (present value interest rate) of 10 percent, an expense escalation rate of 6 percent, and a life cycle of 30 years. The levelizing factor gives a stream of revenues which has the same present worth for a discount rate, i.e., cost of money, of 10 percent as the non-levelized stream. These parameters and equation 5-1 yield a BEC of 4.34¢/kWh. If the higher fixed charge rate used to obtain the results of Figure 5-2 (18 versus 15 percent) is used, the BEC is 5.03¢/kWh. To break even as a fuel saver, the Mod-X must displace fuel with a levelized cost of 4.34¢/kWh. If it is assumed that the cost of fuel in 1978 increases for 30 years at an assumed inflation rate of 6 percent (no real escalation case), the levelizing factor used for O&M expenses may be applied to the BEC for 1978 to obtain the required fuel cost. The break-even cost of fuel in 1978 for the Mod-X at a 6.26 m/s (14 mph) site is thus 2.30¢/kWh, i.e., (4.34/1.886).

The cumulative distribution of the estimated average cost of fossil fuel for 310 utilities in 1978 is illustrated in Figure 5-3. The estimates are based on utility data for 1976 and 1977 which were compiled by the Federal Power Commission. The utilities account for nearly 98 percent of U.S. fossil-fuel generation. Fuel prices for 1977 were increased by 6 percent to obtain estimates in 1978 dollars. Figure 5-3 indicates that approximately 30 percent of the total number of utility companies may find a BEC of 4.34¢/kWh economically attractive for the fuel saver case with no real fuel price escalation. However, sites with an annual mean wind speed of 6.26 m/s may not be available to all utilities. Figure 5-3 shows the locations of busbar energy costs for the Mod-X on the fuel price curve for sites with annual mean wind speeds of 5.36 (12), 6.26 (14), and 8.05 (18) m/s (mph) referenced to 30-ft elevation. For the Mod-2 at a 6.26 m/s (14 mph) site, the break-even cost of fuel in 1978 is 2.0¢/kWh

Table 5-3. Economic parameters* for the determination of the busbar
energy cost for the Mod-X.
($1978)

FCR	15%	
IC	$202,810	($1014/kW)**
LF	1.886	
AOM	$4,000	
AEP (Annual mean wind of 6.26 m/s [14 mph]		
at 30-ft elevation above ground level)	875,000 kWh	
Life cycle	30 years	
Busbar energy cost	4.34¢/kWh	

*The parameters are identified in the text on page 421.
**Estimated mature product cost.

From NASA Lewis Research Center (1979).

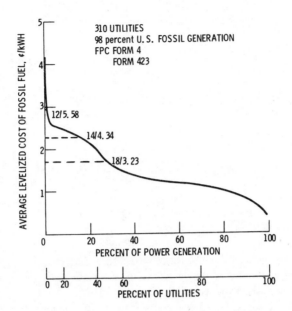

Fig. 5-3. Baseline Mod-X as utility fuel saver. From NASA Lewis Research Center (1979).

and approximately 40 percent of the total number of utility companies may find wind generation attractive for the fuel saver–no price escalation case.

The assumption that fuel prices increase at the general inflation rate is conservative. Table 5-4 shows fuel and electric cost growth rates based on data from the Portsmouth Naval Shipyard and the Public Service Company (PSC) of New Hampshire (Vachon et al., 1979). The annual rate of growth in the average cost of the utility's fuel during the 10-year period 1968-1978 was 18.63 percent; during the 5-year period 1973-1978, the rate was 25.90 percent. The cost of electricity to PSC customers has grown at a lower rate because, until recently, the cost of fuel was a small fraction of the total cost for electric service. Revenues used for fuel purchases amounted to 50 percent in 1977.

The levelizing factor for a discount rate of 10 percent, a fuel cost escalation rate of 12 percent (6 percent above the general inflation rate) and a 30-year stream is 4.26. The break-even cost of fuel in 1978 for the Mod-X at a 6.26 m/s site, based on this scenario and a levelized BEC of 4.34¢/kWh, is 1.02¢/kWh.

Effective Ownership Alternatives

Wind generation developed by the publicly owned sector of the electric utility industry will benefit from financing arrangements which are not available to the private sector. Martin et al. (1979) have reviewed the relationship between financing and various ownership alternatives.

Table 5-4. Fuel and electric cost growth rates.

%

	1973-1978	1968-1978
Utility steam oil [a]	27.26	20.62
Utility all fuels[a]	25.90	18.63
Navy fuel[b]	31.12	20.54
Electric customer service cost[b]	15.72	12.38
General rate of inflation	8	6

Sources: [a] Public Service Company of New Hampshire.
             [b] Portsmouth Naval Shipyard.
Adapted from Vachon et al. (1979).

The public sector includes federal agencies, rural electric cooperatives (REA Co-ops), municipal governments, power districts, and state agencies. Federal projects obtain capital directly from the U.S. Treasury. REA Co-ops obtain part of their capital from the Treasury and the balance from nongovernmental sources and their own banks. Municipals obtain capital at a lower cost because no federal income tax is imposed on municipal bond interest. The tax exemption allows a municipal company to pay its bond owners only the net rate of return they demand, whereas a private utility must pay an interest rate which yields a comparable net return after bondholders have paid personal income tax on their interest. Furthermore, public power systems are not required to pay federal or state corporate income taxes, or property taxes, although payments in lieu of property taxes are sometimes required.

Table 5-5 shows fixed-charge rates for different owners and financing arrangements, including investor-owned utility financing by normal capital sources and by tax-free bonds, public utility financing, and financing for small hydroelectric development. The fixed charge rates vary from 3.58 to 18.54 percent for Federal projects and industrial small hydroelectric development, respectively. The BEC for a Mod-2 at a site with an annual mean wind speed of 6.26 m/s referenced to 30-ft elevation based on the capital and O&M costs of Table 5-2, a levelizing factor of 2.0, an annual energy production of 9,750,000 kWh, and fixed charge rates of 3.58 and 18.54 percent are 0.94 and 3.57¢/kWh, respectively.

COMPREHENSIVE ASSESSMENTS OF THE VALUE OF WIND GENERATION

General Electric Study

The Electric Utility Systems Engineering Department of the General Electric Company developed a methodology for studying the performance and value of wind

Table 5-5. Fixed charge rate for various owners.
(%)

	Investor-Owned Utilities		Federal Projects	State-Owned Projects		Municipally Owned Projects	REA[c] Co-ops	Small Hydroelectric Development	
	Normal Financing	Tax-Free Bonds		SRP[a]	LCRA[b]			Industrial	Private
Required return	10.25	9.50	3.00	6.80	7.50	6.80	4.4[1]	15.00	13.00
Depreciation (sinking fund, 30 years)	0.58	0.58	0.58[d]	0.58	0.58	0.58	0.58	0.58	0.58
Income taxes	3.88	3.49	–	–	–	–	–	0.46	0.45
Local taxes and insurance	2.50	2.50	–	3.60	–	0.10	1.30	2.50	2.50
Total	17.21	16.07	3.58	10.98	8.08	7.48	6.28	18.54	16.53

[a] SRP = Salt River Project.

[b] LCRA = Lower Colorado River Authority.

[c] REA = Rural Electrification Administration.

[d] Depreciation is variable and would be charged at the cost of major repairs in the year of expenditure.

From Martin et al. (1979).

generation in utility systems (Marsh, 1979). The study was performed for the Electric Power Research Institute (EPRI, 1979). Comparisons of the cost of energy from different generating devices based on the busbar energy cost method are meaningful provided that the devices have the requisite ability to start, stop, and follow the utility load curve, the same impact on the network requirement for reserve capacity (equal effective capability), and the same capacity factors throughout their life cycle. New generating devices with similar characteristics may be compared using the BEC method; however, a stochastic device such as a wind turbine may not be compared in a meaningful fashion with conventional devices, e.g., oil, coal, or nuclear generating units. Wind performance, reliability, and production cost models were therefore developed to perform a total generating system analysis.

Wind Performance. Weather data consisting of the wind speed, ambient temperature, and barometric pressure were used in conjunction with power output models for a number of different machines to obtain hourly values of the net electrical output.

Reliability Model. The effective capability of a generating device provides a measure of the unit's ability to contribute to the reliability of the total utility system. "Reliability" refers to the system's ability to supply the demand load. The General Electric researchers used the loss-of-load probability (LOLP) method to evaluate the reliability of systems with and without wind generation. The major contributions to the overall LOLP occur during peak load hours. Therefore, in general industry practice, the reliability (LOLP) is evaluated only for the peak load of the day. The overall LOLP is thus based on 365 calculations per year and is usually expressed as the number of days per year that the total system is unable to satisfy the demand load. A typical industry standard for the annual LOLP is 0.1 days per year.

The effective capability of a generating unit is defined as the amount of additional peak load which the utility system could serve at the reliability (LOLP) which existed before the addition of this unit. The effective capability is often expressed as a percentage of the rated capacity of the generating unit. For example, a 1000-MW nuclear unit with an effective capability of 750 MW permits a utility power system to provide 750 MW of additional load demand with the same reliability; the effective capability of the unit may also be expressed as 75 percent. The effective capability is a function of the unit's size, forced outage rate, maintenance requirements, and the characteristics of the utility system in which it operates.

The General Electric researchers used the wind turbine output to reduce the net load to be served by conventional units. The LOLP method was then applied to evaluate the reliability of the utility system.

Production Cost Model. Standard methods were used to simulate, hour-by-hour, the future operation of the generating system. The costs of fuel, operation, and maintenance for each generating unit is calculated. The simulation includes scheduled maintenance, spinning reserve, startup and shutdown rules, priority commitment of units, incremental cost dispatch, and detailed representation of both conventional and pumped storage hydro.

Procedure. The EPRI study was performed for utilities in three areas: the Kansas Gas & Electric Company in Central Kansas (area 1), the Niagara Mohawk Power Corporation in Northern New York (area 2), and the West Group of the Northwest Power Pool (area 3). Two sites were chosen in the latter area, one on the Southern Oregon coast and the other in the Columbia River Gorge. The annual mean wind speeds referenced to 50-m (165-ft) elevation for the four regimes are 8.3, 6.9, 9.5, and 11.5 m/s (19, 15, 21, and 26 mph), respectively. The performance and cost were calculated for each utility system based on present plans without wind generation (base case); calculations were then performed for substitutions of wind generation; and total system costs were then compared with results for the base case. The LOLP method was used to substitute wind generation in a quantity which produced equal system reliability.

The value of the wind generation was calculated in terms of two components: energy value, the production cost savings, and capacity value, the savings in capital cost which results from omitting some amount of conventional generation. The value of the wind generation expressed as dollars per kilowatt was then compared with cost projections for mass-produced wind turbines.

Results. Calculations were performed for a 1.5-MW, horizontal-axis, fixed-speed wind turbine. The rated wind speed selected for each site was slightly different. Figure 5-4 shows the effective capability and economic value for each of the four wind regimes. Conventional generating units generally have effective capabilities in the range of 70 to 95 percent. The effective capabilities for wind turbines installed in the four regimes range from approximately 5 to 45 percent. For the Niagara Mohawk regime, the value is approximately 38 percent, and 50 MW of wind generation capacity could displace 27 MW of conventional capacity with an effective capability of 70 percent (.38 × 50/70). The effective capability of the Northwest Power Pool Gorge site is very low because the hourly wind power for the study year did not show a high correlation with the hourly peak loads.

Figure 5-5 shows the economic value for each of the four wind regimes. The results are based on the case of wind turbines whose capacity represents 5 percent of the original system capacity. The value, i.e., Putnam's worth of the wind generation, ranges from $480/kW to $1050/kW. Predicted values of the costs of mass-produced MW-scale wind turbines fall within the range between the Kansas and New York sites. The energy value is generally larger than the capacity value and the energy value is greatest for utilities with good winds and a substantial amount of oil-fired generation. The contribution from capacity value to the overall value ranges from approximately 5 to 52 percent.

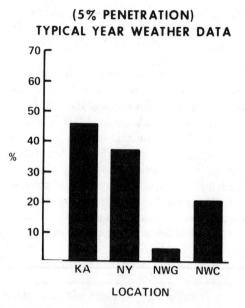

Fig. 5-4. Effective capability for wind turbines in four utility systems. The definition of effective capability is given on page 428. From Marsh (1979).

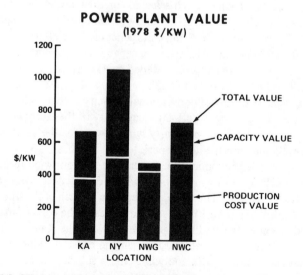

Fig. 5-5. Wind generation value in four utility systems. From Marsh (1979).

JBF Study

The JBF Scientific Corporation developed a similar model for the determination of wind generation value for the U.S. DOE (Johanson and Goldenblatt, 1978, 1979). The economic analysis was based on maximizing the net present value of the savings which accompany wind generation. The results were obtained as marginal savings (or marginal value), i.e., the value of one additional unit at the penetration level of interest. Penetration was defined as the ratio of the total installed wind turbine rated capacity to the utility peak load for the first year of operation, expressed as a percentage. The model permits a utility to determine the economic value of wind generation for the fuel saver case, and also for a reoptimized mix mode case. In the former case, the optimum mix of conventional generating units was selected to satisfy load and reliability requirements at minimum cost without wind generation. Wind turbines were then added to save fuel and obtain capacity credit toward the utility's reserve margin requirement. In the latter case the wind generation was treated as a negative load and the mixture of conventional generating units was optimized to satisfy the remaining load.

The model was applied to two generating systems, the New England Gas and Electric Association (NEGEA), which serves Southeastern Massachusetts, and a privately owned utility which was synthesized from a report prepared by the Jet Propulsion Laboratory (JPL).

Figure 5-6 shows the marginal value expressed as 1984 dollars per kW for NEGEA and the synthetic JPL utility for the reoptimized mix mode case. The results for the fuel saver case were similar up to a penetration of 10 percent; above this, the fuel saver mode has less value. The decreasing trend in the marginal value with increasing penetration is caused by a decrease in the cost of the fuel saved by each successive wind generating unit. The first wind generating units save the most expensive fuels and have the highest value.

The mean wind speed within the NEGEA service area is approximately 5 m/s (11 mph) referenced to 10-m (33-ft) elevation, and does not vary much with location. The calculations were performed for a wind turbine design with a rated power of 500 kW. The value of the first installed unit to NEGEA in 1984, which includes the turbine, transportation to the site, and erection, is $847/kW for both the fuel saver and mixed mode cases. The total installation unit worth, which includes the site and interface to utility, is $976/kW in 1984 dollars for the first installed unit.

Figure 5-7 shows the relative value of the components of NEGEA's 1985 singleyear savings for the fuel saver and mixed mode cases. Results are shown for penetrations at the 10 and 40 percent levels. At low percentages, the total savings are approximately equal for the two cases, since the savings are largely fuel savings. At 40 percent penetration, savings in fixed costs associated with high-cost baseload generating

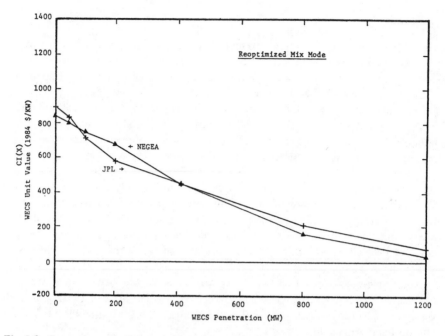

Fig. 5-6. Comparison of wind-generating unit worth for NEGEA and the JPL private utility cases. From Johanson and Goldenblatt (1979).

units are important due to a shift in capacity expansion toward more peaking units and less baseload capacity. The up-front fixed cost savings yield larger total savings for the reoptimized mix mode case. At intermediate and high levels of penetration, reoptimization yields smaller fuel savings and larger capacity credit.

The wind operating reserve (WOR) is the difference between the maximum expected load variation before and after the addition of wind generating units to the generating system. Goldenblatt (1979) has described the method used to calculate the required WOR. For NEGEA, the cost of WOR makes a small (negative) contribution to the value of wind generation.

The JBF researchers found potential sites for approximately 593 500-kW wind turbines in four farms in the NEGEA service area. Using an availability of 90 percent, the on-line rated capacity amounts to 267 MW (593 × .5 × .9). This capacity amounts to approximately 30 percent of the expected peak load of 903 MW in 1985.

The economic values for penetration in excess of 10 percent and thus the advantages of reoptimization may not be relevant due to stability requirements.

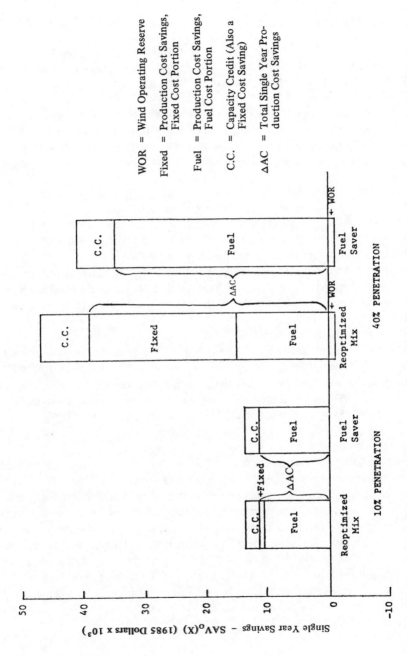

Fig. 5-7. Relative magnitude of the components of NEGEA'S 1985 single-year savings for the fuel saver and reoptimized mix modes. Johanson and Goldenblatt (1979).

Wind-Hydro Studies

Hightower and Watts (1977) have studied the integration of wind generation with a hydroelectric system. Their economic analysis was based on the generation from 49 Mod-1 generators at a site near Medicine Bow, Wyoming. Machine costs were based on production estimates for the General Electric 2-MW unit. Estimates were used for the costs of interconnection, land, site facilities, overhead, contingencies, and operation and maintenance. Repayment of the investment over a 30-year life cycle was based on the Bureau of Reclamation's current 7 percent interest rate and the present Colorado River Storage Project (CRSP) capacity charge for energy storage.

A comparison with the costs of other energy sources and customer load requirements indicated that a marketing method that provides firming energy for peaking capacity, in place of generation from oil-fired turbines, was the most cost-effective application. The wind turbine capacity was sized using the available firm peaking capacity, with energy, available in 1977 within the CRSP.

The installed capacity of the wind generators is 98 MW. The wind-generation system in combination with the CRSP hydroelectric system could provide peaking service in the amount of 277 MW during the winter, and 108 MW in the summer, in addition to the present firm power which is marketed. The energy amounts to 407.5 million kWh, the equivalent of 799,000 barrels of oil per year. The total capital investment was estimated as $77.83 million (1977 dollars). The average costs of capacity and energy for the 49-unit installation was estimated to be 3.01 ¢/kWh, an amount which is significantly less than the cost of generation from oil-fired systems (4.6¢/kWh). The use of the Mod-2 cost and energy production, and a more accurate analysis of the wind resource yields similar costs (Hightower, 1979).

Watts (1979) has discussed two different marketing methods for integrating wind energy into the CRSP system.

Todd et al. (1977) have considered wind generation in conjunction with pumped-storage hydroelectric plants for load leveling and existing transmission lines for interconnecting wind farms and storage sites with load centers at a distance of up to 2000 km. The potential wind generation capacity was estimated to be larger than 100 GW for wind farms in 17 western states, and many times this amount in Arctic North America. For the 100-GW level of development, the busbar cost was estimated to be 1.0¢/kWh. Energy storage required for load leveling and transmission costs add 1.1¢/kWh for a total cost at the load center of 2.1¢/kWh (1976 dollars). This cost is more than competitive with the cost of generation near load centers from new nuclear and oil-fired generating units. The fuel cost of oil at $40/barrel corresponds to approximately 7.1¢/kWh.

6
Legal-Institutional and Environmental Issues

In this chapter, the legal-institutional and environmental issues which will have a significant impact on the development of large-scale wind generation will be described.

LEGAL-INSTITUTIONAL ISSUES

Researchers at George Washington University have studied the legal and institutional implications of both land-based and ocean-based wind turbine generators (Mayo, 1977).

They concluded that the most serious impediments to the implementation of land-based systems, in order of importance, are: (1) difficulties concerning the availability of supplementary power from external sources during windless periods, (2) zoning, (3) the preclusive effect of state utility commission certification procedures, i.e., the inability of some new wind generation entities to operate due to the "regulated monopoly" structure of the electric power industry, (4) the absence of appropriate legal devices for regulating or preventing upwind obstructions to the wind flow, (5) the lack of specific reference to wind systems in a variety of state and federal tax and loan incentive programs which could easily apply to wind generation, (6) state and local building, safety, and housing codes, and (7) a likely lack of statutory authority for Interior Department federal utilities to undertake large-scale wind generation projects.

The researchers found that numerous legal provisions and devices exist for affecting the functions required for the implementation of wind generation. The most important are: (1) institutional vehicles appropriate to different types of large-scale wind generation projects, (2) an electric utility regulatory structure directed to most of the operations undertaken by most such vehicles, (3) legal devices for obtaining the supplementary electricity necessary for wind generation operations to function during windless periods, (4) eminent domain provisions which will facilitate obtaining the land needed for large wind generation operation by certain utilities, and (5) various state and federal tax and loan provisions which should be applied to wind generation ventures.

The conclusions which concern offshore wind turbine arrays are: (1) it is probable that an adequate legal-institutional scheme for offshore pilot and demonstration projects does not exist at this time, (2) while private offshore wind generation must comply with numerous state and federal regulatory schemes, there is no overall statutory structure establishing regularized procedures for offshore wind generation by nonfederal entities, and (3) uncertainty as to the "vessel" status of moored wind generation units could be a substantial constraint, because this affects many design, inspection, liability, and other considerations.

PURPA

Some of the most important legal and institutional issues which confront wind generation implementation are addressed by the Public Utility Regulatory Policies Act of 1978 (PURPA).

PURPA was enacted to assist the development of small power production facilities which use "biomass, waste, renewable resources, or any combination thereof," and have a capacity under 80 MW. PURPA requires that the small facility be connected to available transmission systems and that provision be made for the sale or exchange of electricity by order of the Federal Energy Regulatory Commission (FERC). Another provision requires the utility to provide a wheeling service to the small facility, i.e., the small producer can be remote from his customer and power must be wheeled on the network. This provision includes the enlargement of transmission capacity required to provide the wheeling services ordered by FERC. However, the law prevents FERC from issuing such orders unless certain conditions are met;

> "the order is not likely to result in a reasonable ascertainable uncompensated economic loss for any electric utilities . . . .; will not place an undue burden on an electric utility . . . . . .; will not unreasonably impair the reliability of any electric utility affected . . . . .; will not impair the ability of any electric utility affected to render adequate service to its customers."

FERC must issue rules which set the rates for purchase by the utilities.

> "No rule prescribed . . . . . shall provide for a rate which exceeds the incremental cost to the electric utility of alternative electrical energy."

The interpretation of the term "incremental cost" will determine the viability of wind generation ventures. In 1979, the Public Utilities Commission of New Hampshire issued the first order under the state's so-called "mini-PURPA," the Limited Electrical Energy Producers Act of 1978. Two utilities were ordered to pay small hydroelectric producers 4¢/kWh for run-of-the-river production, and 4.5¢/kWh for dependable production based on storage (capacity credit). The order was apparently based on the opinion of one of the affected utilities (Vachon et al., 1979). The New Hampshire Electric Cooperative "offered the opinion that it [incremental cost] means not only the cost of the next kilowatt-hour to be bought, but also should include the total cost per kilowatt-hour for the next plant that has to be built sometime in the near future." For example, it was stated that electricity from the Seabrook Nuclear Station would cost 4.5¢ to 5.0¢/kWh when it is completed in 1983.

In June, 1980, the Public Utilities Commission of New Hampshire raised the rates to 7.7 and 8.2¢/kWh for energy and reliable capacity, i.e., for hydroelectric generation. The new rates were based on the fuel cost for a new oil-fired plant (6.181¢/kWh) and additional charges. The capacity limit for small producers is limited to 5 MW.

PURPA may initiate a new era in electric generation by providing conditions which facilitate the individual ownership of generating systems which are relatively free from utility regulations. Wind developers are identifying large, investor-owned, high energy cost utilities which are located in promising wind regimes. An attempt is then made to negotiate a binding purchase power agreement for the wind generation. With a market for the wind farm's product established, the task of raising venture capital for what must still be regarded as a high-risk project is easier. Developers seek a purchase aggreement for the generation which is tied to the escalating cost of primary levels, mostly oil. A description of the wind farm development concept and current efforts to arrange and finance projects is given in *Wind Energy Report* (1979f).

U.S. Wind Power was the first company to secure a purchase power agreement (*Wind Energy Report*, 1979g). The agreement stipulates that the California Department of Water Resources will purchase a maximum of 400 million kWh annually, beginning in April, 1983. This generation would be provided by four 25-MW windfarms composed of 50-kW units. This energy production would require an optimistic average capacity factor of 0.46.

Windfarms Ltd. of San Francisco has announced that it has negotiated a 25-year agreement to supply the electricity produced by a 80 MW wind farm to the Hawaii Electric Company (*Wind Energy Report*, 1979h). This capacity would be provided by 20 4MW Hamilton Standard WTS-4 machines installed at Kahuku Point on Oahu. The Pacific Gas and Electric Company (PG&E), the California Department of Water Resources (CDWR), and Windfarms have announced the signing of an agreement to negotiate toward a contract under which PG&E and CDWR would purchase electricity produced by a Windfarms installation in Solano County, California. The development would occur in three stages. At the third stage of development in 1989, a total of 146 machines with a capacity of 350 MW would be installed. Windfarms is developing other projects in Hawaii and California.

ENVIRONMENTAL ISSUES

The issues in this category include ecological effects, health and safety, electromagnetic interference, land use, noise, and aesthetics (Wind Task Force, 1980).

Ecological Effects

Rogers et al. (1976, 1977) have studied the effects on microclimate, flora, and fauna which were produced by the Mod-0 at Plum Brook, Ohio.

Microclimatic effects are produced by the reduction of the downstream wind speed. Turbine-induced changes near ground level were within those caused by natural processes such as turbulence and topography.

Impact on flora occur where the ground cover is disturbed during the installation process. These impacts are similar to those caused by the construction and maintenance of transmission line towers and are site-specific. For example, land-clearing in mountainous terrain can produce significant impacts. (Vermont Ad Hoc Wind Energy Committee report, 1981). The silva culture in regions above 2500 ft in elevation in Vermont consists predominantly of red spruce and balsam fir. The ground cover consists of ferns, mosses, and alpine tundra. The removal of approximately one-half to one acre of forest destroys the integrity of the canopy which buffers the forest against severe elements. High winds cause trees to dessicate and die. This "wind killing" of trees continues until the surrounding terrain provides shelter to the forest.

The impacts on fauna do not appear to be serious. The release of insects upwind of the Mod-0 did not result in significant mortality. Rogers et al. (1977) concluded that the only probable impact to animal populations which was significant enough to warrant detailed studies was nighttime kills of birds at wind turbine towers. During four migratory seasons, three dead birds were found near the Mod-0 meteorological (two) and wind turbine (one) towers. Even during nights with high migration, low cloud ceiling, and fog, birds approaching the wind turbine at night were observed to take evasive action to avoid the blades. Others flew straight through the blades without incident. The probability of a collision under normal operating conditions is approximately 8 percent. The minimum altitude for most nocturnal migration is 150 m (500 ft).

Phillips (1979) has described the impacts during the installation phase of a 1.5-MW wind turbine: production of fugitive dust, added motor vehicle emissions and heavy equipment noise, and possible soil erosion and siltation of surface waters.

Safety

The largest planned wind turbine could throw a blade 300 to 460 m (Solar Program Assessment, 1977). An attempt is being made to develop fail-safe machines which require small exclusion zones for both people and property. If the wind turbine is sited where the blades are subject to icing, detection and shutoff procedures and exclusion zones or both must be established to eliminate the hazard from falling ice.

Electromagnetic Interference

Senior and Sengupta (1978) and Sengupta and Senior (1978) have studied the electromagnetic interference (EMI) produced by large wind turbine generators. The rotating blades reflect the electromagnetic signals in such a way as to cause periodically constructive and destructive interference between the reflected and transmitted signals. Poor siting of large wind turbines could cause unacceptable interference with television, radio, VOR and DVOR navigational systems, microwave links, and other radio frequency equipment. Metalic blades can produce more severe effects than those composed of fiberglass, wood, or composite materials.

The effects associated with TV reception can be alleviated through the use of a directional (fringe) antenna, local TV repeater, cable reception, or circularly polarized transmission.

Effects on AM reception are minimal. Due to the higher broadcast frequencies, FM reception is more subject to interference. However, interference with FM reception will be negligible except within tens of meters of the wind turbine.

EMI effects may be minimized by the proper siting of large wind turbines.

Land Use

It is expected that most, if not all, utility applications of large-scale wind turbines will be in farms (see Chapter 3). Areas remote from human activity will be acceptable if the environmental versus human benefit trade-off weighs in favor of wind generation. Multiple land use may be feasible in regions where grazing and agricultural activities occur.

Noise

On-site observations at the Mod-0 site (Rogers et al., 1976, 1977) did not reveal significant levels of noise. However, the larger Mod-1 machine has produced an intermittent, low frequency ground which has annoyed the residents of homes within 762 m (2500 ft) of the test unit (*Wind Energy Report*, 1980a). The audible component of the sound makes a "swish-swish" noise. The inaudible component (infrasound) causes windows and other objects in the homes to rattle from the vibrations of the sound. Means of attenuating these effects, including a lowered operating speed and the use of fiberglass blades, are under investigation.

Aesthetics

Since most utility applications will involve farms of large wind turbines, public acceptance of the visual impact may become an important issue. At a 10-diameter spacing, the next visible unit in a farm of Mod-2s would be 914 m (3000 ft) away. In level terrain, when viewed at ground level, the remainder of the units recede from visibility in the distance (see Figure 3-5). Respondents in a survey conducted by the University of Illinois (Survey Research Laboratory, 1977) were shown a picture of a farm with a smaller spacing. The participants preferred a scene with wind turbines to one exhibiting transmission towers and power lines. On mountain ridges which are perpendicular to the wind flow, the interval between units can be smaller. The visual impact will be more severe and the public will have to confront a more difficult trade-off situation. The use of remote locations for farms will isolate them from more than occasional view. The University of Illinois researchers concluded that people are basically in favor of the use of wind and solar sources for electric energy production (Feber, 1978).

People who possess knowledge about wind energy are more inclined to react positively to this means of energy production.

Posivite Impacts

Due to the technological simplicity of wind turbines and their ability to offset fuel consumption, significant environmental benefits will accrue from their use.

Net Energy. Calculations have been performed which suggest that large wind turbines could generate 40 to 45 times more energy during their life cycle than the energy utilized in their manufacture (Coty, 1976, Perry et al., 1977a,b). The energy required to manufacture the wind turbine includes the energy consumed from the mining of raw materials through its construction and installation. The high net energy ratio is two to four times better than that for many conventional and alternative generating systems (Perry et al., 1977a, Rotty et al., 1976). Thus, over time, the production of wind generating systems would yield a large cumulative reduction in the energy used for the creation of new generating systems.

Displacement of Environmental Impacts. Wind generation can have a significant impact upon the national level of pollution by displacing fossil fuels. Coty (1976) has categorized the impacts. The fuel-saving impacts are the product of mining, refining, transportation, and utilization. The land affected impact is the product of extraction, processing, and conversion only. The magnitude of the environmental disturbance which could be avoided or negated by wind generation in 1995 is summarized in Table 6-1.

The quality of ambient air downwind from generating stations will be improved through the fuel savings which accompany wind generation. Table 6-2 gives the composition of air pollutants which are involved in the total fuel process from mining through consumption.

Table 6-1. Annual displacement of environmental impacts
by wind generation implementation – 1995.

Impacts	Energy Source Alternatives		
	Imported Fuel Oil	Domestic Oil	Domestic Coal
Land area affected, $mile^2$	82.2	581	596.7
Water required, acre-ft/y	82.4	525.3	442.2
Air pollutants, 10^6 T/y	7.73	8.39	13.9
Solid waste, 10^6 T/y		0.42	822

Pollution and consumption impacts of steam-electric generating systems to produce 1280 billion kWh/y with full implementation of existing environmental controls. Data is presented for domestic oil which is produced on shore and surface extracted domestic coal.

From Coty (1976).

Table 6-2. Composition of air pollution avoided by wind generation.

Pollutant	Energy Source		
	Imported Fuel Oil (%)	Domestic Oil (%)	Domestic Coal (%)
Particulates	2	1	58
SO_x	67	32	35
NO_x	30	7	7
HC	1	45	–
CO	–	14	–
Others	–	1	–

From Coty (1976).

7
The Future of Large-Scale Wind Generation

The results of the studies described in this edition of Putnam's book indicate that large-scale wind turbines can play an important role in the national effort to produce electrical energy by using renewable energy resources.

Political support for the development of wind power technology will increase due to our diminishing supply of native fossil fuels, our dependence on increasingly unreliable foreign fuel supplies, favorable environmental impacts, and socioeconomic benefits. Recently, the President's Domestic Policy Review on Solar Energy estimated that approximately 1.7 quads of energy per year could be supplied by wind systems in the year 2000. This amount corresponds to 6.3 percent of the General Electric Company's estimate of the year 2000 demand for electrical energy.

Labor unions will support plans for high wind turbine production levels since extensive employment without dislocation would be required in many industries. Aircraft workers would make rotors, automobile workers would make gears and shafts, electrical workers would make generators, structural steel workers would make towers, and eventually, shipyard workers would make flotation equipment.

Inglis (1978) developed an estimate for the magnitude of the industrial effort required for a large amount of wind generation. He determined that the construction of enough MW-scale wind turbines to supply the demand for electrical energy in 1978 would require about as much metal and other materials as is used in three years of production by the automobile industry.

Numerous technical problems must be solved. The performance of wind turbines in utility systems must be verified during demonstration projects to establish confidence in the utility industry. The results must confirm predictions concerning energy production, operation and maintenance costs, the effects of gusting and wake interference on the performance of wind farms, and the impact of wind farms on network reliability and stability. Mass production methods must be devised, and materials used, which yield costs that will permit significant levels of penetration in regions with moderate and low wind regimes.

The Wind Energy Systems Act of 1980 provides a research, development, and demonstration program which will accelerate the widespread utilization of wind energy. The objectives of the act are:

1. To reduce the average cost of electricity produced by installed wind energy systems, by the end of fiscal year 1988, to a level competitive with conventional energy sources

2. To reach a total megawatt capacity in the United States from wind energy systems, by the end of fiscal year 1988, of at least eight hundred megawatts, of which at least one hundred megawatts are provided by small ($\leqslant$ 100 kW) wind energy systems

3. To accelerate the growth of a commercially viable and competitive industry to make wind energy systems available to the general public as an option in order to reduce national consumption of fossil fuel

The program management plan will be developed by the DOE with consultation from NASA, the Department of the Interior, other Federal agencies, and appropriate public and private organizations.

References

Agopian, K. G., and S. C. Crow, 1978: *The Effect of Atmospheric Density Stratification on Wind Turbine Siting*, DOE Report No. RLO-2444-78/1, January.

Ai, D. K., 1979: *ALCOA Wind Turbines, Large Wind Turbine Design Characteristics and R&D Requirements Workshop*, NASA Lewis Research Center, Cleveland, Ohio, April 24–26, Seymour Lieblein, Ed., NASA Conference Publication 2106, DOE Publication CONF-7904111, pp. 155–172.

American Society of Civil Engineers, 1961: "Wind Forces on Structures," *Trans.*, **126**, part II, p. 1124.

Anthes, R. A., and T. T. Warner, 1974: *Prediction of Mesoscale Flows over Complex Terrain*, U.S. Army Electronics Command Report ECOM-5532, March.

Approach Fish, Inc., 1979: "Instructions for Hand-Held Wind Measuring Device," 314 Jefferson Street, Clifton Forge, Virginia 24422.

Baker, R. W., and J. P. Hennessey, Jr., 1977: "Estimating Wind Power Potential," *Power Eng.*, March, pp. 56–57.

Baker, R. W., R. L. Whitney, and E. W. Hewson, 1979: "Wind Profile Measurements Using a Tethered Kite Anemometer," Oregon State University, Corvallis, Oregon.

Baynes, C. G., 1974: *The Statistics of Strong Winds for Engineering Applications*, University of Western Ontario report, BLWT-4-1974, September.

Bendat, J. S., and A. G. Piersol, 1971: *Random Data: Analysis and Measurement Procedures*, Wiley, New York.

Benjamin, J. R. and C. A. Cornell, 1970: *Probability, Statistics, and Decision for Civil Engineers*, McGraw-Hill, New York.

Betz, A., 1920: "Das Maximum der Theoretisch Möglichen Ausnutzung des Windes durch Windmotoren," *Z. Gesamte Turbinen Wesen*, **17**, p. 320, Berlin, West Germany.

Bhumralkar, C., 1980: *Estimation of Wind Characteristics at Potential Wind Energy Conversion Sites*, SRI International, Menlo Park, California, NTIS report PNL-3074.

Bodansky, D., 1980: "Electricity Generation Choices for the Near Term," *Science*, **207**, pp. 721–728, February 15.

Boeing Engineering and Construction, 1978: "2500 kW Wind Turbine System for Electric Power Generation, Mod-2 Project," Seattle, Washington, December.

Bortz, S. A., R. A. Budenholzer, R. D. Carlson, I. Fieldhouse, J. Kornfeld, R. S. Norman, R. W. Porter, and W. Rosenkranz, 1979: *Navy-New Hampshire Wind Energy Program*, Naval Material Command, Contract No. N00014-79-C-6503, Washington, D.C., November.

Bouwmeester, R. J. B., R. N. Meroney, V. A. Sandborn, and M. A. Rider, 1978: "The Influence of Hill Shape on Wind Characteristics over Two-Dimensional Hills," Third Wind Energy Workshop, Washington, D.C., September 19–21, 1977, T. R. Kornreich, Ed., DOE CONF-770921/1, 2, pp. 646-653, May.

Braasch, R. H., 1979: *Darrieus Vertical Axis Wind Turbine Program Overview – Fall 1979*, Fourth Wind Energy Workshop, Washington, D.C., October 29–31, JBF Scientific Corporation, Ed., DOE Conference No. 791097, pp. 39–58.

Brandeis, L., 1979: *The Swedish Wind Energy Program Overview*, Fourth Wind Energy Workshop, Washington, D.C., October 29–31, JBF Scientific Corporation, Ed., DOE Conference No. 791097, pp. 457-460.

Brooks, C. E. P., and N. Carruthers, 1953: *Handbook of Statistical Methods in Meterology*, Her Majesty's Stationary Office, London, England.

Changery, M. J., 1975: *Initial Wind Energy Data Assessment Study*, National Climatic Center, Asheville, North Carolina, NSF/NOAA, NSF-RA-N-75-020.

Conrad, V., and L. W. Pollak, 1962: *Methods in Climatology*, Harvard University Press, Cambridge, Massachusetts.

Consulting Engineer, 1976: "Flying in the Face of the Wind," 40, No. 2, pp. 42–43.

Corotis, R. B., 1974: "Statistical Analysis of Continuous Data Records," *Trans. Eng. J.*, ASCE, 100, No. TEI, Proceeding Paper 10362, February, pp. 195–206.

Corotis, R. B., 1976: *Stochastic Modeling of Site Wind Characteristics*, Northwestern University, Chicago, Illinois, Final Report: NSF Grant AER 75-00357, November.

Corotis, R. B., 1977: *Stochastic Modeling of Site Wind Characteristics*, Northwestern University, Chicago, Illinois, ERDA report RLO/2342-77/2, September.

Corotis, R. B., A. B. Sigl, and M. P. Cohen, 1977: "Variance Analysis of Wind Characteristics for Energy Conversion," *J. Appl. Meteor.*, 16, pp. 1149–1157.

Corotis, R. B., A. B. Sigl, and J. Klein, 1978: "Probability Models of Wind Velocity Magnitude and Persistence," *Solar Energy*, 20, pp. 483–493.

Corotis, R. B., 1979: *Statistical Reliability of Wind Power Assessments*, Conference and Workshop on Wind Characteristics and Wind Energy Siting, sponsored by Battelle Memorial Institute's Pacific Northwest Laboratory and the American Meteorological Society, June 19–21, Portland Oregon, published by the American Meteorological Society, 1980.

Coty, U. A., 1976: Wind Energy Mission Analysis, Lockheed California Company, Burbank, California, ERDA Reports: Executive Summary, SAN/1075-1/3; Final Report, SAN/1075-1/1; Appendix, SAN/1075-1/2.

Court, A., 1974: "Wind Shear Extremes," Proceedings of the Initial Wind Energy Data Assessment Study, Asheville, North Carolina, July 29–31, M. J. Changery, Ed., published as report NSF-RA-N-75-020, May, 1975.

Crafoord, C., 1975: "An Estimate of the Interaction of a Limited Array of Windmills," University of Stockholm, DM-16.

Crutcher, H. L., 1959: *Upper Wind Statistics of the Northern Hemisphere*, 1, NAVAER 50-IC-535, Washington, D.C.

Crutcher, H. L. and L. Baer, 1962: "Computations from Elliptical Wind Distribution Statistics," *J. Appl. Meteor.*, 14, pp. 1512–1520.

Davenport, A. G., 1963: *The Relationship of Wind Structure to Wind Loading*, Proceedings of the Conference on Wind Effects on Structures, National Physics Lab., London, England, pp. 19–82.

Davison, G. N., 1979: *The Boeing Mod-2 Wind Turbine System Rotor*, Large Wind Turbine Design Characteristics and R&D Workshop, NASA Lewis Research Center, Cleveland, Ohio, April 24–26, Seymour Lieblein, Ed., NASA Conference Publication 2106, DOE Publication CONF-7904111, pp. 343–354.

DEFU, 1979: *The Wind Power Program of the Ministry of Commerce and The Electric Utilities in Denmark*, Lundtoftevej 100, DK-2800 Lyngby, Denmark, February.

Department of Energy, 1980: *Wind Energy Systems Program Summary*, DOE/CS-XXXX.

Derickson, R. G., and R. N. Meroney, 1977: *A Simplified Physics Airflow Model for Evaluating Wind Power Sites in Complex Terrain*, Summer Computer Simulation Conference, July

18-20, Hyatt Regency, Chicago, Colorado State University, Fort Collins, Colorado, CEP, 76-77-31.

Dickerson, M. H., and R. C. Orphon, 1975: *Atmospheric Release Advisory Capability (ARAC) Development and Plans for Implementation*, Lawrence Livermore Laboratory, Report No. UCRL-51839.

Donovon, R., 1978: Private communication.

Doran, J. C., J. A. Bates, P. J. Liddell, and T. D. Fox, 1977: *Accuracy of Wind Power Estimates*, Battelle Memorial Institute, Pacific Northwest Laboratories, Richland, Washington, PNL-2442/UC60.

Dornberg, J., 1979: "Danish Amateurs Build the World's Biggest Windmill," *Popular Science*, January, pp. 81-84.

Dornbirer, S. D., and C. J. Wilcox, 1945: "The Future of Large Scale Wind Power – An Analysis," a report prepared for Beauchamp E. Smith of the S. Morgan Smith Co., October.

Douglas, R. R., 1979: *The Boeing Mod-2 Wind Turbine System Rated at 2.5 MW*, Large Wind Turbine Design Characteristics and R&D Requirements Workshop, NASA Lewis Research Center, Cleveland, Ohio, April 24-26, Seymour Lieblein, Ed., NASA Conference Publication 2106, DOE Publication CONF-7904111, pp. 61-78.

Eagan, B. A., 1975: "Turbulent Diffusion in Complex Terrain," Chap. 4 of *Lectures on Air Pollution and Environmental Impact Analysis*, American Meteorological Society, pp.112-135.

Electric Power Research Institute, 1979: *Requirements Assessment of Wind Power Plants in Electric Utility Systems*, Palo Alto, California, Summary Report ER-978-SY, January.

Elliott, D. L., 1977: *Synthesis of National Wind Energy Assessments*, Battelle Memorial Institute, Pacific Northwest Laboratories, Richland, Washington, BNWL-2220 WIND-5, UC-60, July.

Elliott, D. L., 1978: *An Overview of the National Wind Energy Potential*, Preprint Volume, Conference on Climate and Energy: Climatological Aspects and Industrial Operations, Asheville, North Carolina, May 8-12, published by the American Meteorological Society, Boston, Massachussetts.

Essenwanger, O. M., 1959: "Probleme der Windstatistik," *Meteorologische Rundschau*, 12, pp. 37-47, Berlin, West Germany.

Essenwanger, O. M., 1968a: *On Deriving 90-99% Wind and Wind Shear Thresholds from Statistical Parameters*, Preprints Third National Conference on Aerospace Meteorology, New Orleans, American Meteorological Society, pp. 145-153.

Essenwanger, O. M., 1968b: *On Fitting the Weibull Distribution with Nonzero Location Parameter and Some Applications*, Proceedings of the 13th Conference on Design Experiment in Army Research Development and Testing, ARO-D Rep. 68-2, pp. 179-194.

Essenwanger, O. M., 1976: *Applied Statistics in Atmospheric Science*, Part A, Frequencies and Curve Fitting, Elsevier, New York.

Fales, E. N., 1967: "Windmills," *Standard Handbook for Mechanical Engineers*, 7th Ed., Baumeister and Marks, Eds., McGraw-Hill, New York, pp. 9-13.

Feber, R., 1978: "Public Reactions to Wind Energy and Windmill Designs," Third Wind Energy Workshop, Washington, D.C., September 19-21, 1977, 1, T. R. Kornreich, Ed., DOE CONF-770921/1, May.

Feller, W., 1966: *An Introduction to Probability Theory and Its Applications*, 2, Wiley, New York.

Fichtl, G. H., and O. E. Smith, 1977: "Wind," *Terrestrial Environment (Climatic) Criteria, Guidelines for Use in Aerospace Vehicle Development*, 1977 Revision, J. W. Kaufman, Ed., NASA TM 78118, 8.15-8.17.

Fosberg, M. A., W. E. Marlatt, and L. Krupnak, 1976: *Estimation of Airflow Patterns Over Complex Terrain*, U.S. Forest Service Research Paper RM 162.

Freeman, B. E., 1976: "Discussion of the Role of Meteorological Modeling in Selecting Wind Energy Sites," Science Applications, Inc., LaJolla, California.

Freeman, B. E., P. C. Patnaik, and G. T. Phillips, 1976: *Development of a Wind Energy Site Selection Methodology*, ERDA Report No. RLO/2440-76/4, December.

Frenkiel, J., 1964: *Wind Flow Over Hills In Relation to Wind Power Utilization*, Proceedings of the United Nations Conference on New Sources of Energy, 7, Wind Power, Rome, Italy, August 21–31, 1961, pp. 85–111.

Frenkiel, J., 1962: "Wind Profiles Over Hills in Relation to Wind Power Utilization," *Quart. J. Roy. Meteorol. Soc.*, 88, pp. 156–169.

Frost, W., J. R. Maus, and W. R. Simpson, 1973: *A Boundary Layer Approach to the Analysis of Atmospheric Motion over a Surface Obstruction*, NASA CR-2182.

Frost, W. G., 1974: "Wind Fields Over Terrain Irregularities," Initial Wind Energy Data Assessment Study, National Climatic Center, Asheville, North Carolina, NSF/NOAA, NSF-RA-N-75-020.

Frost, W., 1977: "Analysis of the Effect of Turbulence on Wind Turbine Generator Rotational Fluctuations," Proceedings of the International Conference on Alternative Energy Sources, Miami, Florida.

Frost, W., B. H. Long, and R. E. Turner, 1979: *Engineering Handbook on the Atmospheric Environmental Guidelines for Use in Wind Turbine Generator Development*, NASA TP 1359, 3.15.

Frost, W., and D. K. Nowak, 1979: *Summary of Guidelines for Siting Wind Turbine Generators Relative to Small-Scale, Two-Dimensional Terrain Features*, FWG Associates, Inc., DOE Report RLO/2443-77/1, March.

Garate, J. A., 1977: *Wind Energy Mission Analysis*, General Electric, Space Division, Valley Forge, Pennsylvania, ERDA Reports: Executive Summary, COO/2578-1/1, Final Report, COO/2578-1/2, Appendices A–J, COO/2578-1/3, February.

General Electric Company, 1976: *Design Study of Wind Turbines 50kW-3000 kW for Electric Utility Applications*. Summary Report, September, NASA CR-134934.

General Electric Company, 1979: Executive Summary, *Mod-1 Wind Turbine Generator Analysis and Design Report*, DOE/NASA/0058-79/3, March.

Glascow, J. C., and A. G. Birchenough, 1978: *Design and Operating Experience on the U.S. Department of Energy Experimental Mod-0 100 kW Wind Turbine*, DOE/NASA/1028-78/18, August.

Glascow, J. C., and W. H. Robbins, 1979: *Utility Operational Experience on the NASA/DOE Mod-0A 200 kW Wind Turbine*, Proceedings of the Workshop on Economic and Operational Requirements and Status of Large Scale Wind Systems, Monterey, California, March 28–30, Ed. Atlas Corporation, DOE Conference No. 790352, Electric Power Research Institute No. ER-1110-SR, pp. 215–247.

Glascow, J. C., 1979: NASA Lewis Research Center, Cleveland, Ohio, private communication.

Goldenblatt, M., 1979: "A Method for Estimating the Impact of WECS on Utility Operating Reserve Requirements," JBF Scientific Corporation.

Golding, E. W., 1955: *The Generation of Electricity by Wind Power*, Philosophical Library, New York. Revised by R. I. Harris, 1976, Wiley, New York.

Grastrup, H., 1979: *Design and Construction of Two 630 kW Wind Turbines*, Fourth Wind Energy Workshop, Washington, D.C., October 29–31, JBF Scientific Corporation, Ed., DOE Conference No. 791097, pp. 419–435.

Gustavson, M. R. 1979: "Limits to Wind Power Utilization," *Science*, 204, pp. 13–17, April 6.

Hamilton Standard, 1979a: *Wind Turbine System for the 80's*, Advanced Energy Products, Windsor Locks, Connecticut.

Hamilton Standard, 1979b: *Wind Energy Now*, Advanced Energy Products, Windsor Locks, Connecticut, December.

Hardy, D. M., 1976: Wind Power Studies: *Initial Data and Numerical Calculations*, Lawrence Livermore Laboratory Report UCRL-50034-76-1, January.

Hardy, D. M., and J. J. Walton, 1977: "Principal Components Analysis of Vector Wind Measurements," *J. Appl. Meteor.*, 17, pp. 1153-1162.

Heald, R. C., 1979: "Boundary Layer Wind Shear," Conference and Workshop on Wind Characteristics and Wind Energy Siting, sponsored by Battelle Memorial Institute's Pacific Northwest Laboratory and the American Meteorological Society, June 19-21, Portland, Oregon, published by the American Meteorological Society, 1980.

Hellman, G., 1916: "Uber die Bewegung der Luft in den Untersten Schichten der Atmosphäre," *Meteorologische Zeitschrift*, Bd. 34, p. 273, Berlin, West Germany.

Hennessey, J. P., 1977: "Some Aspects of Wind Power Statistics," *J. Appl. Meteor.*, 16, pp. 119-128.

Hennessey, J. P., 1980: A Critique of "Trees as a Local Climatic Wind Indicator," *J. Appl. Meteor.*, 19, pp. 1020-1023.

Heronemus, W. E., 1972: "Pollution Free Energy from Offshore Winds," Reprint, 8th Annual Conference and Exposition, Marine Technology Society, Washington, D.C.

Heronemus, W. E., 1976: "Oceanic Windpower," a paper prepared for and delivered to the conference convened by North Carolina State University under the title: "Energy from the Oceans: Fact or Fantasy," January.

Hewson, E. W., J. E. Wade, and R. W. Baker, 1977: *Vegetation as an Indicator of High Wind Velocity*, RLO/2227-T24-77/2, July.

Hewson, E. W., J. E. Wade, and R. W. Baker, 1979a: *A Handbook on the Use of Trees as an Indicator of Wind Power Potential*, RLO/227-79-3, June.

Hightower, S. J., 1979: United States Department of the Interior, Bureau of Reclamation, Denver, Colorado, private communication, August.

Hightower, S. J. and A. W. Watts, 1977: "A Proposed Conceptual Plan for Integration of Wind Turbine Generators with a Hydroelectric System," United States Department of the Interior, Bureau of Reclamation, Denver, Colorado, March.

Hino, M., 1968: "Computer Experiment on Smoke Diffusion over a Complicated Topography," *Atmospheric Environment*, 2, p. 541.

Hugosson, S., 1979a: *Specification, Siting, and Selection of Large WECS Prototypes*, Large Wind Turbine Design Characteristics and R&D Requirements Workshop, NASA Lewis Research Center, Cleveland, Ohio, April 24-26. Seymour Lieblein, Ed., NASA Conference Publication 2106, DOE Publication CONF-7904111, pp. 89-102.

Hugosson, S., 1979b: *Swedish Large-Scale Prototypes – Design and Siting*, Fourth Wind Energy Workshop, Washington, D.C., JBF Scientific Corporation, Ed., DOE Conference No. 791097, pp. 461-470.

Hütter, U., 1964: *The Aerodynamic Layout of Wind Blades of Wind Turbines with High Tip-Speed Ratio*, Proceedings of the United Nations Conference on New Sources of Energy, 7, Wind Power, Rome, Italy, August 21-31, 1961, pp. 217-228.

Hütter, U. 1973: "Past Developments of Large Wind Generators in Europe," Wind Energy Conversion Systems, Workshop Proceedings, J. M. Savino, Ed., NASA Lewis Research Center, Cleveland, Ohio, PB231341, pp. 19-22, December.

Inglis, D. R., 1978: *Wind Power and Other Energy Options*, University of Michigan Press, Ann Arbor, Michigan.

Inglis, D. R., 1979: "A Windmill's Theoretical Maximum Extraction of Power from the Wind," *Am. J. Phys.*, 47, pp. 416-420.

Jackson, P. S., and T. C. R. Hunt, 1975: "Turbulent Wind Flow over a Low Hill," *Quart. J. Roy. Meteor, Soc.,* 101, pp. 929-955.

Jackson, P. S., 1975: "A Theory for Flow over Escarpments," Ministry of Works and Development, New Zealand.

Johanson, E. E., and M. K. Goldenblatt, 1978: "An Economic Model to Establish the Value of WECS to a Utility System," a paper presented at the 1978 annual meeting of the American Section of the International Solar Energy Society, Denver, Colorado, August 28-31.

Johanson, E. E., and M. K. Goldenblatt, 1979: *Wind Energy Systems Application to Regional Utilities,* Proceedings of the Workshop on Economic and Operational Requirements and Status of Large Scale Wind Systems, Monterey, California, March 28-30, Atlas Corporation, Ed., EPRI Report ER-1110-SR, DOE Conference No. 790352, pp. 127-157.

Justus, C. G., 1974: *Wind Data Collection and Assessment,* Proc. of the Initial Wind Energy Data Assessment Study, Asheville, NC, July 29-31, M. J. Changery, Ed., published as report NSF-RA-N-75-020, May, 1975.

Justus, C. G., W. R. Hargraves, and A. Yalcin, 1976a: "Nationwide Assessment of Potential Output from Wind Powered Generators," *J. Appl. Meteor.,* 15, pp. 673-678.

Justus, C. G., W. R. Hargraves, and A. Mikhail, 1976b: *Reference Wind Speed Distributions and Height Profiles for Wind Turbine Design and Performance Evaluation Applications,* ORO/5108-76/4, August.

Justus, C. G., and A. Mikhail, 1976: "Height Variation of Wind Speed and Wind Distribution Statistics," *Geophys. Res. Lett.,* 3, p. 261.

Justus, C. G., and W. R. Hargraves, 1977: *Wind Energy Statistics for Large Arrays of Wind Turbines (Great Lakes and Pacific Coast Regions),* Georgia Institute of Technology, DOE Report RLO/2439-77/2.

Justus, C. G., W. R. Hargraves, A. Mikhail, and D. Graber, 1978: "Methods for Estimating Wind Speed Frequency Distributions," *J. Appl. Meteor.,* 17, pp. 350-353.

Justus, C. G., 1978a: *Winds and Wind System Performance,* Franklin Institute Press, Philadelphia, Pennsylvania.

Justus, C. G. and A. Mikhail, 1978a: "Generic Power Performance Estimates for Wind Turbines," *Wind Tech. J.,* 2, p. 45.

Justus, C. G., and A. Mikhail, 1978b: "Energy Statistics for Large Wind Turbine Arrays," *Wind Engineering,* 2 pp. 184-202.

Justus, C. G., and A. S. Mikhail, 1978c: *Energy Statistics for Large Wind Turbine Arrays,* Georgia Institute of Technology, DOE Report RLO/2439-78/3, May.

Justus, C. G., 1978b: "Wind Energy Statistics for Large Arrays of Wind Turbines (New England and Central U.S. Regions)," *Solar Energy,* 20, pp. 379-386.

Justus, C. G., K. Mani, and A. S. Mikhail, 1979: "Interannual and Month-to-Month Variations of Wind Speed," *J. Appl. Meteor.,* 18, pp. 913-920.

Kao, J. H. K., 1958: "Computer Methods for Estimating Parameters in Reliability Studies," *IRE Trans. Reliability Qual. Control,* PG-RQC-03, pp. 15-22.

Kamen Aerospace Corporation, 1976: *Design Study of Wind Turbines 50 kW-3000 kW for Electric Utility Applications, Analysis and Design,* Final Report, February, NASA CR-134934.

Kasahara, A., E. Isaacson, and J. Stoker, 1965: "Numerical Studies of Frontal Motion in the Atmosphere," *Tellus,* 17, p. 261.

Kilar, L. A., 1979: *Design Study and Economic Assessment of Multi-Unit Offshore Wind Energy Conversion Systems Applications,* Four Volumes, Westinghouse Electric Corporation, Advanced Systems Technology, East Pittsburgh, Pennsylvania, Final Report, WASH/2330-78/4.

Kirchhoff, R. H., 1979: *Turbulence and WTG Performance,* Conference and Workshop on Wind Energy Siting, sponsored by Battelle Memorial Institute's Pacific Northwest Laboratories and the American Meteorological Society, June 19–21, 1979, Portland, Oregon, published by the American Meteorological Society.

Knox, J. B., D. M. Hardy, C. A. Sherman, and T. J. Sullivan, 1976: *Status Report: Lawrence Livermore Laboratory Wind Energy Studies*, Lawrence Livermore Laboratory, Report UCIP-17157-1, 18 pp.

Koeppl, G. W., 1979a: "Information and Analysis to Support a Candidate Site Proposal for the Installation of a Mod-2 WECS at a Site Near Scrag Mountain in Response to DOE PON 06-79RL10081.000- First Selection," a report prepared for the Green Mountain Power Corporation, Burlington, Vermont, August.

Koeppl, G. W., 1979b: "The Wind Characteristics of the Lincoln Ridge, Part II and Part III, Sections A.1 and A.2 of DOE PON 06-79RL10081.000- Second Selection," a report prepared for the Green Mountain Power Corporation, Burlington, Vermont, October.

Koeppl, G. W., 1979c: "The Wind Characteristics at the Summit of Stratton Mountain, Part II and Part III of DOE PON 06-79RL10081.000- Second Selection," a report prepared for the Vermont Electric Cooperative, Johnson, Vermont, October.

Koeppl, G. W., 1980: *Further Analysis of Smith-Putnam Data*, to be published.

Körber, F., and H. A. Thiele, 1979: *Large Wind Energy Converter – Growian 3MW*, Large Wind Turbine Design Characteristics and R&D Requirements Workshop, NASA Lewis Research Center, Cleveland, Ohio, April 24–26, Seymour Lieblein, Ed., NASA Conference Publication 2106, DOE Publication CONF-7904111, pp. 121–131.

Lange, K. O., 1964: *Some Aspects of Site Selection for Wind Power Plants on Mountainous Terrain*, Proceedings of the United Nations Conference on New Sources of Energy, 7, Wind Power, Rome, Italy, August 21–31, 1961, published by the United Nations in 1964.

Lavoie, R. L., 1972: "A Mesoscale Numerical Model of Lake Effect Storms, *J. Atmos. Sci.*, 29, p. 1025.

Lowe, J. E., 1978: Boeing Engineering and Construction, Seattle, Washington, private communication, December.

Lowe, J. E., and W. W. Engle, 1979: "The Mod-2 Wind Turbine – Now a Viable Option for the Power System Planner," Preprint 3738, ASCE Convention and Exhibition, Atlanta, Georgia, October 23–25.

MacCready, P. B., 1977: *Assessing the Local Wind Field for Siting of Wind Power Systems*, DOE Report RLO-2441-76/16.

Maschinenfabrik Augsburg-Nurnberg Aktiengesellschaft, 1979a: *Large Wind Energy Converter – Growian*, summary prepared by F. Körber under the auspices of KFA Julich GmbH for the Minister of Research and Technology of the Federal Republic of Germany, p. 17, April.

Maschinenfabrik Augsburg-Nurnberg Aktiengesellschaft, 1979b: *Grosse Windenergie-Anlage-Growian*, Dachauer Strasse 667, D8000 München 50, West Germany.

Marrs, R. W., and J. Marwitz, 1978: *Locating Areas of High Wind Energy Potential by Remote Observation of Eolian Geomorphology and Topography*, Third Wind Energy Workshop, Washington, D.C., September 19–21, 1977, by T. R. Kornreich, Ed., DOE CONF-770921/1, 1, pp. 307–320, May.

Marrs, R. W., and S. Kopriva, 1978: *Regions of the Continental United States Susceptible to Eolian Action*, RLO-2343-78/2, May.

Marsh, W. D., 1979: *Requirements Assessment of Wind Power Plants in Electric Utility Systems*, Proceedings of the Workshop on Economic and Operational Requirements and Status of Large Scale Wind Systems, Monterey, California, March 28–30, Atlas Corporation, Ed., EPRI Report ER-1110-SR, DOE Conference No. 790352, pp. 110–126.

Martin, P., M. Balma, G. Crooks, R. Endlich, C. Maxwell, W. Schubert, J. Witwer, L. Manfield, B. Stevens, 1979: *A Feasibility Study of Windpower for the New England Area*, a report prepared by SRI International for Navy Material Command, Washington, D.C., Contract No. N00014-79-C-0539.

Mayo, L. H., 1977: *Legal-Institutional Implications of Wind Energy Conversion Systems*, Final Report, NSF/RA-770204.

Meroney, R. N., V. A. Sandborn, R. J. B. Bouwmeester, and M. A. Rider, 1976a: *Sites for Wind Power Installations: Wind Tunnel Simulation of the Influence of Two-Dimensional Ridges on Wind Speed and Turbulence*, ERDA Report No. ERDA/NSF/00702-75/1, July.

Meroney, R. N., V. A. Sandborn, R. Bouwmeester, and M. Rider, 1976b: "Wind Tunnel Simulation of the Influence of Two-Dimensional Ridges on Wind Speed and Turbulence," International Symposium on Wind Energy Systems, BHRA Fluid Engineering, Cranfield, Bedford, England, Paper A6, September.

Meroney, R. N., A. J. Bowen, D. Lindley, J. R. Pearse, 1978: *Wind Characteristics Over Complex Terrain: Laboratory Simulation and Field Measurements at Rakaia Gorge, New Zealand*, DOE Report No. RLO/2438-77/2.

Meroney, R. N., 1979: *WECS Site Screening by Physical Modeling*, Conference and Workshop on Wind Characteristics and Wind Energy Siting, sponsored by Battelle Memorial Institute's Pacific Northwest Laboratories and the American Meteorological Society, June 19-21, Portland, Oregon, published by American Meteorological Society, 1980.

Messerschmitt-Bölkow-Blohm GmbH, 1979: *Wind-energieanlage WEA 5000 (Growian II)*, München, West Germany.

Mikhail, A. S., 1977: "Atmospheric Boundary Layer Similarity Theory for Application in Wind Energy Fields," Ph.D. Thesis, Georgia Institute of Technology, September.

Monin, A. S., and M. A. Obukhov, 1954: "Dimentionless Characteristics of Turbulence in the Surface Layer," *Akad, Nauk. SSSR* Geofizicheskii Institut-Trudy No. 24, pp. 163-187, Moscow, Russia.

NASA Lewis Research Center, 1978: *200 kW Wind Turbine Project*, a part of the U.S. Department of Energy Federal Wind Energy Program, July.

NASA Lewis Research Center, 1979: *200 kW Wind Turbine Generator Conceptual Design Study*, DOE/NASA/1028-79/1, January.

Neumann, R., and R. Windheim, 1978: *Wind Energy R&D Program of the Federal Republic of Germany and Main Lines of Development*, Third Wind Energy Workshop, Washington, D.C., September 19-21, 1977, T. R. Kornreich, Ed., DOE CONF-770921/1, **1**, pp. 438-443, May.

Neustadter, H. E., 1979: *The Use of Wind Data with an Operational Wind Turbine in a Research and Development Environment*, NASA Lewis Research Center, Cleveland, Ohio, DOE/NASA/1004-79/16.

O'brien, J. J. and H. E. Hurlburt, 1972: "A Numerical Model of Coastal Upwelling," *J. Phys. Oceangr.*, **2**, January, p. 14.

Orville, H. D., 1968: "Ambient Wind Effects on the Initiation of Cumulus Clouds over Mountains," *J. Atmos. Sci.*, **25**, pp. 385-403.

Ossenbrugen, P. J., G. P. Pregent, L. D. Meeker, 1979: "Offshore Wind Power Potential," *J. Energy Div.*, Proceedings of the American Society of Civil Engineers, **105**, No. EYI, pp. 81-92, January.

Pederson, B. M., 1979: *The Danish Large Wind Turbine Program*, Large Wind Turbine Design Characteristics and R&D Requirements Workshop, NASA Lewis Research Center, Cleveland, Ohio, April 24-26, Seymour Lieblein, Ed., NASA Conference Publication 2106, DOE Publication CONF-7904111, pp. 103-120.

Perry, A. M., et al., 1977a: *Net Energy Analysis of Five Energy Systems*, Oak Ridge Associated Universities, Institute for Energy Analysis, ORAU-IEAR-77-12, September.

Perry, A. M., et al., 1977b: *The Energy Cost of Energy – Guidelines for Net Energy Analysis*, Oak Ridge Associated Universities, Institute for Energy Analysis, ORAU/IEA(R)-77-14.

Petterssen, S., 1964: *Some Aspects of Wind Profiles*, Proceedings of the United Nations Conference on New Sources of Energy, 7, Wind Power, Rome, Italy, August 21–31, 1961, pp. 133–136, published by the United Nations in 1964.

Phillips, P. D., 1979: "NEPA and Alternative Energy: Wind as a Case Study," *Solar Law Reporter*, 1, No. 1, pp. 29–54.

Poor, R. H. and R. B. Hobbs, 1979a: *The General Electric Mod-1 Wind Turbine Generator Program*, Large Wind Turbine Design Characteristics and R&D Requirements Workshop, NASA Lewis Research Center, Cleveland, Ohio, April 24–26, Seymour Lieblein, Ed. NASA Conference Publication 2106, DOE Publication CONF-7904111, pp. 35–59.

Poor, R. H., and R. B. Hobbs, 1979b: "The General Electric Mod-1 Wind Turbine Generator Program," General Electric Company, Space Division, Valley Forge, Pennsylvania.

Pristov, J., 1959: "Abweichungen des Windes aufden Alpine Beabachtungs-stationen im begug aufdie Stromung under Frein Atmosphere," *Berichte des Deutschen Wittendienstes*, Bd. 8, 5, p. 241, Bad Kissingen.

Proceedings of the United Nations Conference on New Sources of Energy, 7, Wind Power, Rome, Italy, August 21–31, 1961, published in 1964 by the United Nations.

Putnam, P. C., 1948: *Power From the Wind*, VanNostrand Reinhold, New York.

Reed, J. W., 1974: "Wind Power Climatology," *Weatherwise*, 27, p. 237.

Reed, J. W., 1975: *Wind Power Climatology of the U.S.*, Sandia Laboratories, Albuquerque, New Mexico, SAND 74-0348, July.

Reed, J. W., 1976: *Predicting Wind Power at Turbine Level from an Anemometer Record at Arbitrary Height*, Sandia Laboratories, Albuquerque, New Mexico, SAND 76-5397.

Robbins, W. H., and J. E. Sholes, 1978: *ERDA/NASA 200 kW Mod-0A Wind Turbine Program*, Third Wind Energy Workshop, Washington, D.C., September 19–21, 1977, T. R. Kornreich, Ed., DOE CONF-770921/1, 1, pp. 59–75, May.

Robbins, W. H., and R. L. Thomas, 1979: "Large Horizontal-Axis Wind Turbine Development," prepared for the Wind Energy Innovative Systems Conference sponsored by the Solar Energy Research Institute, Colorado Springs, Colorado, May 23–25, DOE/NASA/1059-79/2.

Rogers, S. E., M. A. Duffy, J. G. Jefferis, P. R. Sticksel, D. A. Tolle, 1976: *Evaluation of the Potential Environmental Effects of Wind Energy System Development*, Battelle Columbus Laboratories, Columbus, Ohio, ERDA/NSF/07378-75/1, August.

Rogers, S. E., B. W. Cornaby, C. W. Rodman, P. R. Sticksel, and D. A. Tolle, 1977: *Environmental Studies Related to the Operation of Wind Energy Conversion Systems*, Battelle Columbus Laboratories, Columbus, Ohio, COO-0092-77/2, December.

Rose, M., 1979: *Cuttyhunk Island Installation*, WTG Energy Systems, Inc., MP1-200 Control System Design, Proceedings of the Workshop on Economic and Operational Requirements and Status of Large Scale Wind Systems, Monterey, California, March 28–30, DOE Conference No. 790352, Electric Power Research Institute No. ER-1110-SR, pp. 283–296.

Rotty, R. M., et al., 1976: *Net Energy from Nuclear Power*, Oak Ridge Associated Universities, Institute for Energy Analysis, PB-254 059.

Savino, J., Ed., 1973: "Wind Energy Conversion Systems," Workshop Proceedings, NSF/RA/W-73-006, December.

Scheffler, R. L., 1979a: *Status of the Southern California Edison Company 3 MW Wind Turbine Generator Development Project*, Large Wind Turbine Design Characteristics and R&D Requirements Workshop, NASA Lewis Research Center, Cleveland, Ohio, April 24–26, Seymour Lieblein, Ed., NASA Conference Publication 2106, DOE Publication CONF-7904111, pp. 355–362.

Scheffler, R. L., 1979b: Private communication.

Scheffler, R. L., 1980: Private communication.

Schlicting, H., 1968: *Boundary Layer Theory*, 6th Ed., McGraw-Hill, New York, p. 564.

Seltari, A., and R. B. Lantz, 1974: "A Turbulent Flow Model for Use in Numerical Evaluation of Air Quality," *J. of Can. Petrol. Ind.*, October-December, Montreal, Canada.

Sengupta, D. L., and T. B. A. Senior, 1978: *Electromagnetic Interference by Wind Turbine Generators*, University of Michigan, Ann Arbor, Michigan, Radiation Laboratory, COO/2846-2, March.

Senior, T. B. A., and D. L. Sengupta, 1978: *Wind Turbine Generator Siting and T. V. Reception Handbook*, University of Michigan, Ann Arbor, Michigan, Radiation Laboratory, COO/2846-1, January.

Shepherd, D. G., 1978: *Wind Power, Advances in Energy Systems and Technology*, Vol. 1, P. Auer, Ed., Academic Press, New York.

Sherman, C. A., 1975: "A Mass Consistent Model for Wind Fields over Complex Terrain, Lawrence Livermore Laboratory," Report UCRL-76171; *J. Appl. Meteor.*, 17, pp. 312-319.

Sigl, A. B., R. B. Corotis, and D. J. Won, 1979: "Run Duration Analysis of Surface Wind Speeds for Wind Energy Application," *J. Appl. Meteor.*, 18, No. 2, pp. 156-166.

Sisterson, D. L., and B. B. Hicks, 1979: "On the Application of Power Laws for Wind Energy Assessment," Conference and Workshop on Wind Characteristics and Wind Energy Siting, sponsored by Battelle Memorial Institute's Pacific Northwest Laboratory and the American Meteorological Society, June 19-21, Portland, Oregon, published by the American Meteorological Society, 1980.

Smith, M. E., 1968: *Recommended Guide for the Prediction of the Dispersion of Airborne Effluents*, American Society of Mechanical Engineers, New York.

Smith, O. E., 1971: *An Application of Distributions Derived from the Bivariate Normal Density Function*, Proceedings of the International Symposium on Probability and Statistics in the Atmospheric Sciences, Honolulu, Hawaii, pp. 162-168.

Smith, O. E., 1976: *Vector Wind and Vector Shear Models, 0 to 27 km Altitude for Cape Kennedy, Florida, and Vandenberg AFB*, California, NASA TM-X-73319, July.

Solar Program Assessment, 1977: *Environmental Factors*, Wind Energy Conversion, ERDA 77-47/6, March.

Spaulding, A. P., 1979: *WTG Energy Systems MP1-200 − 200 kW Wind Turbine Generator*, Large Wind Turbine Design Characteristics and R&D Requirements Workshop, NASA Lewis Research Center, Cleveland, Ohio, April 24-26, NASA Conference Publication 2106, DOE Publication CONF-7904111, pp. 79-88.

Spera, D. A., 1977: *Comparison of Computer Codes for Calculating Dynamic Loads in Wind Turbines*, DOE/NASA/1028-78/16, September.

Spera, D. A., L. A. Viterna, T. R. Richards, and H. E. Neustadter, 1979: *Mod-1 Wind Turbine Generator*, Fourth Wind Energy Workshop, Washington, D.C., October 29-31, JBF Scientific Corporation, Ed., DOE Conference No. 791097, pp. 99-117.

Spera, D. A., and T. R. Richards, 1979: *Modified Power Law Equations for Vertical Wind Profiles*, Conference on Wind Characteristics and Wind Energy Siting, sponsored by Battelle Memorial Institute's Pacific Northwest Laboratories and the American Meteorological Society, June 19-21, Portland, Oregon, published by American Meteorological Society, 1980.

Stewart, D. A., and O. M. Essenwanger, 1978: "Frequency Distribution of Wind Speed Near the Surface," *J. Appl. Meteor.*, 17, pp. 1633-1642.

Szostak, J., 1980: "DAF," *Wind Power Digest*, 109 E. Lexington, Elkhart, Indiana, pp. 30-32, Spring.

Survey Research Laboratory, 1977: *Public Reactions to Wind Energy Devices*, University of Illinois, Urbana-Champaign, Illinois, Final Report, NSF/RA-77-0026, October.

Takle, E. S., and J. M. Brown, 1977: "Note on the Use of Weibull Statistics to Characterize Wind Speed Data," *J. Appl. Meteor.*, 17, pp. 556-559.

Templin, R. M., 1974: "An Estimate of the Interaction of Windmills in Widespread Arrays," National Research Council of Canada, LTR-LA-171.

Thomas, R. L., and R. M. Donovon, 1978: *Large Wind Turbine Generators*, DOE/NASA/1059-78/1, February.

Thomas, R. L., and T. R. Richards, 1978: *ERDA/NASA 100 kW Mod-0 Wind Turbine Operations and Performance*, Third Wind Energy Workshop, Washington, D.C., September 19–21, 1977, T. R. Kornreich, Ed., DOE CONF-770921/1, 1, pp. 35–58, May.

Thomas, R. L. and W. H. Robbins, 1979: *Large Wind Turbine Projects*, Fourth Wind Energy Workshop, Washington, D.C., October 29–31, JBF Scientific Corporation, Ed., DOE Conference No. 791097, pp. 75–97.

Todd, C. J., R. L. Eddy, R. C. James, and W. E. Howell, 1977: *Cost-Effective Electric Power Generation from the Wind*, U.S. Department of the Interior, Bureau of Reclamation, Denver, Colorado, August.

Traci, R. M., G. T. Phillips, P. C. Patnaik, and B. E. Freeman, 1977: *Development of a Wind Energy Site Selection Methodology*, DOE Report No. RLO/2440-11, June.

Traci, R. M., G. T. Phillips, and P. C. Patnaik, 1978: *Wind Energy Site Selection Methodology Development*, DOE Report No. RLO/2440-18/2, September.

Traci, R. M., G. T. Phillips, and K. C. Rock, 1979: *The Utility and Verification of Mathematical Windfield Models for Wind Energy Regional Screening and Site Selection*, Conference and Workshop on Wind Characteristics and Wind Energy Siting, sponsored by Battelle Memorial Institute's Pacific Northwest Laboratory and the American Meteorological Society, June 19–21, Portland, Oregon, published by the American Meteorological Society, 1980.

United Nations Conference on New Sources of Energy, 1964: "Proceedings of the United Nations Conference on New Sources of Energy," 7, Wind Power, Rome, Italy, August 21–31, 1961 published by the United Nations in 1964.

Vachon, W. A., W. T. Downey, F. March, F. R. Madio, G. R. Schimke, and J. E. Wade, 1979: *Wind Energy in the Mountains of New Hampshire as a Potential Energy Source for the Portsmouth Shipyard*, a report prepared by A. D. Little, Inc. for Navy Material Command, Washington, D.C., No. NR521-710.

Van Bronkhorst, J., 1979: *The Mod-1 Steel Blade*, Large Wind Turbine Design Characteristics and R&D Requirements Workshop, NASA Lewis Research Center, Cleveland, Ohio, April 24–26, Seymour Lieblein, Ed., NASA Conference Publication 2106, DOE Publication CONF-7904111, pp. 325–342.

Vermont Ad Hoc Wind Energy Committee, 1981: "Wind Energy, The Prospects for Vermont," Vermont State Energy Office, Montpelier, Vermont.

von Kármán, T., 1921: *Z. Angew. Math. Mech.*, Bd. 1, p. 239, Berlin, West Germany.

Vukovich, F. M., and C. A. Clayton, 1977: *On a Technique to Determine Wind Statistics in Remote Locations*, DOE Report No. RLO/2445-78/1.

Wade, J. E., and E. W. Hewson, 1979b: "Trees as a Local Climatic Wind Indicator," *J. Appl. Meteor.*, 18, pp. 1182–1187.

Wahl, F. W., 1966: *Windspeed on Mountains*, Airforce Cambridge Research Laboratories, Bedford, Massachusetts, Final Report AFCRL-66-280, January.

Watts, A. W., 1979: "Methods of Integrating Wind Energy with Hydroelectric Systems," United States Department of the Interior, Bureau of Reclamation, Denver, Colorado, June.

Wegley, H. L., M. M. Orgill, and R. L. Drake, 1978: *A Siting Handbook for Small Wind Energy Conversion Systems*, Battelle Memorial Institute, Pacific Northwest Laboratory, PNL-2521/UC60, May.

Wendell, L. L., 1979: *Overview of the Wind Characteristics Program*, Fourth Wind Energy Workshop, October 29–31, Washington, D.C., JBF Scientific Corporation, Ed., DOE Conference No. 791097, pp. 209–244.

Wentink, T., Jr., 1976: *Study of Alaskan Wind Power and Its Possible Application*, NSF/RANN/SE/AER-74-00239/FR-76/1.

Widger, W. K., 1976: "Estimating Wind Power Feasibility," *Power Engineering*, August, pp. 58–61.

Widger, W. K., 1977: "Estimations of Wind Speed Frequency Distributions Using Only the Monthly Average and Fastest Mile Data," *J. Appl. Meteor.*, **16**, pp. 244–247.

Wilson, R. E., and P. B. S. Lissaman, 1974: *Applied Aerodynamics of Wind Power Machines*, National Science Foundation Report PB-238-595.

Wind Energy Report, 1978: "California Utility to Test 3 MW Wind Turbine," 104 S. Village Avenue, Rockville Center, N.Y., July.

_____ , 1979a: "Canadian Utility Buying WTG Unit," November.

_____ , 1979b: "Bendix Acquires Rights to Schachle WECS," August.

_____ , 1979c: "SCE To Buy ALCOA 500 kW VAWT," September.

_____ , 1979d: "Oregon Utility Buying ALCOA 500 kW VAWT," June.

_____ , 1979e: "Offshore Wind Energy, Is It An Opportunity?" October.

_____ , 1979f: "PURPA, Windfarms and the Financing of Wind Power," December.

_____ , 1979g: "SWECS Wind Farm Announced for California," April.

_____ , 1979h: "80 MW Wind Farm Planned for Hawaii," November.

_____ , 1980a: "Mod-1 Emits Low Frequency Sound, Vibrations," April.

_____ , 1980b: "Fourth and Last, Mod-0A on Line In Hawaii," July.

Wind Power Group, 1979: *Development of Large Wind Turbine Generators*, a design feasibility and cost study undertaken by a group consisting of British Aerospace Dynamics Group, Cleveland Bridge & Engineering Company, Electrical Research Association, North Scotland Electricity Board, Taylor Woodrow Construction, Department of Energy, Thames House South, Millbank, London SW1P 4QJ, March.

Wind Task Force, 1980: Preliminary Draft, "Wind Task Force Report," Oregon Department of Energy, Salem, Oregon.

Windheim, R., and R. Neumann, 1979: *Wind Energy R&D Program of the Federal Republic of Germany and Current Wind Energy Projects*, Fourth Wind Energy Workshop, Washington, D.C., October 29–31, JBF Scientific Corporation, Ed., DOE Conference No. 791097, pp. 249–267.

Index

costs (continued)
   energy, 139–140, 144–145, 189
   engineering, 178, 180
   fuel, 237–240
   gears, 139–140, 154, 158
   generators, 139–140, 154, 158
   installation, 185–187
   installed, 178, 180–181, 187
   kilowatt hour, 7, 119–120, 419, 423–425, 434
   manufacturing, 180–181
   mass production, 162–163, 212–213, 217–218
   multiple units, 118, 139
   nonstandard parts, 183–185, 205–206
   offshore wind power systems, 418
   preproduction model, 160, 162–163
   reduction, 205–213
   standard parts, 181–183, 205
   and unit size, 118, 139–140, 216–217
Coty, U. A., 225, 281, 440
couplings, 118, 207–208
   costs, 154, 158
   electric, 181, 207–208, 211–212
   elimination, 211–212
   flexible, 182
Court, A., 242
Crafoord, C., 338, 340
Cree, Albert, 8
Crow, S. C., 288–295
Crown Point, Vt., 44, 47, 74–75, 77
Crutcher, H. L., 227–228, 242
cubic models, 316–317
Culebra, Puerto Rico, 351
Cuttyhunk Island, Mass., 385–386
Czechoslovakia, 114

DAF Indal, 415
daily variations, in velocity, 72, 95, 97, 268–277, 280–281
damping elimination, 209
Danish Electrical Supply Undertakings (DEFU), 382
Darrieus, 4, 102, 105, 107, 109
Darrieus vertical-axis wind turbine (VAWT), 415
Davenport, A. G., 257
Davis, Aaron, 4
Davison, G. N., 365
deformation, of trees, 54–59, 62–63, 86–91, 281–286

Denmark, 215, 378–384
density
   air, 17, 36–37, 315–316, 326
   of wind power, 225–233
Department of Energy (Britain), 393–395
Department of Energy (DOE) (U.S.), 262, 276, 282, 307, 315, 340, 347, 364, 388, 421, 431, 443
Department of Interior, 435, 443
Department of the Treasury, 426
Department of Water Resources (Calif.), 437
Derickson, R. G., 287
design
   costs, 205–213
   development of, 5–7, 113, 162–163, 168
   dimensions, 119–120
     large turbines, 139–140, 144–145
     preproduction model, 163, 168
     test unit, 120, 144
   efficiency, 93, 309–311
   icing considerations, 37, 111, 168
   loading assumptions, 13–14, 124, 126, 168, 172–174
   other than Smith-Putnam, 102–112
   preproduction model, 162–163, 168
   and structure of wind, 35
   test unit, 120–124, 131–133
   and wind generation capacity, 234, 237
Dickerson, M. H., 296
direction probability distribution functions, 241–256, 318–320
distributions, probability, 241–256, 318–320
diurnal variations, in velocity, 72, 95, 97, 268–277, 280–281
Donovan, R., 348
Doran, J. C., 254
Dornberg, J., 378
Dornbirer, Stanton D., 11, 15, 174, 184–186, 209, 262, 282
Douglas, R. R., 314, 364
Douglas firs, 282
Drzewiecki, 102
DuPont DeNemours, E. I., 183
Durgin, Harold L., 8
Dutch designs, 102, 117, 417
Dynamatic Corp., 181

Eagan, B. A., 287
East Boston, Mass., 27, 38, 73, 76
East Mountain, 41–43, 47, 94–95, 97